D1698884

Edited by
Andreas Öchsner and Waqar Ahmed

Biomechanics of Hard Tissues

Further Reading

Kumar, C. S. S. R. (ed.)

Biomimetic and Bioinspired Nanomaterials

2010

ISBN: 978-3-527-32167-4

Ghosh, S. K. (ed.)

Self-healing Materials
Fundamentals, Design Strategies, and Applications

2009

ISBN: 978-3-527-31829-2

Öchsner, A., Murch, G. E., de Lemos, M. J. S. (eds.)

Cellular and Porous Materials
Thermal Properties Simulation and Prediction

2008

ISBN: 978-3-527-31938-1

Ruiz-Hitzky, E., Ariga, K., Lvov, Y. M. (eds.)

Bio-inorganic Hybrid Nanomaterials
Strategies, Syntheses, Characterization and Applications

2008

ISBN: 978-3-527-31718-9

Breme, J., Kirkpatrick, C. J., Thull, R. (eds.)

Metallic Biomaterial Interfaces

2008

ISBN: 978-3-527-31860-5

Brito, M. E.

Developments in Porous, Biological and Geopolymer Ceramics
Ceramic Engineering and Science Proceedings, Volume 28, Issue 9

2008

ISBN: 978-0-470-19640-3

Kumar, C. S. S. R. (ed.)

Nanomaterials for Medical Diagnosis and Therapy

2007

ISBN: 978-3-527-31390-7

Edited by Andreas Öchsner and Waqar Ahmed

Biomechanics of Hard Tissues

Modeling, Testing, and Materials

WILEY-VCH Verlag GmbH & Co. KGaA

The Editors

Prof. Dr. Andreas Öchsner
Technical University of Malaysia
Faculty of Mechanical Engineering
81310 UTM Skudai
Johor
Malaysia

Prof. Waqar Ahmed
University of Lancashire
Institute of Advanced Manufacturing
Preston PR1 2HE
United Kingdom

Library of Congress Card No.: applied for

British Library Cataloguing-in-Publication Data
A catalogue record for this book is available from the British Library.

Bibliographic information published by the Deutsche Nationalbibliothek
The Deutsche Nationalbibliothek lists this publication in the Deutsche Nationalbibliografie; detailed bibliographic data are available on the Internet at <http://dnb.d-nb.de>.

Typesetting Laserwords Private Limited, Chennai, India
Printing and Binding Fabulous Printers Pte Ltd., Singapore
Cover Design Formgeber, Eppelheim

Printed in Singapore
Printed on acid-free paper

ISBN: 978-3-527-32431-6

Contents

Biomechanics of Hard Tissues: Modeling, Testing, and Materials.
Edited by Andreas Öchsner and Waqar Ahmed

ISBN: 978-3-527-32431-6

Preface

Biomechanics, the application of mechanical methods to biological systems, is a rapidly growing area of immense importance. The ability to influence the "lifetime" of parts of the human body or to offer adequate replacements in the case of failure has a direct influence on our entire well-being. This becomes increasingly important during old age when joints must be replaced in order to guarantee an adequate mobility of the various components of the human body. To adopt the mechanical performance of structural parts of our body or to offer alternatives if they do not function properly any more in order to meet the biological life expectancy is a major challenge that requires coordinated efforts of a number of academic disciplines.

The application of analytical, numerical, and experimental characterization methods, which were originally applied to engineering structures, allows us to model, analyze, and understand the behavior and the performance of biological structures. However, this area is much more complicated than engineering structures. For example, classical test specimens, as in the case of metals, are difficult to manufacture from biological materials. Biological materials are in many cases not as isotropic and homogeneous as traditional engineering materials and their mechanical properties depend on many factors and may be subjected to a significant variation during the entire lifetime of the structure.

This monograph focuses on hard tissues, that is, tissues having a firm intercellular structure such as bone or cartilage. Their structure and physical properties are described in detail. Modeling approaches on different length scales are presented in order to predict the mechanical properties. The influence of different biological, mechanical, and other physical factors and stimuli on the performance and regeneration ability is discussed in several chapters. Other chapters are related to the topic of bone replacement using implants. Different types of implants are characterized, tribological aspects covered, and the bone–implant interaction modeled and simulated numerically. Finally, the design of bone implants based on mathematical criteria is presented.

The editors wish to thank all the chapter authors for their participation, cooperation, and patience, which has made this monograph possible.

Biomechanics of Hard Tissues: Modeling, Testing, and Materials.
Edited by Andreas Öchsner and Waqar Ahmed

ISBN: 978-3-527-32431-6

Finally, we would like to thank the team at Wiley-VCH, especially Dr. Heike Nöthe, Dr. Martin Preuss, and Dr. Martin Ottmar, for their excellent cooperation during this important project.

August 2010

Andreas Öchsner
Waqar Ahmed

List of Contributors

Stephen C. Cowin
The City College of New York
Department of Biomedical
Engineering
138th Street and Convent Avenue
New York NY 10031
USA

Darlan Dallacosta
Federal University of
Santa Catarina
Department of Mechanical
Engineering
Group of Analysis and Design
88040-900
Florianópolis
Brazil

Manuel Doblaré
University of Zaragoza
Aragón Institute of
Engineering Research (I3A)
Group of Structural Mechanics
and Material Modelling
Betancourt Bldg.
María de Luna s/n
50018 Zaragoza
Spain

and

Centro de Investigación
Biomédica en Red en
Bioingeniería
Biomateriales y Nanomedicina
I+D+i Bldg. Mariano Esquillor
s/n - 50018 Zaragoza
Spain

Eduardo A. Fancello
Federal University of Santa
Catarina
Department of Mechanical
Engineering
Group of Analysis and Design
88040-900
Florianópolis
Brazil

Virginia L. Ferguson
University of Colorado
Department of Mechanical
Engineering
UCB 427
Boulder, CO 80309
USA

Biomechanics of Hard Tissues: Modeling, Testing, and Materials.
Edited by Andreas Öchsner and Waqar Ahmed

ISBN: 978-3-527-32431-6

Paulo R. Fernandes
TU Lisbon
Instituto Superior Técnico
Mechanical Engineering
Department
IDMEC-IST
Av. Rovisco Pais
1049-001 Lisbon
Portugal

Joao Folgado
TU Lisbon
Instituto Superior Técnico
Mechanical Engineering
Department
IDMEC-IST
Av. Rovisco Pais
1049-001 Lisbon
Portugal

José Manuel García-Aznar
University of Zaragoza
Aragón Institute of
Engineering Research (I3A)
Group of Structural
Mechanics and
Material Modelling
Betancourt Bldg.
María de Luna s/n
50018 Zaragoza
Spain

and

Centro de Investigación
Biomédica en Red en
Bioingeniería
Biomateriales y Nanomedicina
I+D+i Bldg. Mariano Esquillor
s/n - 50018 Zaragoza
Spain

María José Gómez-Benito
University of Zaragoza
Aragón Institute of
Engineering Research (I3A)
Group of Structural
Mechanics and
Material Modelling
Betancourt Bldg.
María de Luna s/n
50018 Zaragoza
Spain

and

Centro de Investigación
Biomédica en Red en
Bioingeniería
Biomateriales y Nanomedicina
I+D+i Bldg. Mariano Esquillor
s/n - 50018 Zaragoza
Spain

Seyed Mohammad Hossein Hosseini
Technical University of Malaysia
Department of Applied
Mechanics
Faculty of Mechanical
Engineering
81310 UTM Skudai
Johor
Malaysia

Kuo-Kang Liu
University of Warwick
School of Engineering
Library Road
Coventry
CV4 7AL
UK

Yoshitaka Nakanishi
Kumamoto University
Department of Advanced
Mechanical Systems
Graduate School of
Science and Technology
2-39-1 Kurokami
Kumamoto 860-8555
Japan

Andreas Öchsner
Technical University of Malaysia
Faculty of Mechanical
Engineering
Department of Applied
Mechanics
81310 UTM Skudai
Johor
Malaysia

and

The University of Newcastle
University Centre for Mass and
Thermal Transport in
Engineering Materials
School of Engineering
Callaghan
2308 New South Wales
Australia

Michelle L. Oyen
University of Cambridge
Department of Engineering
Trumpington Street
Cambridge
CB2 1PZ
UK

María Ángeles Pérez
University of Zaragoza
Aragón Institute of
Engineering Research (I3A)
Group of Structural
Mechanics and Material
Modelling
Betancourt Bldg.
María de Luna s/n
50018 Zaragoza
Spain

and

Centro de Investigación
Biomédica en Red en
Bioingeniería
Biomateriales y Nanomedicina
I+D+i Bldg. Mariano Esquillor
s/n - 50018 Zaragoza
Spain

Carlos R. M. Roesler
Biomechanical Engineering
Laboratory
University Hospital
Federal University of
Santa Catarina
88040-900
Florianópolis
Brazil

Rui B. Ruben
ESTG, CDRSP
Polytechnic Institute of Leiria
P-2411-901 Leiria
Portugal

Duncan E. T. Shepherd
University of Birmingham
School of Mechanical
Engineering
Edgbaston
Birmingham
B15 2TT
UK

Humphrey Hak Ping Yiu
Heriot-Watt University
School of Engineering and
Physical Sciences
Chemical Engineering
Edinburgh
EH14 4AD
UK

Ryszard Wojnar
IPPT PAN
ul. Pawińskiego 5B 02-106
Warsaw
Poland

1
Bone and Cartilage – its Structure and Physical Properties

Ryszard Wojnar

1.1
Introduction

Here we describe the morphology and biology of bone, analyzing its components. The chapter is divided into two sections:

1) Macroscopic structure of bone
   a. Growth of bone
   b. Structure of body
   c. Structure of bone
2) Microscopic structure
   a. Osteons
   b. Bone innervations
   c. Bone cells
   d. OPG/RANK/RANKL signaling system
   e. Proteins and amino acids
   f. Collagen and its properties
   g. Polymer thermodynamics
   h. Architecture of biological fibers

All organs of the body are made up of four basic tissues: epithelia, connective tissue (CT), muscle tissue, and nervous tissue. Blood, cartilage, and bone are usually regarded as CTs. All tissues of an organism are subject to different stimuli, among others, to mechanical forces. These forces arise from various reasons, such as blood circulation, inertial forces created during motion, and gravity forces that act ceaselessly in normal conditions.

Bone is a specialized form of CT that serves as both a tissue and an organ within higher vertebrates. Its basic functions include mineral homeostasis, locomotion, and protection. Bone has cellular and extracellular parts. The extracellular matrix (ECM) of the bone comprises approximately 9/10 of its volume, with remaining 1/10 comprising cells and blood vessels. The ECM is composed of both organic and inorganic components.

Biomechanics of Hard Tissues: Modeling, Testing, and Materials.
Edited by Andreas Öchsner and Waqar Ahmed

ISBN: 978-3-527-32431-6

Greek philosopher and scientist Aristotle of Stageira maintained that "Nature, like a good householder, is not in the habit of throwing away anything from which it is possible to make anything useful. Now in a household the best part of the food that comes in is set apart for the free men, the inferior and the residue of the best for the slaves, and the worst is given to the animals that live with them. Just as the intellect acts thus in the outside world with a view to the growth of the persons concerned, so in the case of the embryo itself does Nature form from the purest material the flesh and the body of the other sense-organs, and from the residues thereof bones, sinews, hair, and also nails and hoofs and the like; hence these are last to assume their form, for they have to wait till the time when Nature has some residue to spare" [1].

About two-thirds of the weight of a bone, or half of its volume, comes from an inorganic material known as the *bone salt* that conforms, so to say, the bone to a nonliving world. It is an example of biomineralization: the process by which living organisms produce minerals. Owing to the inorganic architecture of the bone its biological properties may often be assumed as time independent, and the bone may be described by the methods of mathematics and mechanics developed for inanimate materials. However, by treating the bone as a live tissue, we observe that its biological activity is essentially directed toward keeping the whole organism in a state of well-being. The functionality of a bone is closely related to that of a cartilage tissue. In embryogenesis, a skeletal system is derived from a mesoderm, and chondrification (or chondrogenesis) is a process by which the cartilage is formed from a condensed mesenchymal tissue, which differentiates into chondrocytes and begins secreting the molecules that form an ECM. Cartilage is a dense CT and, along with collagen type 1, can be mineralized in the bone.

The high stiffness and toughness of biomineralized tissues of a bone are explained by the material deformation mechanisms at different levels of organization, from trabeculae and osteons at the micrometer level to the mineralized collagen fibrils at the nanometer length scale. Thus, inorganic crystals and organic molecules are intertwined in the complex composite of the bone material [2].

Bone, like every living tissue, cannot be described completely in terms of a nonanimate matter description. It breaks as a lifeless stick if overloaded, but if set up it recovers after some time. Under some loads, microcracks can appear; these are *ad hoc* healed, and the bone undergoes reinforcement. These properties of a bone are due to a complicated but coordinated structure, as it is seen during remodeling. Living bone could be treated as a solid-state fluid composite with circulating blood and living cells, while a bone skeleton has hierarchical structure and variable biomechanical properties. In addition, the blood flows through bones according to the rhythm of the heart beat.

Following Erwin Schrödinger, one sees that most physical laws on a large scale are due to stochasticity on a small scale ("order-from-disorder" principle). For example, the diffusion, in macroscopic description, is an ordered process, but in microscopic view it is caused by random movement of particles. If the number of atoms in the particle increases, the behavior of the system becomes less and less random. The life greatly depends on order and the master code of a living organism has to consist of a large number of atoms. The living organism seems to be a macroscopic system, the behavior of which approaches the purely mechanical (as contrasted with thermodynamical) conduct to which all systems tend, as the temperature approaches absolute zero and the molecular disorder is removed. The life is based on "order-from-order" principle. Schrödinger indicates that a periodic crystal is the material carrier of life, in contrast to periodic crystal of classical (inanimate) physics [3].

The asymmetry of living bodies was emphasized first by Louis Pasteur, who widely used examples with spiral structures. Upon examination of the minuscule crystals of sodium ammonium tartrate, Pasteur noticed that the crystals came in two asymmetric forms that were mirror images of one another (1849): solutions of one form rotated polarized light clockwise, while the other form rotated light counterclockwise. As a result, he devoted himself to the study of what he called *dissymmetry*, pointing out that inorganic substances are not dissymmetrical in their crystallization, while all the products of vegetable and animal life are dissymmetric. He concluded that there was some great biological principle underlying this: "All artificial products of the laboratory and all mineral species are superposable on their images. On the other hand, most natural organic products (I might even say all, if I were to name only those which play an essential part in the phenomena of vegetable and animal life), the essential products of life, are asymmetric and possess such asymmetry that they are not superposable on their images" [4, 5]. Geometry itself makes distinction between living and inanimate bodies, and this difference protects the autonomy of life. In the bone, both seemingly opposite substances, living and unliving, meet together and cooperate toward creating bone tissue.

This is the problem of three-dimensional growth through spiral forms, discussed in 1884 by Lord Kelvin [6], who earlier proposed (1873) the term *chirality*, cf. [7] and [8]. The repetition of such a way of development is observed from molecular level until macroscopic forms. In nature, helical structures arise when identical structural subunits combine sequentially, the orientational and translational relation between each unit and its predecessor remaining constant. A helical structure is thus generated by the repeated action of a screw transformation acting on a subunit. A plane hexagonal lattice wrapped around a cylinder provides a useful starting point for describing the helical conformations of protein molecules, investigating at the same time the geometrical properties of carbon nanotubes, and certain types of dense packings of equal spheres.

Essential for life proteins are organic compounds made of amino acids that are arranged in a linear chain and joined together by peptide bonds between the carboxyl and amino groups of adjacent amino-acid residues. The sequence of amino acids in a protein is defined by the sequence of genes that is encoded in the genetic code, which specifies 20 standard amino acids. The collagen is the main protein of CT in animals and the most abundant protein in mammals.

Among the substances, some are ungenerated and imperishable, while others partake in generation and perishing. Linus Pauling in 1970 indicated several attributes that distinguish a living organism from an inanimate object [9]. One is the ability to reproduce – the power of having progeny belonging to the same species, being sufficiently similar to the parental organism. Another attribute is the ability to ingest certain materials and subject them to own metabolism. Also a capacity to respond to environment, possibility of healing, as well as the memory, and the capacity to learn are typical for living organisms. The complicacy of chemical processes and the constancy of biological systems that we recognize so easily in the heredity phenomenon appear to contradict our intuitions.

The cell theory states that all living beings are composed of cells, which can be regarded as the basic units of life, and that cells come from other cells [10]. Botanist Matthias Jakob Schleiden (1804–1881) was a cofounder of the cell theory, along with Theodor Schwann and Rudolf Virchow. The growth of an organism is effected by consecutive cell divisions (such a cell division is called *mitosis*). In biology, another division process, namely, meiosis is dealt with. Meiosis is a process of reductional division in which the number of chromosomes per cell is halved. In animals, meiosis always results in the formation of gametes, while in other organisms it can give rise to spores.

Many important substances in cells and tissues occur as thin, of the order of 20 Å, and highly elongate particles. (1 angström $\equiv 1$ Å $= 0.1$ nm $= 10^{-10}$ m.) Proteins (such as myosin, collagen, and nerve-axon protein), nucleic acids (DNA and RNA), and polysaccharides (cellulose and hyaluronic acid) are examples. Many of these substances are themselves polymers (as the protein macromolecules are polymers of many amino-acid residues) but, as monomers, these elongated macromolecules polymerize end-to-end and laterally to form fibrous structures. Schmitt *et al.* [12] suggested that tropocollagen (TC), the macromolecule of collagen, has dimensions of about 14×2800 Å [11–15].

The nucleic acids can be longer. DNA polymers contain millions of repeating units called *nucleotides*. One nucleotide unit is 3.3 Å (0.33 nm) long. Two nucleotides on opposite complementary DNA (or RNA) strands that are connected *via* hydrogen bonds are called a *base pair* (bp). In DNA, adenine forms a base pair with thymine, as does guanine with cytosine, and the DNA chain is 22–26 Å wide. The largest human chromosome, chromosome number 1, is approximately 220 million bp long. This gives a length of $0.33 \times 10^{-9} \times 220 \times 10^{6} = 72.6$ mm [16–18].

In Figures 1.1 and 1.2, one sees results of interactive computer animation – simulation of the famous Bragg–Nye bubble raft experiment. Two-dimensional crystallization is realized by equalizing distribution of atoms on torus, not only globally as enforced by Descartes–Euler Law, but also locally. Consecutive iterations repel atoms and shift them to centroids of Voronoi polygons. In consequence, the fraction of pentagon–heptagon pairs of defects (disclinations, curvatures, vortices, . . .) among prevailing crystalline hexagons is gradually diminishing. A progressive coalescence and coarsening of crystal grains are occurring during rotations and growth of circular inclusions while five to seven edge dislocations align into migrating and rearranging grain boundaries [19]. In such systems, a local order cannot propagate throughout space. Contradiction

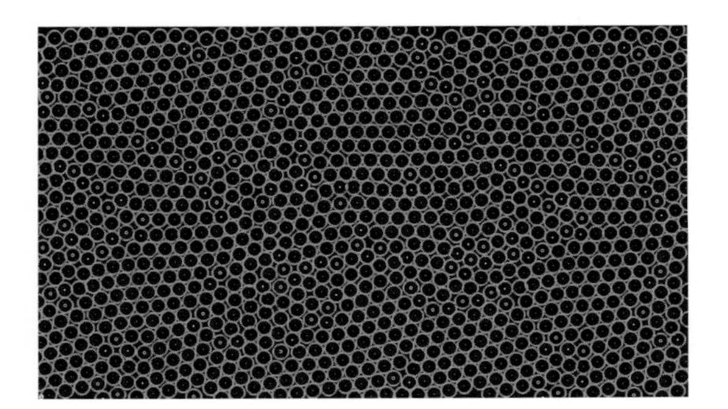

Figure 1.1 Frustrated matter – a prototype of biological medium: numerical simulation of Bragg–Nye bubble raft experiment [21], with Centroidal Voronoi network. Experiment produced image of a bubble raft showing vacancies and five to seven edge dislocations. Disclination 5 (with five neighbors) is denoted by white point, while disclination 7 (with seven neighbors) is denoted by white circle. Courtesy of Andrzej Lissowski.

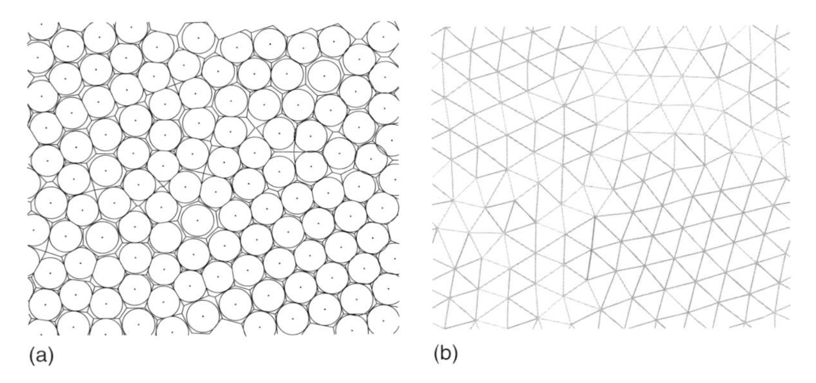

Figure 1.2 Pentagon–heptagon dipoles of dislocations in hexagonal lattice. (a) Numerical simulation of Bragg–Nye bubble raft experiment with Centroidal Voronoi network. (b) The same experiment in dual (triangular) representation. Courtesy of Andrzej Lissowski.

between local and global configurations is known as *geometrical frustration* [20].

From microorganisms to man, from the smallest cell organelle to the human brain, life presents us with examples of highly ordered cellular matter, precisely organized and shaped to perform coordinated functions [22].

A topological distribution of approximately hexagonal lattice of osteocytes in osteon and osteons in a compact bone is observed (cf. Figure 1.16). Creation of one new osteon leads to disturbance of the lattice and is equivalent to the lattice defects – two pentagon–heptagon (5–7) dislocations – analogous to those observed in crystal lattice, cf. Figure 1.1.

The problem of cell, in particular bone cell, communication is important. Paracrine signaling is a form of cell signaling in which the target cell is near ("para" = near) the signal-releasing cell. Two neurons would be an example of a paracrine signal. Only neighbor–neighbor interactions of osteocytes would permit them to create an image of the bone state. This would be in the spirit of the research by David G. Kendall, who in the 1970s discussed recovery of a structure from fragmentary information [23, 24]. Such a behavior was observed in a colony of bacteria or in another collective response to external disturbances by Howard Bloom in 2007, cf. [25]. The bone marrow, distributed in many parts, acts and reacts as one organ [26].

A quantitative experimentation is still needed for understanding the behavior of regulatory actions that manifest themselves in many biological systems, explaining principles of cellular information processing, and to advance predictive modeling of cellular regulation.

The body is made up of cells that communicate with each other and with external cues *via* receptors at their surfaces. To generate cellular responses, signaling pathways are activated, which initiate movement of proteins to specific locations inside the cells, notably the nucleus, where DNA is situated.

Enzymes called *kinases* are widely used to transmit signals and control processes in cells. One particular pathway is the extracellular signal-regulated kinase (ERK) pathway. ERKs act as messenger molecules by relaying signals that are received from outside the cell to the administrative core, the nucleus. To do so, ERK must move from the intracellular fluid to the nucleus of the cell and turn on several genes while turning off others, which in turn finally makes the cell to divide or differentiate. ERK's entry in the nucleus is unconventional, because the protein lacks the ability to bind to the known nuclear import proteins. Recently, Lidke and her colleagues showed that protein pairing – known as *dimer formation* – is not necessary for ERK to move into the nucleus after all. Instead, the process was found to be dependent solely on the rate at which stimuli activate the ERK. A delay in activation triggers a delay in nuclear entry of ERK, indicating that ERK entry in the nucleus is a direct consequence of activation [27].

Shekhter has applied the systems approach to the analysis of mechanisms whereby CT is integrated into one functional system. The primary CT functions in health and in disease (biomechanical, trophic, protective, reparative, and morphogenetic) are carried out by means of cell–cell, cell–matrix, and intertissue interactions based on feedback between these components. Both CT as a whole and its cellular and extracellular components exhibit structural and functional heterogeneity, which increases the capacity of CT for adaptation. Shekhter's study supports the concept of internal network regulation of the CT composition, functions, and growth through intercellular interactions at different levels of structure [28].

There is another analogy in which a bone structure resembles the mechanical structure of a tree trunk. In plants, the carbohydrate cellulose is the most important constituent of the wall cells, while in bones the hard skeleton in the exterior of bone cells is created by osteocytes. The resulting mechanical effect – a strong sponge-like structure – is similar. In biology, convergence of analogous (but without apparent common origin) structures is known as *homoplasy*.

1.1.1 The Structure of Living Organisms

The structure of the organism is realized by tissues. Malpighi (1628–1694), physician and biologist, studied subdivisions of the bone, liver, brain, spleen, kidneys, and skin layers, concluding that even the largest organs are composed of minute glands [29]. In 1835, before the final cell theory (which regards cells as the basic unit of life) was developed, Jan Evangelista Purkyně observed small "granules" while looking at the plant tissue through a microscope [30, 31].

As remarked by Pauling, chemical investigation of the plant viruses has shown that they consist of the materials called *proteins and nucleic acids.* Molecular weight of the enzyme urease is 483 000. The viruses (looked at sometimes as giant molecules) with a molecular weight of the order of magnitude of 10 000 000 may be described as aggregates of smaller molecules [9]. Viruses vary from simple helical and icosahedral shapes to more complex structures. Most viruses are about 100 times smaller than an average bacterium. Tobacco mosaic virus (TMV) is a rod-shaped virus of length 3000 Å, diameter 150 Å, and molecular weight 50 million. Franklin has shown that the TMV protein is in the form of structural subunits of molecular weight about 29 000 that are arranged on a helix of pitch 23 Å and the axial repeat period 69 Å [32, 33].

Many microorganisms, such as molds, bacteria, protozoa, consist of single cells, cf. Figure 1.3. These cells may just be big enough to be seen with an ordinary microscope, having diameter around 10 000 Å ($=1000$ nm $= 1$ μm $= 1 \times 10^{-6}$ m), or they may be much bigger – as large as a millimeter or more in diameter. For comparison, atomic diameters range between 1 and 2 Å. The cells have a structure, consisting of a cell wall, a few hundred angströms in thickness, within which is enclosed a semifluid material called *cytoplasm*, and other components. Other plants and animals consist largely of tissues – aggregates of cells, which may be of many different kinds in one organism. The muscles, blood vessels and lymph vessel walls, tendons, CTs, nerves, skin, and other parts of the body of a man consist of cells attached to each other to constitute a well-defined structure. There are also cells that are not attached to this structure, but float around in the body fluids. Most numerous among these cells are the red corpuscles of the blood. The red corpuscles in man are flattened disks, about 7500 nm in diameter and 2000 nm thick. There are about 5 million red cells per cubic millimeter of blood, and a man contains about 5 l of blood. Some cells are smaller, like the red cells, and some larger – single nerve cell may be about 1 μm in diameter and 100 cm long – extending from the toe to the spinal cord. A typical cell size is 10 μm, and a typical cell mass is 1 ng. The total number of cells in the adult human body is about 5×10^{14} [9]. Groups of cells combine and form tissue, which combines to form organs, which work together to form organ systems. The study of tissues is known as *histology*.

Some foreign cells, often nocive, also can dwell in bone. Osteomyelitis (osteo- from the Greek word osteon, meaning bone, myelo- meaning marrow, and its meaning is inflammation) means an infection of the bone or bone marrow. The infection is often caused by bacteria called *Staphylococcus aureus*, a member of the

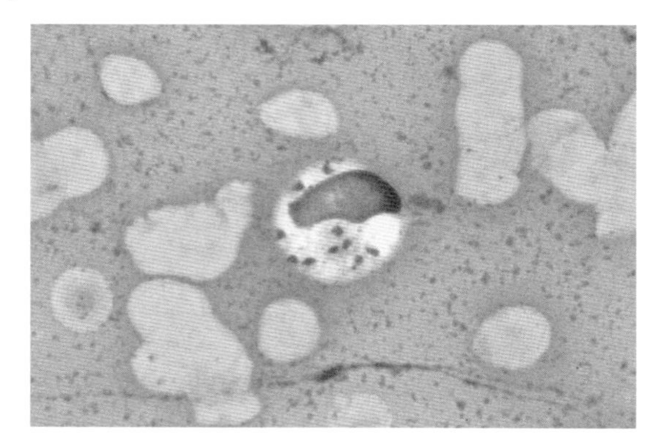

Figure 1.3 A protozoan *Leishmania donovani* in a bone marrow cell. Leishmania is a genus of parasites that are the etiologic agents of diseases of humans, such as leishmaniasis. After [34].

normal flora found on the skin and mucous membranes. In children, osteomyelitis usually affects the long bones of the arms and legs. Osteomyelitis often requires prolonged antibiotic therapy, and may require surgical debridement. Severe cases may lead to the loss of a limb [35, 36].

Leishmania is a genus of trypanosome protozoa and is the parasite responsible for the disease leishmaniasis, Figure 1.3. It is spread through sandflies. *Leishmania* commonly infects vertebrates: hyraxes, canids, rodents, and humans [34].

The body also contains the body fluids such as blood and lymph, as well as fluids that are secreted by other organs [9]. The bones are laid down as excretions of bone-making cells, called *osteocytes*. The quantitative proportion of cells in the mass of cartilage and bone is very low, while the major part is taken by extracellular substances. In the hollow interior of bones, bone marrow tissue is found. In adults, marrow in large bones produces new blood cells. It constitutes 4% of total body weight. The bone marrow is composed of stroma and parenchyma parts. Hematopoiesis is performed by parenchymal cells, while the stroma provides hematopoietic microenvironment.

Interaction of huge biomolecules with precision and infallibility is the essence of the living state. Looking at cell metabolism in relation to the intricate structure of a cell Peters (1930) stated that in the cell "extreme order has to be reconciled with a fluid anatomy (. . .). The cell must be considered as a reflex entity, structurally organized so far as even its chemistry is concerned, with chains of chemical substances acting as it were as reflex arcs (. . .). It is perfectly possible to appreciate how a coordinated structure may be maintained in a medium which is apparently liquid. This theory is all, that is, needed to enable us to understand how substances can reach a special site in the cell. Between the chains of molecules, fixed by their radiating webs, there will exist paths from the external to the internal surface, the capillaries of the cell" [37].

Wheatley observes that "as many as 4000 reactions may be occurring simultaneously in a cell (...) and every one has to be harmoniously controlled. There is no factory on earth that comes anywhere near this complexity and, at the same time, gives the fidelity or replicative performances while remaining flexible and adaptable to its environment" [38].

1.1.2
Growth of Living Organisms

It is supposed from the works of Braun, Schimper, brothers Bravais, Schwendener, Wulff, and Lewis that the crystallization and growth of a living tissue are similar [39–46]. The dislocations' gliding and climbing is the basis for such similarity. In two-dimensional packing, it is realized by the motion of pentagons and heptagons (five to seven) among crystalline hexagons.

One of the most striking aspects of symmetry in plants is in phyllotaxis – the arrangement of leaves on a stem or of flowers in the inflorescences. It is an interdisciplinary study involving mathematics, botany, and crystallography among others. The phyllotaxis should be properly studied at the shoot apical meristem (SAM). It is at the meristemic apex that the organs of shoot such as primordia of leaves, buds, or flowers originate, cf. also [47]. A *primordium*, in embryology, is defined as an organ or tissue in its earliest recognizable stage of development.

Biological systems are the best prototypes of genuine smart structures. A unique example is provided by considering the important and mysterious phenomenon of spiral phyllotaxis – leaf primordia packing with Fibonacci differences between nearest neighbors, Figure 1.4. For the Fibonacci spiral, in polar coordinates r and ϕ, the nth primordium has the position

$$r_n = A\sqrt{n}, \quad \varphi_n = C$$

where A is a constant, $C = 360^\circ \cdot u \approx 222.5^\circ$, and $u = (\sqrt{5} - 1)/2$, or, what is equivalent for structural form $C = 360^\circ/(2 + u) \approx 137.5^\circ$. The golden symmetry ratio u is typical for quasicrystals, and for the icosahedron dimensions.

As they grow, older primordia are displaced radially away from the center of the circular meristem. The newest primordium initiates in the least crowded space at the edge of the meristem. The growth process is accomplished in an exceptional order. Phyllotaxis compromises local interactions giving rise to long range order and assures the best way of optimal close packing.

Meristems are classified according to their location in the plant as apical (located at the root and shoot tips), lateral (in the vascular and cork cambia), and intercalary (at internodes, or stem regions between the places at which leaves attach, and leaf bases). Lateral meristems, found in all woody plants and in some herbaceous ones, consist of the vascular cambium and the cork cambium. They produce secondary tissues from a ring of vascular cambium in stems and roots. The lateral meristems surround the stem of a plant and cause it to grow laterally, cf. Figure 1.5. Nature uses the same pattern to place seeds on a seedhead, to arrange petals around the edge of a flower, and to place leaves around a stem.

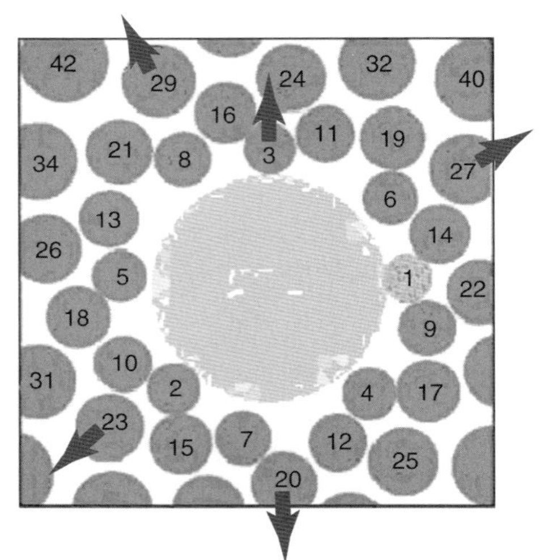

Figure 1.4 Norway Spruce (*Picea abies*): spiral phyllotaxis of needle primordia emerging from central SAM. The primordial numbers are in reverse order of their appearance. The newest primordium initiates at the periphery of the meristem where there is the largest free space. As they grow, older primordia are displaced radially away from the center of the circular meristem. Then, the older the primordium, the farther it is from the center. For example, the contact on the parastichy spiral with difference 5 (between 8 and 13 spirals) changes after 5–7 flip to the contact on the spiral with difference 21. After [48]. With kind permission of the author Pau Atela.

1.1.2.1 **Ring-Shaped Grain Boundary**

D'Arcy Thompson emphasized the deep correlation between mathematical statements, physical laws, and fundamental phenomena of organic growth of biological structures. At the end of "On Growth and Form" we read: " ... something of the use and beauty of mathematics I think I am able to understand. I know that in the study of material things number, order, and position are the threefold clue to exact knowledge" [49].

Occurrence of defects begins a process of destruction of a crystal. Destruction is therefore necessary for the crystal growth, as shown by Rivier and Lissowski [50]. In a similar manner, appearance of one new osteon leads osteon lattice defects to arise – a pair of pentagon–heptagon (five to seven) dislocations. Hence, the bone turnover is realized by the defects of osteonic structure, cf. Figures 1.15 and 1.16 of compact bone structure.

In 1868, from his microscopic study of plant meristems, botanist Hofmeister [51] proposed that a new primordium always forms in the least crowded spot along the meristem ring, at the periphery of SAM.

In a manner analogous to the propagation of defects during crystallization, the growth of a tissue stress leads to buckling and undulation down to the order of the cell diameter. Structural control in tissue development is accomplished by wavelike 5–7 dislocation rearrangement. The oriented cell divisions as 5–7 climbing can

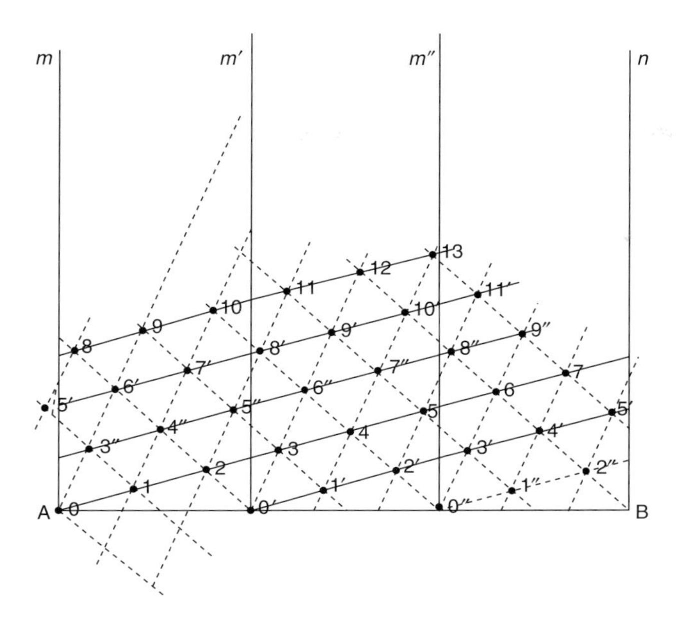

Figure 1.5 Spiral distribution of primordia around the cylindrical stem, after development on a plane, as observed by Bravais in 1837. The neighboring primordia differ by Fibonacci numbers: 1 along the solid lines, 3 along the dashed lines. After [42].

be explained in a similar manner. A vortex interpretation of the 5–7 pair in SAM growth was given in [52].

In a trial to explain the phyllotaxis pattern, Newell, Shipman, and Sun assume that the auxin-produced growth is proportional, in a first approximation, to how much average tensile stress the local elemental volume (which will contain many cells) feels. This is best measured by the trace of the stress tensor at that location. Fluctuations in auxin concentration influence the mechanical forces in the tunica by creating uneven growth and are manifested by an additional strain contribution in the stress–strain relationships. On the other hand, inhomogeneities in the stress distribution are assumed to lead to changes in auxin concentration. The exact way in which stresses influence biological tissue growth (weight-bearing bones and fruit stems become stronger) is still an open challenge to biologists [53].

Gebhardt in 1911 found a similarity between bone formation and the chemistry of colloids [54, 55]. He found that collagen fibrils are organized into distinct lamellae, the molecular orientations being parallel in each lamella. Much later, in 1988, it was pointed out by Giraud-Guille that together with normal (i.e., Gebhardt's) plywood architecture, a twisted plywood distribution of collagen fibrils in human compact bone osteons is observed, comparable with a liquid crystalline self-assembly [56]. Geometrical resemblance of long fibrils and long molecules leads to a similar arrangement of these objects.

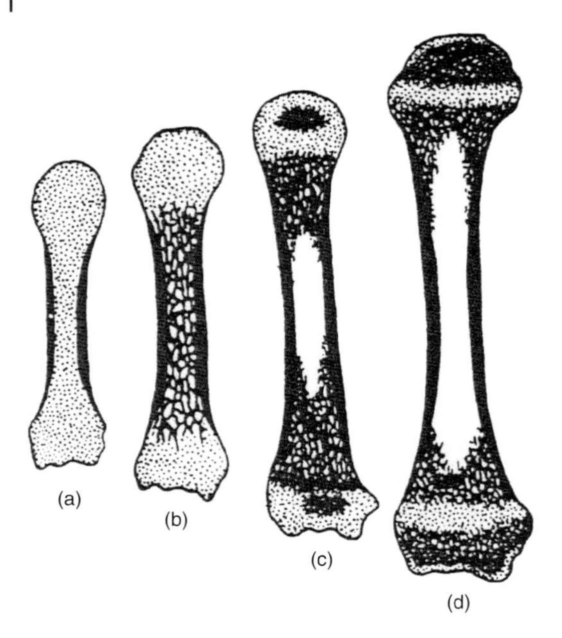

Figure 1.6 Ossification of a long bone: (a) hyaline cartilage model with surface ossification only; (b) development of trabecular bone network; (c) development of medullary cavity; (d) after achievement of bone maturity, only articular cartilage and epiphyseal plate are left from the cartilage tissue, while the medullary cavity is enlarged. Cartilage part is denoted by dots and the bone part is denoted by black color. After [66].

Remarkable symmetry characterizes crystals and organic forms, because both are subdued primarily to the topological laws of close packing by the Descartes–Euler theorem.

1.1.3
Planarity of Biological Structures

The development of living world is accomplished in two-dimensional structures. Such structures are easily accessible to environment and external influence. But it seems that the fundamental reason is geometrical topology. The Descartes–Euler theorem on polyhedra in three-dimensional space assures that, for a polyhedron, the following relation among the number of vertices N_V, number of edges N_E, and number of faces N_F is satisfied $N_V - N_E + N_F = 2$.

In two dimensions, this relation becomes more sharp, as only two independent numbers are left. Therefore, all processes with phase transformation are accomplished more easily in two dimensions. The Descartes–Euler Law forces the compensation of positive and negative curvatures in planar tissue: $3 \times p3 + 2 \times p4 + p5 = p7 + 2 \times p8 + 3 \times p9 +$ sum of $(N - 6) \times pN$ for $N > 9$ where pN are percentages of N-sided cells and each cell gives 6-N units of curvature.

The predominantly hexagonal cell pattern of simple epithelia was noted in the earliest microscopic analyses of animal tissues: a topology commonly thought to reflect cell sorting into optimally packed honeycomb arrays. The development of specific packing geometries is tightly controlled. For example, in the *Drosophila* wing epithelium, cells convert from an irregular to a hexagonal array shortly before hair formation. Packing geometry is determined by developmental mechanisms that likely control the biophysical properties of cells and their interactions, cf. [57–60].

1.2 Macroscopic Structure of the Bone

Bone tissue (or osseous tissue) is the main structural and supportive tissue of the body. The basic elements of the bone tissue are cells (osteocytes, osteoblasts, osteoclasts) and extracellular substance (ECM). The bone matrix consists of an organic part (collagen of type I and other proteins) and a nonorganic one (mainly hydroxyapatite).

CT consists of cells and extracellular materials secreted by some of those cells. Thus, the cells in CT may be separated from one another within the ECM. The ECM consists of ground substance and fibers. In many types of CTs, the matrix-secreting cells are called *fibroblasts*. Frequently, other cell types (e.g., macrophages, mast cells, and lymphoid cells) may also be present. Ground substance is a term for the noncellular components of ECM containing the fibers. This substance is gel like, amorphous, and is primarily composed of glycosaminoglycans (mostly hyaluronan), proteoglycans, and glycoproteins. CT is the most diverse of the four tissue types and fulfills different functions. Its consistency ranges from the gel-like softness of areolar CT to the hardness of bone.

Cells are surrounded by ECM in tissues, which acts as a support for the cells. Ground substance does not include collagen but does include all the other proteinaceous components, such as proteoglycans, matrix proteins, and, most prevalent, water. The noncollagenous components of ECM vary depending on the tissue, cf. monographs by Ross *et al.* [61, 62] and Gray's anatomy [63, 64].

1.2.1 Growth of the Bone

The development of all large complex animals and human beings is accomplished in a two-dimensional, layered way. The mesoderm germ layer is formed in the embryos of triploblastic animals. During gastrulation, some of the cells migrating inward contribute to the mesoderm (middle layer), an additional layer between the endoderm and the ectoderm. From the mesoderm, skeletal muscle, the skeleton, the dermis of skin, CT, the urogenital system, the heart, blood (lymph cells), and the spleen are formed [63].

There are two different methods of the ossification process: intramembranous ossification is bone formation from an organic matrix membrane, whereas endochondral ossification occurs within a cartilaginous model. However, there is only one mechanism of bone formation: the laying down of the osteoid matrix by osteoblasts, followed by the deposition of crystalline apatite [65].

In embryogenesis, the skeletal system is derived from the mesoderm germ layer. Chondrification (or chondrogenesis) is the process by which cartilage is formed from the condensed mesenchymal tissue, which differentiates into chondrocytes and begins secreting the molecules that form the ECM. Early in the fetal development, the greater part of the skeleton is cartilaginous. It is the *temporary* cartilage that is gradually replaced by the bone (endochondral ossification), a process that ends at puberty. The cartilage in the joints is *permanent* – it remains unossified during the whole of life.

During the fetal stage of development the bone can be formed by two processes: intramembranous or endochondral ossification. Intramembranous ossification mainly occurs during the formation of the flat bones of the skull; the bone is then formed from the mesenchymal tissue.

Endochondral (intracartilaginous) ossification occurs in long bones. In this process, the bone is formed from cartilage, which is gradually replaced by the bone as the embryo grows. The steps of endochondral ossification are visible in Figure 1.6.

Adult hyaline articular cartilage is progressively mineralized at the junction between cartilage and bone. It is termed *articular calcified cartilage*. A mineralization front advances through the base of the hyaline articular cartilage at a rate dependent on cartilage load and shear stress. Adult articular calcified cartilage is penetrated by vascular buds, and the new bone produced in the vascular space in a process similar to endochondral ossification at the physis. A cement line separates the articular calcified cartilage from the subchondral bone.

Both the bone and the cartilage are classified as supportive CT.

- Bone (osseous tissue) makes up the skeleton in adult vertebrates.
- Cartilage makes up the skeleton in chondrichthyes (known also as *cartilaginous fishes*). In most other adult vertebrates, the cartilage is primarily found in joints, where it provides bearing and cushioning.

Bone-forming cells called *osteoblasts* deposit a matrix of collagen, but they also release calcium, magnesium, and phosphate ions, which chemically combine and harden within the matrix into the mineral hydroxyapatite. The combination of hard mineral and flexible collagen makes the bone harder than cartilage without being brittle.

Bone marrow can be found in almost any bone that holds cancellous tissue. In newborns, all such bones are filled with red marrow only, but as the child ages it is mostly replaced by yellow, or fatty, marrow. In adults, red marrow is mostly found in the flat bones of the skull, the ribs, the vertebrae, and pelvic bones, cf. [67–73].

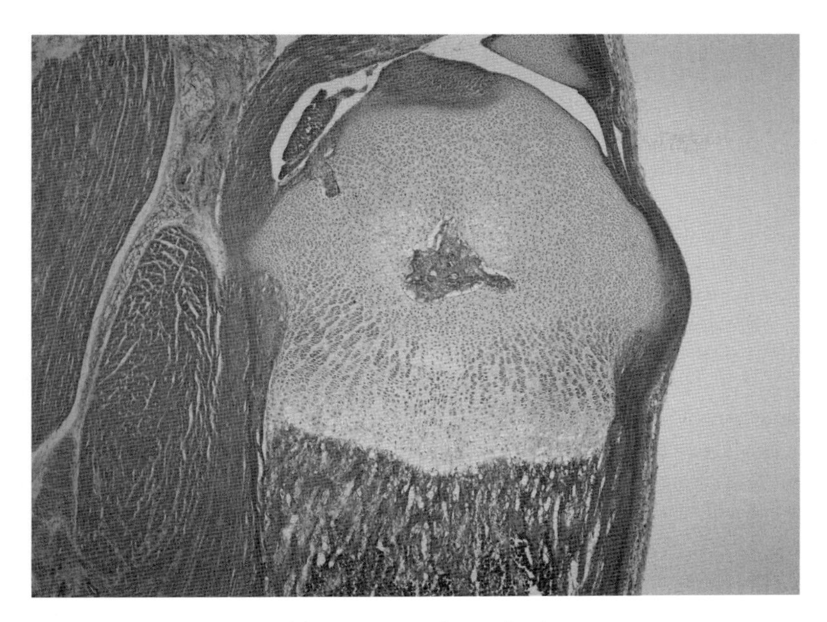

Figure 1.7 Cross section of the young rat tibia in development. The trabecular bone network with medullary cavity in the center is exposed. To the left, a smaller cross section of fibula is seen. Courtesy of Litwin and Gajda.

Endochondral ossification begins with points in the cartilage called "*primary ossification centers.*" They mostly appear during fetal development, though a few short bones begin their primary ossification after birth. They are responsible for the formation of the diaphyses of long bones, short bones, and certain parts of irregular bones. Secondary ossification occurs after birth, and forms the epiphyses of long bones and the extremities of irregular and flat bones. The diaphysis and both epiphyses of a long bone are separated by a growing zone of cartilage (the epiphyseal plate), cf. Figures 1.6 and 1.7.

Epiphyseal plates (growth plates) are located in the metaphysis and are responsible for growth in the length of the bone, cf. Figure 1.10. Because of their rich blood supply, metaphysis of long bones are prone to hematogenous spread of *Osteomyelitis* in children.

When the child reaches skeletal maturity, all of the cartilage is replaced by the bone, fusing the diaphysis and both epiphyses together (epiphyseal closure).

Exterior shape of the bone is characteristic of every species and is revealed by different roughnesses, spikes, spicules, openings, and holes; it is an effect of modulating influence from the side of the soft components of the organism. This *paradoxal* observation is explained by the fact that bones develop relatively late when soft parts are formed, and the growing bone has to match its form to the shape of soft components.

1.2.2
Structure of the Body

The bones of vertebrates compose the internal skeleton of these organisms. Bones are divisible into four classes: long, short, flat, and irregular. The number of bones in the organism is variable and depends on the age.

There are 206 bones in the adult human body and about 270 in an infant. A human adult skeleton consists of the following distinct bones: skull (22): cranium (8), face (14); spine and vertebral column (26), hyoid bone, sternum and ribs (26), upper extremities (64), lower extremities (62), and auditory ossicles (6). The patellae are included in this enumeration, but the smaller sesamoid bones are not taken into account, cf. [66, 71–73, 77].

In particular, the metatarsal bones are a group of five long bones in the foot that are located between the tarsal bones of the hind- and mid-foot and the phalanges of the toes, see Figure 1.8. The metatarsal bones are numbered from the medial side (side of the big toe): the first, second, third, fourth, and fifth metatarsal. The metatarsals are analogous to the metacarpal bones of the hand. In human anatomy, the metacarpus is the intermediate part of the hand skeleton that is located between the phalanges (bones of the fingers) distally and the carpus, which forms the connection to the forearm.

The bone fulfills three essential roles in the organism:

- Mechanical (constructional) – being a scaffold of the body and being responsible together with skeletal muscles for the movement and locomotion of the organism;
- protective – shielding internal organs against external hurts;
- metabolic – hematopoietic processes of blood production by red and yellow marrow, within the medullary cavity of long bones and interstices of cancellous bone, storage of fat as yellow bone marrow, storage of minerals such as calcium and phosphorus, assuring acid–base balance by absorbing or releasing alkaline salts, necessary for holding the ionic homeostasis in the organism.

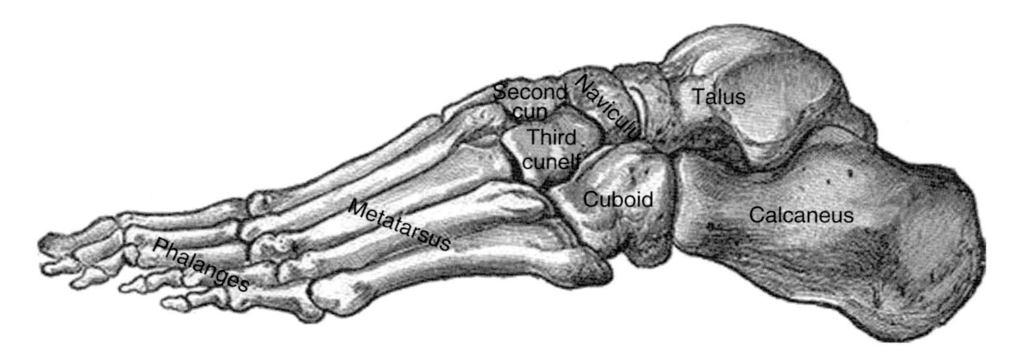

Figure 1.8 Skeleton of human foot with metatarsus, in lateral aspect. After [63].

The skeleton supports soft parts of the body, enables the movements of body and its members, and protects its internal organs (e.g., the skull protects the brain and the ribs protect the heart and lungs). Bones together with tendons, ligaments, joints, and skeletal muscles, steered by the nervous system, create forces and motion of the entire body or its parts only. Another mechanical function of the bone is linked to sound transduction in the ear and hearing. These mechanical functions are the subject of study in biomechanics.

The bone owes its hardness to the osseous tissue, which can be regarded as a composite material (mineralized organic matrix) formed of a mineral – hydroxyapatite and a protein – collagen. The living cells are embedded in the osseous tissue.

Bones are organs made up of bone tissue as well as marrow, blood vessels, epithelium, and nerves, while the term *bone tissue* specifically refers to the mineral matrix that forms the rigid sections of the organ.

1.2.3
Macroscopic Structure of Skeleton

The bones of the skeleton are joined to one another at different parts of their surfaces, and such connections are termed *joints* or *articulations*. It happens that the joints are *immovable*, as in the articulations between practically all the bones of the skull; the adjacent margins of the bones are almost in contact, being separated merely by a thin layer of fibrous membrane (sutural ligament). In certain regions at the base of the skull, this fibrous membrane is replaced by a layer of cartilage. In the *freely movable* joints, the surfaces are completely separated; the bones forming the articulation are expanded for greater convenience of mutual connection, covered by cartilage and enveloped by capsules of fibrous tissue. The cells lining the interior of the fibrous capsule form an imperfect membrane – the synovial membrane – which secretes a lubricating fluid. The joints are strengthened by strong fibrous bands called *ligaments*, which extend between the bones forming the joint.

Bones constitute the main elements of all the joints. In the long bones, the extremities are the parts that form the articulations; they are built of spongy cancellous tissues with a thin coating of compact substance. The layer of compact bone that forms the joint surface, and to which the articular cartilage is attached, is called the *articular lamella*. It differs from ordinary bone tissue in that it contains no Haversian canals, and its lacunae are larger and have no canaliculi. The vessels of the cancellous tissue, as they approach the articular lamella, turn back in loops, and do not perforate it; this layer is consequently denser and firmer than ordinary bone, and forms an unyielding support for the articular cartilage.

Cartilage is a nonvascular structure found in various parts of the body: in adult life, chiefly in the joints, in the parietes of the thorax, and in various tubes, such as the trachea, bronchi, nose, and ears, which require to be kept permanently open [63–73, 77].

1.2.4
Apatite in the Bone

The bone tissue is a mineralized CT. The bones consist of inorganic constituents, calcium hydroxyphosphate, $Ca_5(PO_4)_3OH$, also known as the *mineral apatite* (or hydroxyapatite, abbreviated sometimes as HA or HAP), and calcium carbonate $CaCO_3$, and an organic constituent, collagen, which is a protein. Nature has evolved sophisticated strategies for developing hard tissues through the interaction of cells and, ultimately, proteins with inorganic mineral phases.

Hydroxyapatite (or hydroxylapatite – according to the nomenclature accepted by the International Mineralogical Association) is a naturally occurring form of calcium apatite with the formula $Ca_{10}(PO_4)_6(OH)_2$ (it is written in this form to denote that the crystal unit cell comprises two entities). We have $Ca_{10}(PO_4)_6(OH)_2 \leftrightarrow 10Ca^{2+} + 6PO_4^{3-} + 2OH^-$. It has relatively high compressive strength but low tensile strength of the order of 100 MPa. It has a specific gravity of 3.08 and is 5 on Mohs hardness scale. It crystallizes in the hexagonal system.

Pure hydroxylapatite powder is white. Naturally occurring apatites can, however, also have brown, yellow, or green colorations, comparable to the discolorations of dental fluorosis. It is estimated that a modified form of the inorganic mineral hydroxylapatite (known as *bone mineral*) accounts for about 50% of the dry weight of bone.

A calcium phosphate mineral found in the bone is similar in composition and structure to minerals within the apatite group. It belongs to biominerals – minerals produced by living organisms. Apatites are widely distributed as accessory minerals in different rocks and are important for the study of geological thermal history [78–81].

Apatites have the general formula $Ca_{10}(PO_4)_6\,X_2$ where X denotes F (fluorapatite, abbreviated as FAp), OH (hydroxyapatite, abbreviated OHAp), or Cl (chlorapatite, ClAp). The apatite lattice is tolerant of substitutions, vacancies, and solid solutions; for example, X can be replaced by $1/2CO_3$ or $1/2O$; Ca by Sr, Ba, Pb, Na, or vacancies; and PO_4 by HPO_4, AsO_4, VO_4, SiO_4, or CO_3.

The mineral of bones and teeth is an impure form of OHAp, the major departures in composition being a variable Ca/P mol ratio (1.6–1.7, OHAp is 1.66), and a few percent CO_3 and water. The mineral is microcrystalline. The crystals are approximately 15 nm wide by 40 nm long in bone and dentine, and 40 nm wide by 100 nm to 5 μm or more long in dental enamel. They are much thinner compared to their width. The mineral in the bone comprises crystals that are smaller than those in dental enamel, so that many of the constituent ions occupy surface, or near-surface, positions. The result is that there are greater uncertainties about the crystal structure of bone mineral, compared with that of dental enamel. Apatite OHAp is also used as a biomaterial, for bone replacement, and for coating metal prostheses to improve their biocompatibility. The osseous tissue without collagen would be hard and brittle, and its fairly large elasticity is contributed by collagen. Biological apatites present in the natural bone, dentin, and enamel contain different amounts of carbonate: 7.4, 5.6, and 3.5 wt % (weight percent), respectively [78–80].

In 1945, Beevers and McIntyre, as a result of X-ray crystal analysis, discovered the (nano)porous structure of hydroxyapatite, recognized then as an essential mineral constituent of bone and of the enamel and dentin of teeth [81]. They have shown that the unit cell of the apatite structure has two equal edges inclined at 120° to one another. These edges are of length 9.37 Å in the case of fluorapatite and 9.41 Å in the case of hydroxyapatite. The third edge is at right angles to these and has a length of 6.88 Å in both fluor- and hydroxyapatites. They also indicated the three important properties of apatite structure: (i) apatites have a tunnel structure with walls composed of corner-connected CaO_6 and PO_4 polyhedra as relatively invariant units; (ii) filling of these tunnels by Ca and anions (OH, F) leads to adjustments that best satisfy bond-length requirements; and (iii) even slight changes in the ionic radii of the tunnel atoms lead to expansion or contraction of the tunnel. It was proposed that the "very critical fit" of the fluorine and hydroxyl ions was responsible for the greater stability of fluorapatite, consistent with the fact that bone could take up fluorine selectively even from dilute solutions.

The size of the hexagonal channels is mainly determined by the calcium and phosphate arrangement. Another feature is the planar arrangement of three Ca atoms around each F in fluorapatite or OH in hydroxyapatite. From these two features, it results that the structure is selective in its choice of ions to occupy the position of the F ions in fluorapatite.

The only ions known to occupy these positions are the ions F^- and OH^-, and these two are nearly of the same size. Each has two K and eight L electrons, but F^- has one nucleus with charge +9, while OH^- possesses two nuclear charges +8 and +1, respectively. This makes OH^- just a little larger than F^-. The hydroxyapatite structure is a little expanded as compared with the fluorapatite one.

The critical fit of the F or OH ion is responsible for the difference in the stability of the two apatites. The fluorapatite is more stable. It is shown by the well-known facts that the fossil bone becomes gradually transformed from hydroxyapatite to fluorapatite, and that bone will take up fluorine selectively even from media very dilute in fluorine. The sensitivity of the macroscopic structure of teeth to fluorine and the relation between the incidence of dental caries and fluorine content are the other observations that have practical implications.

As is seen in Figure 1.9, there are four different types of crystallographic positions in the apatitic unit cell: (i) tetrahedral sites for six P^{5+} ions, each in fourfold coordination with oxygen, (ii) Ca [1] sites for four of the Ca^{2+} ions, (iii) Ca [2] sites for the six other Ca^{2+} ions (arranged in such a way that they form a channel along the *c*-axis, the so-called anion-channel), and (iv) the channel site, which is typically occupied by two monovalent anions (most commonly OH^-, F^-, and Cl^-) per unit cell. Among these anions, the one that best fits into the channel site is F^-. Its ionic radius is small enough to permit F^- in the most symmetric position in the channel (i.e., on mirror planes perpendicular to the *c*-axis), and thus fluorapatite is the apatite with the highest symmetry. Because the OH^- ion is not spherical, the two mirror planes normal to the *c*-axis channel cannot be

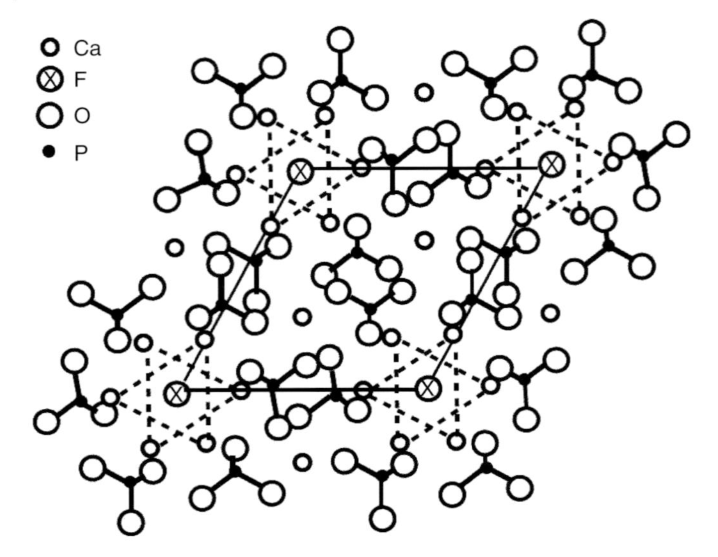

Figure 1.9 Three-dimensional structure of fluorapatite. View down the *c*-axis showing PO_4 tetrahedral ionic groups, Ca-ions, and channel ions. The parallelogram indicates the outline of the unit cell. The unit cell consists of two triangular prismatic subcells forming a rhombic prism. Six of the Ca^{2+} atoms form a sixfold site (indicated by dashed lines) in which the channel ions reside (F^- in the case of fluorapatite). These channels are oriented perpendicular to the page. Every crystallographic site (including the channel site) has a certain size, and thus not every atom or ionic group will fit into each site (the sizes of atoms are not drawn to scale). After [82].

preserved in hydroxylapatite. Thus, it has a lower symmetry than fluorapatite. Such differences in symmetry impact the growth morphology of the crystals, important to the mechanical properties of a composite material like bone.

Wopenka and Pasteris commented in 2005 that in the contemporary biomedical, orthopedic, and biomaterials literature, the mineral component of bone is still usually referred to as *hydroxy(l)apatite* or *carbonated hydroxy(l)apatite*, as if biological apatite were a well defined and well understood material. Whereas the Raman spectra of apatite in enamel, just like those of both geologic OHAp and synthetic OHAp, show the O–H modes for hydroxyl within the apatite structure, the spectra for apatite in bone do not. This is the property of all cortical bones of different mammals that were analyzed in [82].

The crystallographic structure of bone apatite is similar to that of OHAp, but there are important differences. The Raman spectra of synthetic OHAp, geologic OHAp, human enamel apatite, and cortical mouse bone apatite provide several differences between OHAp and biological apatites.

The bone apatite does not have a high concentration of OH groups, which is the feature of the mineral hydroxylapatite. Some bone apatites may not contain any OH groups at all. There is growing evidence for the lack of OH in bone apatite based not only on the results obtained *via* Raman spectroscopy but also on results of infrared spectroscopy, inelastic neutron scattering, and nuclear magnetic resonance spectroscopy, cf. [82–86].

1.2.5
Structure of the Bone

The structure of bone is nonhomogeneous. The bone tissue contains two main types of tissues: dense cortical bone and porous trabecular bone, cf. Figure 1.10. The tissues have similar biological activity; the difference is in geometry – in the arrangement of the microstructure. The outer layer of bone tissue is hard and is called the *compact bone* (known also as *cortical* or *dense bone*). This part of the tissue gives bones their smooth, white, and solid appearance, and accounts for 80% of the total bone mass of an adult skeleton. Compact bone tissue is called so because of its very small gaps and spaces in comparison with the inner trabecular bone.

Trabecular (cancellous or spongy) bone accounts for approximately 15% of the total bone mass. The vertebrae and pelvic bones contain relatively high amounts of trabecular tissue and are common sites of osteoporotic fractures, whereas the long bones (e.g., femoral neck) contain a relatively high amount of cortical bone.

The metaphysis is the wider portion of a long bone adjacent to the epiphyseal plate, cf. Figure 1.10b. It is this part of the bone that grows during childhood; as

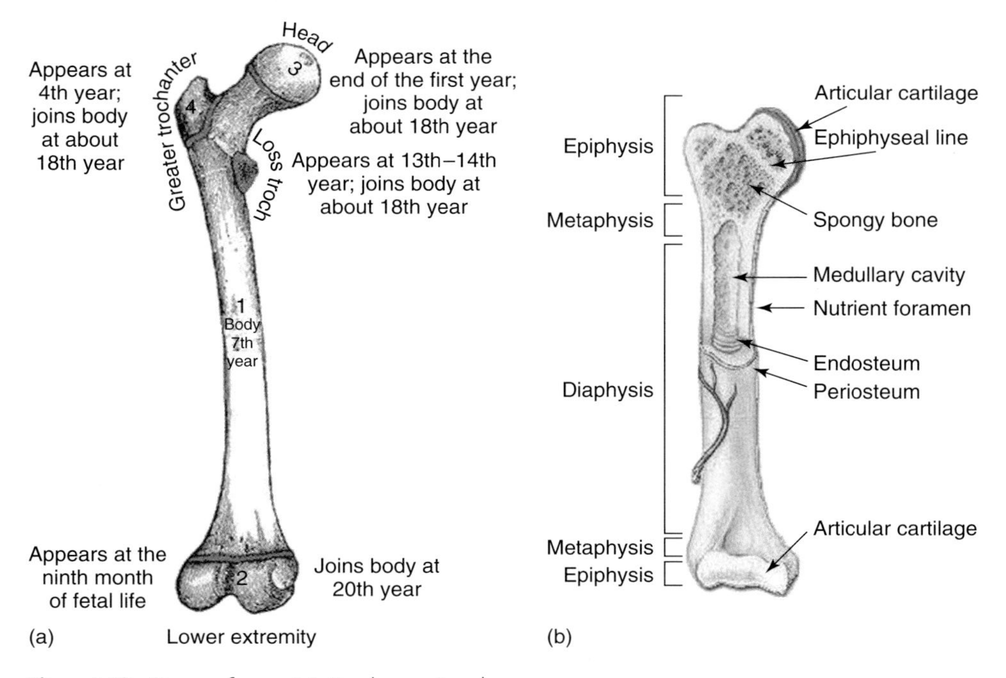

Figure 1.10 Human femur. (a) Development and (b) anatomical structure. It is noted that the metaphysis is essential for the growth of long bone. After [63, 74–76].

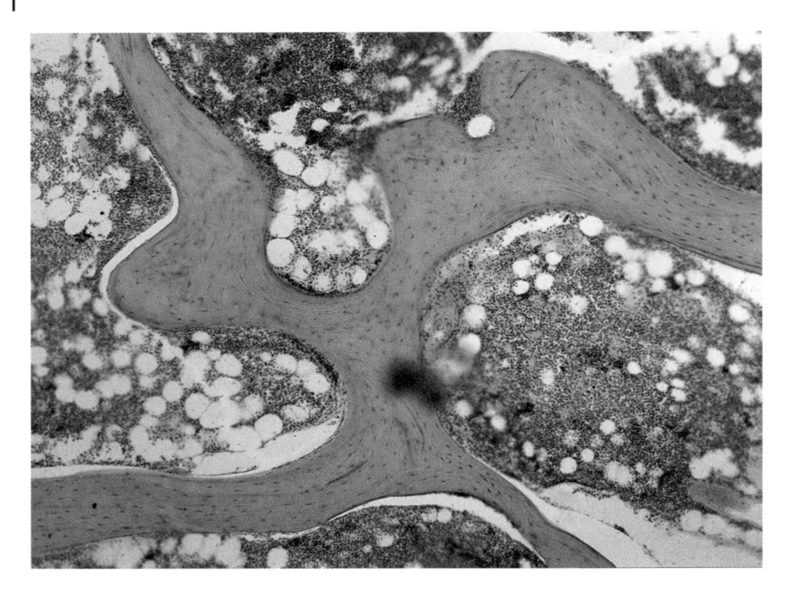

Figure 1.11 The spongy bone in histological section. The trabecules are surrounded by the marrow, and white spheres of fat in the marrow are seen. Courtesy of Litwin and Gajda.

it grows, it ossifies near the diaphysis and the epiphyses. At about 18–25 years of age, the metaphysis stops growing and completely ossifies into a solid bone.

The interior of the bone is known as the *spongy bone* (or cancellous or trabecular bone), cf. Figures 1.10b and 1.11. Spongy tissue has porous appearance and is composed of a network of trabecules, and rod- and plate-like elements that make the tissue lighter and allow space for blood vessels and marrow. Spongy bone accounts for 20% of the total bone mass, but (from macroscopic point of view) has nearly 10 times the surface area of compact bone.

1.3 Microscopic Structure of the Bone

1.3.1 General

Bone is composed of three major components: (i) small plate-shaped crystals of carbonate apatite, (ii) water, and (iii) macromolecules, of which type I collagen is the major constituent. The manner in which the crystals and the collagen fibrils are organized in a bone has not been resolved yet. In mineralized collagen fibrils of turkey tendon, the crystals are arranged in parallel layers across the fibrils, with the crystal *c* axes aligned with the fibril lengths, and in rat bones the plate-shaped crystals are also arranged in parallel layers within individual lamellae [76].

There are two methods of preparing sections for microscopy. One method provides the dry bone section. A piece of dead bone is broken or sawed from the main bone. Next, it is ground and polished to be very thin (about 15–45 μm thick). That polished piece is placed on a microscope slide and viewed directly. The bone cells are missing from the dead, polished bone specimen, and in a humorous sense one is looking at a skeleton of the skeleton. A second method for obtaining bone for histology is to soak a piece of bone in an acid solution for some time. The acid treatment dissolves the bone salts from the tissue in a process called *demineralization*. With this method, the cells stay behind and can be stained before observation under the microscope, cf. [87].

It is seen in Figure 1.12 that the bone is porous at two scales: containing macropores measuring 100 μm or more (Haversian and Volkmann's canals, lacunae), and micropores measuring up to 0.02 μm (= 20 nm) in diameter (canaliculi). The double porosity and interconnectedness of pores enable the bone to fulfill two vital

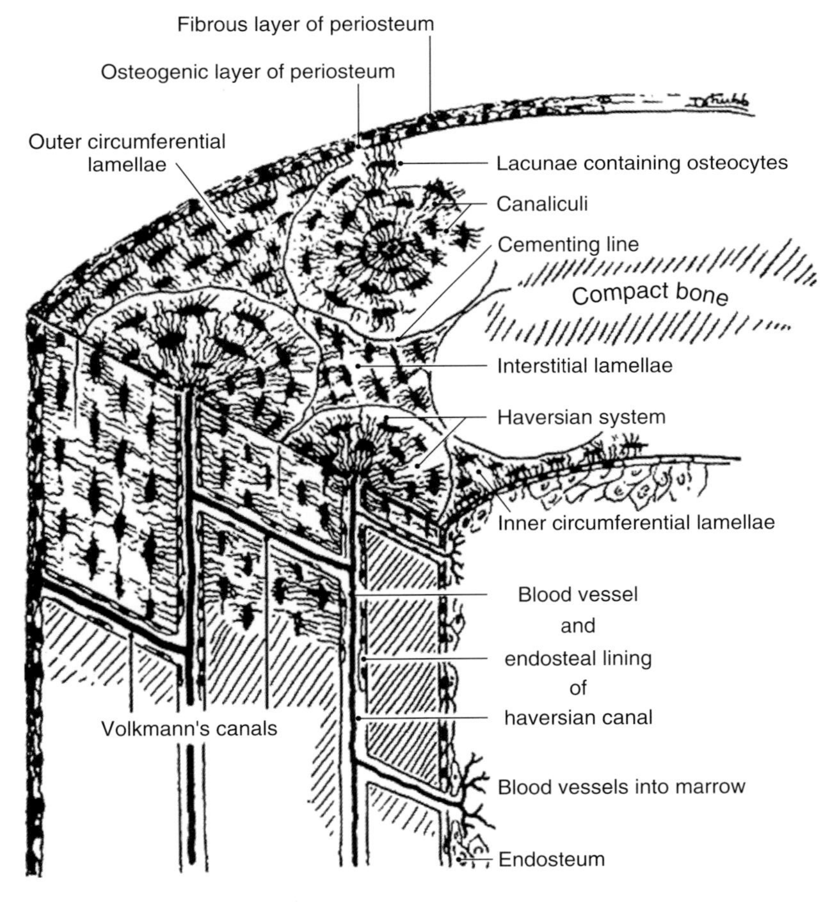

Figure 1.12 Design of the microstructure of cortical bone. After Piekarski and Munro [88].

functions. The macropores give space to permit bone cells to grow and allow blood to circulate, and the micropores facilitate the cell adhesion and crystallization of bone structure.

1.3.2 Osteon

The principal organizing unit of the compact bone is the osteon. A synonym for osteon is Haversian system. The osteon can be approximated as a long narrow cylinder that is 0.2 mm (200 μm) wide and 10 mm long. Osteons are found in the bones of mammals, birds, reptiles, and amphibians, running in a meandering way but generally parallel to the long axis of bones. Morphology of the osteon, obtained by electron microscopic techniques, for the study of compact bone is given in [89].

When the compact bone osteons are being formed, collagen fibers are laid down first. The collagen patterns are reflected in the structure known as a *lamella*. Osteons have between 4 and 20 lamellae with each measuring between 3 and 7 μm in width.

Leeuwenhoek, the father of microbiology, had reported his observations of the canal system in bones to the Royal Society of London in 1678. He called the canals *pipes*, and pointed out that they run both longitudinally and transversely in bones [90–92]. Thirteen years later, Havers did provide a more extensive description of the canal system in bones, linking it with his ideas of the lamellar nature of the bone tissue [93]; see also [94–96].

The group of cells functioning as an organized unit was called *basic multicellular unit* or bone multicellular unit (BMU) by Frost [97–100]. Remodeling process occurs with a specific sequence of events in the BMU.

The microscopic structure of a mammalian compact bone consists of repeating units called *osteons* or *Haversian systems*. Each system has concentric layers of mineralized matrix, called *concentric lamellae*, which are deposited around a central canal, also known as the *Haversian canal*, containing blood vessels and nerves that service the bone. By the longitudinal axis of the osteon runs a central canal, called the *Haversian canal* (synonyms: *Canalis nutricius, Canalis nutriens*, Haversian space, nutrient canal of bone) [101–103]. The elements of the osteon are shown in Figure 1.13.

The central canal is surrounded by concentric layers of matrix called *lamellae*. The lamellae are laid down one after the other over time, each successive one inside

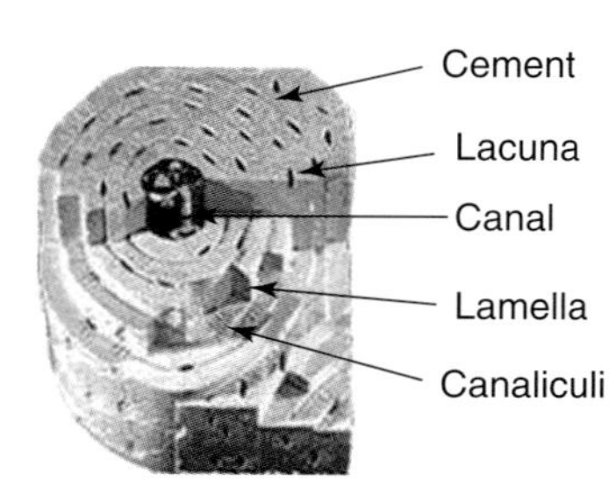

Figure 1.13 Elements of an osteon. After [87]. With permission of the author Blystone.

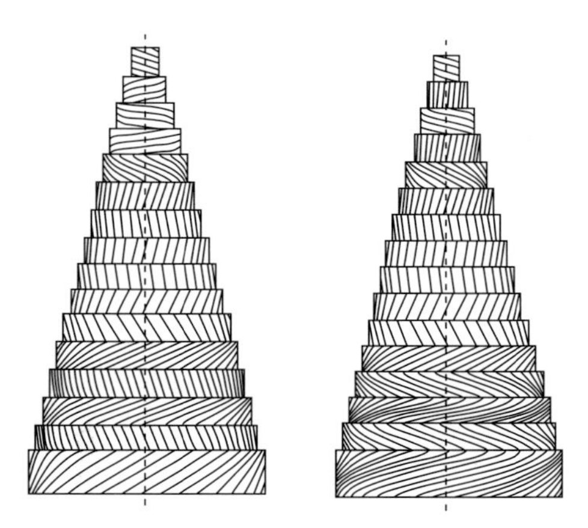

Figure 1.14 Examples of the osteon structure according to Gebhardt, with different collagen fiber orientations in the osteon lamellae. Helical course of fibers are noted at the successive lamellae. After [104].

the preceding one. Collagen fibers in a lamellae run parallel to each other but the orientation of collagen fibers across separate lamellae is oblique, cf. Figure 1.14. The fiber density is also lower at the border between adjacent lamellae, which accounts for the distinctive appearance of an osteon. In addition to blood vessels, Haversian canals contain nerve fibers and bone cells called *bone lining cells.* Bone lining cells are actually osteoblasts that have taken on a different shape following the period in which they have formed bone.

In 1905, Gebhardt [104, 105] performed research on bone structure, in particular on osteon architecture, using optical polarized microscope. Observations under polarized light indicate preferable directions of fibers in the lamellae of osteon. As a result, Gebhardt found that osteons are composed of a number of lamellae in which collagen fibers lay in different directions, cf. Figure 1.14.

These observations were repeated half a century later by Ascenzi and Bonucci [106–108]. They suggested that osteons that appear bright under polarized light are composed of lamellae in which collagen fibers lay (in prevailing number) parallel to the plane and perpendicular to the Haversian canal. The dark osteons under polarized light in their model consist of lamellae in which collagen fibers are oriented parallel to the long axis of the bone. In intermediate (alternating) osteons, collagen fibers should in this classification alternate orientation from one lamella to the other, having some lamella in which collagen fibers are orientated parallel (dark bands) and some orientated perpendicular (bright bands) to the long axis of the bone.

Ascenzi and Bonucci examined the mechanical properties of these three classes of osteons. Dark osteons were found to be the strongest under tensile loading. Bright osteons were stronger under compression. Intermediate osteons possess intermediate properties between bright and dark osteons, cf. [106–112] (also [113]).

The structure used by Gebhardt (as well as by Ascenzi and Benucci) is known as the *orthogonal plywood model*: only two fibril directions exist, making an angle 90°. Next, in 1988, Giraud-Guille presented the twisted plywood model of collagen fibril orientation within cortical bone lamellae. The twisted plywood model allows for parallel collagen fibrils, which continuously rotate from one plane to another in a helical structure [56].

Lying between or within the lamellae are special holes known as *lacunae*. Each lacuna has an oblong ellipsoidal form and provides enough space for an individual bone cell (osteocyte) to reside. In a microscopic section, viewed by transmitted light, lacunae appear as fusiform opaque spots. Lacunae are connected to one another by small canals called *canaliculi*. The osteocyte inside the lacuna is responsible for secreting the bone salts surrounding it. Osteocytes are found between concentric lamellae, within their cavelike lacunae, and connected to each other and the central canal by cytoplasmic processes through the canals called *canaliculi*. Osteocytes communicate with each other, and their network permits the exchange of nutrients and metabolic waste. The human osteocyte under normal conditions lives for about 25 years. Thus, in the lifetime of a person there would be about four generations of osteocytes.

Osteons are separated from each other by cement lines. Collagen fibers and canaliculi do not cross cement lines. The space between separate osteons is occupied by interstitial lamellae, which were formed by preexisting osteons that have since been reabsorbed. Osteons are connected to each other and the periosteum by oblique channels called *Volkmann's canals*.

Figures 1.15 and 1.16 are images of a sectioned bone. Cross section of a real osteon is not perfectly circular and the lamellae are not perfectly concentric.

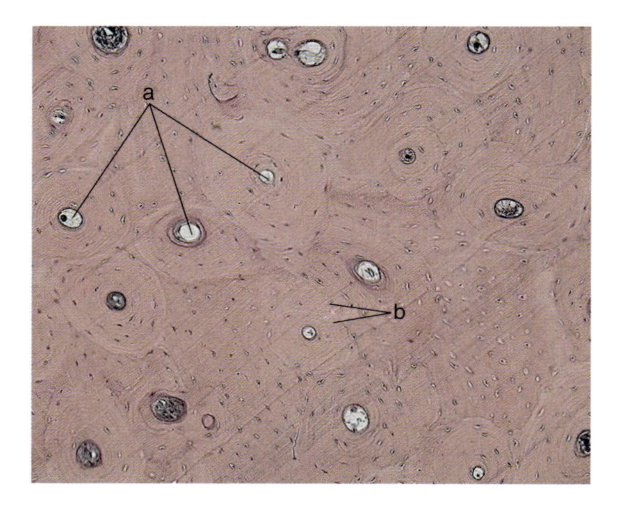

Figure 1.15 Compact bone – decalcified cross section. Osteonic structure is seen in the magnification: a – Haversian canals; b – lacunar spaces. Courtesy of Litwin and Gajda.

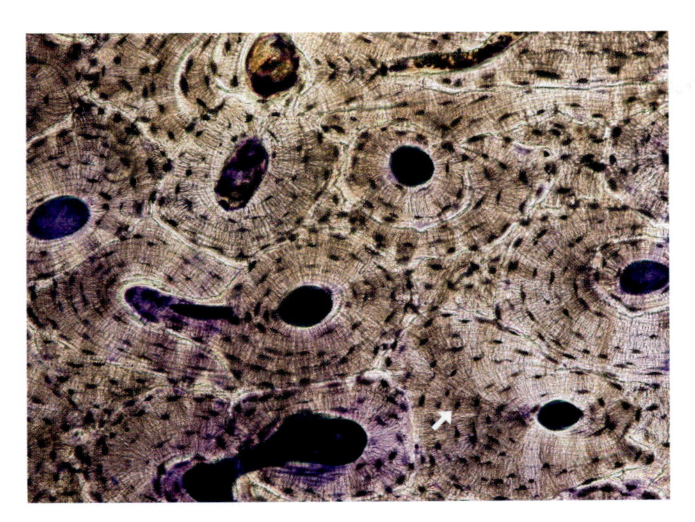

Figure 1.16 Compact bone – ground cross section. System of osteons that is visible in transverse histological section of the cortical bone. Haversian canals (large dark circular holes) are surrounded by the rings of lamellae. The Haversian canal in the center of the osteon has a diameter ranging between 50 and 90 μm. The smaller dark circles or ellipses (one is indicated by white arrow) are lacunar spaces within the osteon. In lacunae, the bone cells – osteocytes – are sheltered. Volkmann's canals, linking Haversian ones, are also seen in the lower part of the figure. Courtesy of Litwin and Gajda.

Impressive microphotographs, being the images of osteon in large magnification, obtained by means of scanning electron microscopy (SEM), were provided by Frasca *et al.* [114]. The arrangements of collagen fibers in lamellae are shown here, for example, decalcified osteon sample with exposed lamellar interfaces (×200), coexisting longitudinal and transverse fibers in one lamella (×1000), and complex fiber arrangement in one lamella, with partial stripping of two sequential lamellae.

The manner in which the crystals and the collagen fibrils are organized in the osteon has still not been resolved, even though it is known that they are intimately related.

Weiner, Arad, and Traub observed in 1991 that the plate-shaped crystals of rat bone are arranged in parallel layers that form coherent structures up to the level of individual lamellae. The crystal layers of the thin lamellae are parallel to the lamellar boundary, whereas those of the thicker lamellae are oblique to the boundary. The basic structure of rat bone can be described as *rotated plywood* [115].

Enlow in his microscopic study of the bone at the tissue level considered that a bone section is always a slice at the time of ontogeny. The actual tissue types express the succession of events that took place at that very level during bone development [116].

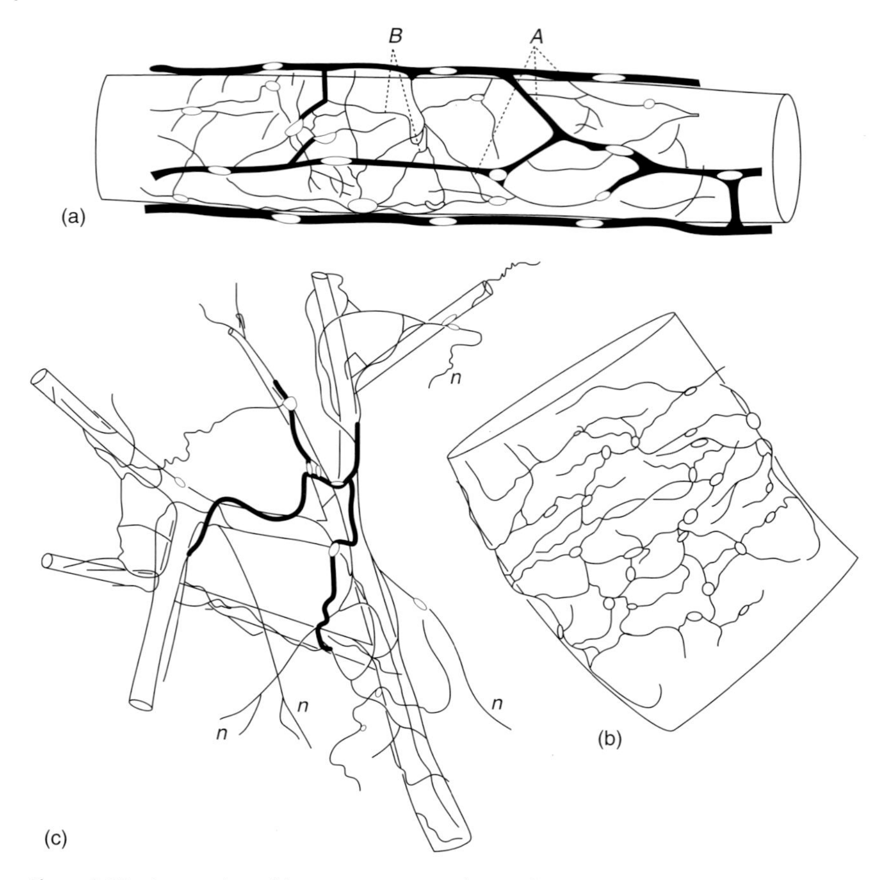

Figure 1.17 Innervation of bone marrow. (a) Plexus of nerve fibers around a vein in the marrow of rabbit tibia. (b) Portion of nerve plexus around an artery in the marrow of rabbit tibia. (c) Plexus of nerve fibers touching arteriola in the marrow of chicken tibia. The thinnest nerve fibers, denoted by *n*, penetrate in the pulp of marrow. After [118].

1.3.3 Bone Innervation

Bone is not only richly supplied with blood but is also abundantly innervated. The study of bone innervation dates back to the first half of the nineteenth century when Gros described the distribution of nerves in the femur of a horse [117]. Even early morphological studies applying classic histological methods, such as methylene blue staining and silver impregnation, revealed an intense innervation pattern of the bone in mature animals and humans, cf. also [118–120]. That the bone marrow is innervated is known since 1901 when Ottolenghi discussed the presence of nerves surrounding marrow arteries with fibers passing into the parenchyma, cf. Figure 1.17. According to Ottolenghi, the nerve fibers within the marrow cavity fall into three main groups: (i) those that penetrate the walls of arterioles and form delicate plexiform networks between the adventitia and the media, (ii) those that surround the capillaries, and (iii) those that terminate between the cells of parenchyma.

Fliedner *et al.* [26] studied the question concerning the mechanisms that allow the bone marrow hemopoiesis to act as one cell renewal system although the bone marrow units are distributed throughout more than 100 bone marrow areas or units in the skeleton. The effect that "the bone marrow" acts and reacts as "one organ" is due to the regulatory mechanisms: the humeral factors (such as erythropoietins, granulopoietins, thrombopoietins, etc.), the nerval factors (central nervous regulation), and the cellular factors (continuous migration of stem cells through the blood to assure a sufficient stem cell pool size in each bone marrow "subunit").

The nervous system is differentiated into efferent nerves and afferent nerves. Efferent nerves – otherwise known as *autonomic or motor or effector neurons* – carry nerve impulses *away* from the central nervous system to effectors such as muscles or glands. The opposite activity of direction or flow is afferent (sensory) [121].

The majority of the skeletal innervation system is composed of sensory fibers originating from primary afferent neurons located in the dorsal root and some cranial nerve ganglia, whereas the other nerve fiber populations are adrenergic and cholinergic in nature and originate from paravertebral sympathetic ganglia. The

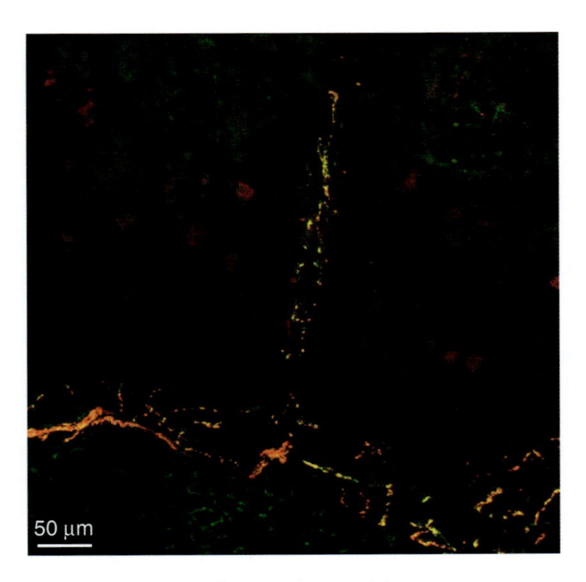

Figure 1.18 Nerve fibers in the canal between periosteum and proximal mataphysis of four-week rat tibia. Growth-associated protein (GAP-43) and protein gene product (PGP) 9.5 are visible. Courtesy of Litwin and Gajda.

sensory fibers were detected in the periosteum, bone marrow cavity, and vascular canals in long bones of mature and developing animals.

The blood vessels in the bone marrow are abundantly innervated, through both sympathetic and afferent nerve fibers. Afferent nerve fibers are also connected with receptors imbedded in the parenchyma of marrow [121]. In Figure 1.18, an example of the innervations is given.

Growth-associated protein (GAP-43) is expressed in conditions of embryonic growth, during axonal regeneration, and even at maturity in certain areas of the brain known to exhibit synaptic plasticity. Protein gene product (PGP) 9.5 is a cytoplasmic protein specific for neurites, neurons, and cells of the diffuse neuroendocrine system. GAP-43 and PGP 9.5 are often used as neuronal markers, cf. [122].

In the field of neuroscience, tachykinin peptides are one of the largest families of neuropeptides, found from amphibians to mammals. They are named so because of their ability to rapidly induce contraction of gut tissue. Tachykinins are widely distributed in the body and function as neurotransmitters and neuromodulators. Five tachykinin subtypes: substance P (SP), neurokinin A, neurokinin B, neuropeptide K, and neuropeptide Y; and three receptor subtypes: neurokinin-1, -2, and -3 receptors, have been identified. SP was the first peptide of the tachykinin family to be identified. It is considered to be an important neuropeptide, and to function in the nervous system and intestine. However, recent studies in the analysis of SP receptors, particularly neurokinin-1 receptors (NK_1-Rs) that have high affinity for SP, have demonstrated that NK_1-Rs are distributed not only in neurons and immune cells but also in other peripheral cells, including bone cells [123]. The distribution of tachykinin-immunoreactive axons and neurokinin receptors suggests that tachykinins may directly modulate bone metabolism through neurokinin receptors, cf. survey paper on the bone innervations by Goto [124], and also [125].

SP is an undecapeptide with multiple effects on the cardiovascular, gastrointestinal, and urinary systems as well as complex central nervous system functions such as learning and memory. SP is released from the terminals of specific sensory nerves; it is found in the brain and spinal cord, and is associated with inflammatory processes and pain [126, 127].

1.3.3.1 Anatomy of Bone Innervation

An extensive plexus of nerve fibers investing the periosteum and joints gives bone the lowest pain threshold of any of the deep tissues. A-delta (small myelinated) fibers and C (small unmyelinated) fibers contain deep somatic nociceptors with free nerve endings. Deep somatic pain is usually described more as aching than sharp, and is less well localized than cutaneous somatic pain. In the human femur, all cortical Volkmann's and Haversian canals contain unmyelinated fibers, and some contain both myelinated and unmyelinated fibers. SP, which mediates pain sensation, is attached to these fibers.

In rat and dog models, nerves in bone marrow have been found to be associated with venous sinuses. These are single fiber nerves, independent of the blood vessels in the marrow, that enter the Haversian canals from both periosteum and the marrow.

Gajda *et al.* [128] investigated the development of sensory innervation in long bones, see Figure 1.18. Their model was rat tibia in fetuses and in juvenile individuals on postnatal days. A double immunostaining method was applied to study the co-localization of the neuronal growth marker GAP-43 and the pan-neuronal marker PGP 9.5 (9.5) as well as that of two sensory fiber-associated neuropeptides, calcitonin gene-related peptide (CGRP) and SP. The earliest, not yet chemically coded, nerve fibers were observed in the perichondrium of the proximal epiphysis. Further development of the innervation was characterized by the successive appearance of nerve fibers in the perichondrium/periosteum of the shaft, the bone marrow cavity and intercondylar eminence, the metaphyses, the cartilage canals penetrating into the epiphyses, and finally in the secondary ossification centers and epiphyseal bone marrow. Maturation of the fibers, manifested by their immunoreactivity for CGRP and SP, was investigated in these cases also.

1.3.4 Bone Cells

1.3.4.1 Cells

The living cells are divided into two types: prokaryotic and eukaryotic. The prokaryotes are organisms that lack a cell nucleus (karyon). Prokaryotes are divided into the bacteria and archea. Animals, plants, fungi, and protists are eukaryotes – organisms whose cells are organized into complex structures enclosed within membranes. The defining membrane-bound structure that differentiates eukaryotic cells from prokaryotic cells is the nucleus. The cells of protozoa, higher plants, and animals are highly structured. These cells tend to be larger than the cells of bacteria, and have developed special packaging and transport mechanisms that are appropriate to their larger size. There are many different cell types: approximately 210 distinct cell types in the adult human body.

In Figures 1.19 and 1.20, the animal and plant cells may be compared. It is seen that a cell wall – a thick, rigid membrane – surrounds a plant cell. This layer of cellulose fiber gives the cell most of its support and structure. The cell wall also bonds with other cell walls to form the structure of the plant.

Existence of wall in the plant cell provides the main difference between plant and animal body from a mechanical point of view. The wall in plant cell gives the plant support and structure. The animal body whose cells have no walls should be supported by special tissues – the bones in the case of vertebrates.

The cytoskeleton (a cellular skeleton or cell scaffolding) is present in all cells, being contained within the cytoplasm. It is a dynamic structure made out of protein molecules that protects the cell, maintains the cell shape, enables cellular motion (using structures such as flagella, cilia, and lamellipodia), and plays important roles in both intracellular transport (such as the movement of vesicles and organelles) and cellular division. Microfilaments (or actin filaments) are the thinnest filaments of the cytoskeleton found in the cytoplasm of all eukaryotic cells [130–135].

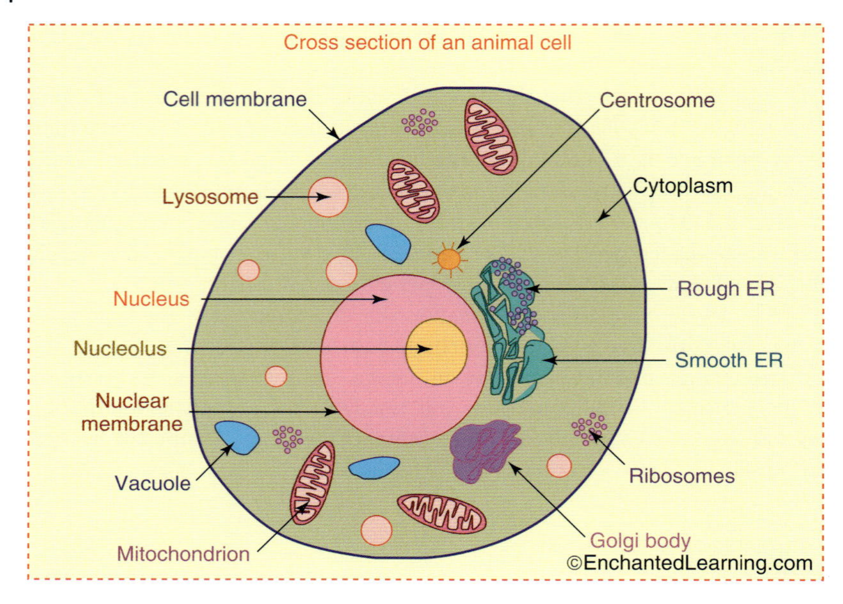

Figure 1.19 An animal cell. Peptide chains, alpha-1 and alpha-2 chains, known as pre-procollagen, are formed during translation on ribosomes along the rough endoplasmic reticulum (Rough ER, RER) inside the cell. Triple helical structure is formed inside the endoplasmic reticulum from each two alpha-1 chains and one alpha-2 chain. Procollagen is shipped to the Golgi apparatus, where it is packaged and secreted by exocytosis. After [129].

Leeuwenhoek [91] pointed out an analogy between the structures of bone and wood. This analogy is almost apparent and shows how different methods are developed by the nature to reach the same goal, in our case: resistance and moderate stiffness. Similar opinion was expressed by Monceau (1700–1782), a botanist and agronomist.

The bones have to support the body of vertebrates and the tree trunk with branches has to support the weight of a plant, cf. [136]. These similar mechanical requirements lead at first sight to similar solutions, known as *homoplasy*. The stem of a plant is built from large cells. Every cell of the plant has a rigid cover – the cellulose wall, in distinction from animal cell. Therefore, the skeleton of a plant itself is composed of cells. The internal geometry of the bone resembles the geometry of the trunk. It is built from a rigid material and it is partitioned into (cellular in general meaning) structures, but this hydroxyapatite structure does not belong to the bone cells (osteocytes). Moreover, while all parts of the bone are active, in the stem of a vascular plant, only the phloem with cambium situated under the bark is the living tissue.

1.3.4.2 Cell Membrane

Amphiphile (Gr. $\alpha\mu\phi\iota\varsigma$, amphis: both and $\phi\iota\lambda\acute{\iota}\alpha$, philia: love, friendship) is a term describing a chemical compound possessing both hydrophilic (*water-loving)*

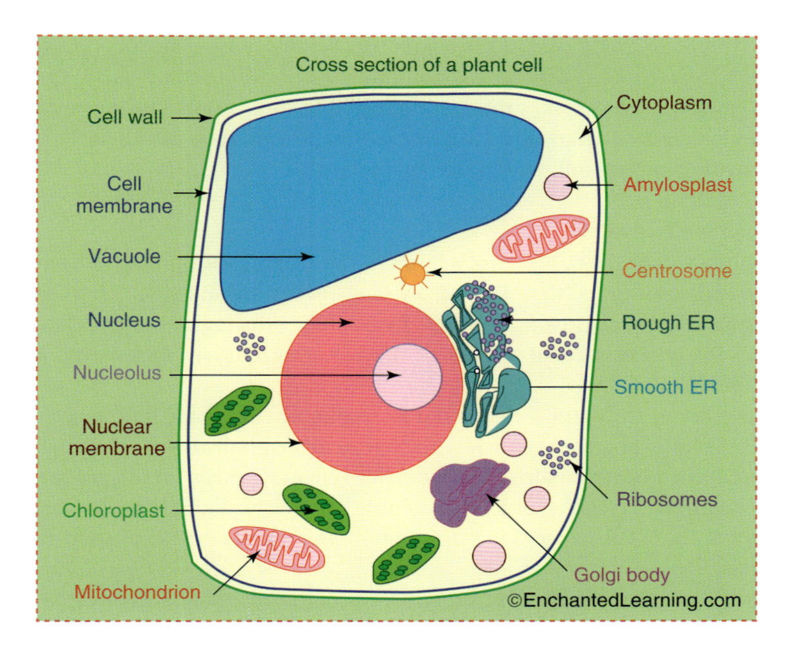

Figure 1.20 A plant cell. In vascular plants cellulose is synthesized at the cell membrane (also called the plasma membrane) by rosette terminal complexes. After [129].

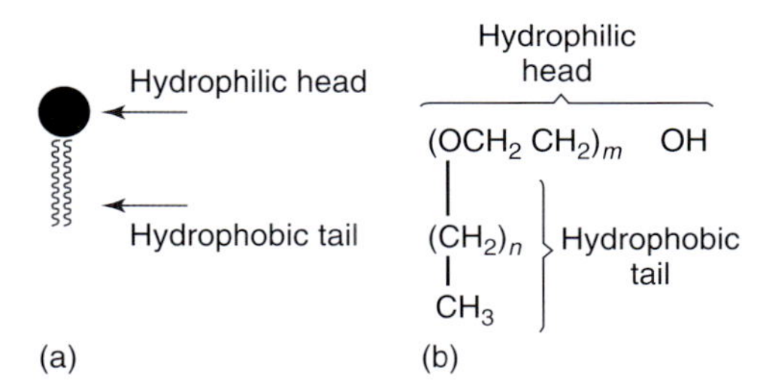

Figure 1.21 An example of an amphiphilic molecule: (a) symbolic representation and (b) chemical structure (alkyl polyoxyethylene).

and lipophilic (*fat-liking*) properties. Such a compound is called *amphiphilic* or *amphipathic*. Common amphiphilic substances are soaps and detergents.

An amphiphilic molecule is composed of two different parts: hydrophobic tail and hydrophilic head. The tail is composed of one or more hydrocarbon chains, and the head is composed of chemical groups with high affinity to water, Figure 1.21.

Phospholipids, a class of amphiphilic molecules, are the main components of biological membranes. When the phospholipids are immersed in an aqueous solution, they arrange themselves into bilayers, by positioning their polar groups toward the surrounding aqueous medium, and their lipophilic chains toward the inside of the bilayer, defining a nonpolar region between the two polar ones.

There are three different major classes of lipid molecules – phospholipids, cholesterol, and glycolipids. Different membranes have different ratios of the three lipids. However, what makes the membrane truly special is the presence of different proteins on the surface that fulfill various functions such as cell surface receptors, enzymes, surface antigens, and transporters [137–141].

1.3.4.3 **Membrane Transport**

The cell membrane primarily consists of a thin layer of amphipathic phospholipids that spontaneously arrange so that the hydrophobic *tail* regions are shielded from the surrounding polar fluid, causing the more hydrophilic *head* regions to associate with the cytosolic and extracellular faces of the resulting bilayer. This forms a continuous lipid bilayer.

The arrangement of hydrophilic heads and hydrophobic tails of the lipid bilayer prevents polar solutes (e.g., amino acids, nucleic acids, carbohydrates, proteins, and ions) from diffusing across the membrane, but generally allows for the passive diffusion of hydrophobic molecules. This affords the cell the ability to control the movement of these substances *via* transmembrane protein complexes such as pores and gates.

Thus, the cell is surrounded by a semipermeable membrane that separates the cell interior from exterior and controls the movement of substances in and out of the cell. Efflux pumps are proteinaceous transporters localized in the cell membranes. They are called *active transporters*, and they require a source of chemical energy to perform their function, cf. [142]. In eukaryotic cells, the existence of efflux pumps has been known since the discovery of p-glycoprotein in 1976 by Juliano and Ling [143]. They remarked that the relative amount of surface-labeled p-glycoprotein correlates with the degree of drug resistance in a number of independent mutant and revertant clones. A similar high-molecular-weight glycoprotein is also present in drug-resistant mutants from another cell line. Observations on pleiotropic drug resistance are interpreted in terms of a model wherein certain surface glycoproteins control drug permeation by modulating the properties of hydrophobic membrane regions. (Pleiotropy means the influence of a single gene on multiple phenotypic traits.)

Enzymes are proteins that catalyze (i.e., increase the rates of) chemical reactions. Almost all processes in a biological cell need enzymes to occur at proper rates. In enzymatic reactions, the molecules at the beginning of the process are called *substrates*, and the enzyme converts them into different molecules, called the *products*. Enzyme kinetics is the investigation of how enzymes bind substrates and turn them into products.

Diastase catalyzes the hydrolysis (breakdown) of carbohydrates [144] and a protease (or proteinase) catalyzes the hydrolysis of proteins. Sucrase is the name given to a number of enzymes that catalyze the hydrolysis of sucrose to fructose and glucose. Invertase is a sucrase enzyme. It breaks down sucrose (table sugar) to fructose and glucose, usually in the form of inverted sugar syrup.

In 1902, Henri investigated the action of diastases and proposed a quantitative theory of enzyme kinetics [145, 146]. In 1913, Michaelis and his postdoc Menten performed experiments on kinetics of invertase activity [147] and confirmed Henri's equation, which is referred now to as *Henri–Michaelis–Menten kinetics* (sometimes as Michaelis–Menten kinetics only). The important contribution of Henri was to consider enzyme reactions in two stages. In the first, the substrate binds reversibly to the enzyme, forming the enzyme–substrate complex. The enzyme then catalyzes the chemical step in the reaction and releases the product.

A membrane transport protein is a protein involved in the movement of ions, small molecules, or macromolecules, such as another protein across a biological membrane [148]. Transport proteins are integral membrane proteins; that is, they exist within and span the membrane across which they transport substances. The proteins may assist in the movement of substances by facilitated diffusion or active transport. The permeases are membrane transport proteins, a class of multipass transmembrane proteins that facilitate the diffusion of a specific molecule in or out of the cell. Many membrane transporters behave as permeases and have several characteristics in common with enzymes. For example, both have binding sites on their surfaces that bind substrate (enzymes) and solute (transporters), both lower the activation energy, both exhibit saturation with increases in substrate or solute concentration, and both exhibit kinetic constants, K_M and v_{max}. The Michaelis–Menten enzyme kinetics is the principal analytical method used to characterize the kinetic properties of enzymes and also that of membrane transport proteins [149, 150]. The Michaelis–Menten equation relates the initial reaction rate v_0 to the substrate concentration S

$$v_0 = \frac{v_{max} S}{K_M + S}$$

The corresponding graph is a hyperbolic function; the maximum rate is described as v_{max}. Constant K_M is the substrate concentration at which the reaction rate v_0 is one-half of v_{max}. The limitation for the Henri–Michaelis–Menten equation is that it relies upon the law of mass action, which is derived from the assumptions of Fickian diffusion. However, many biochemical or cellular processes deviate significantly from such conditions. Voituriez *et al.* have shown that the state attained by reversible diffusion-limited reactions at time $t = \infty$ is generally *not a true thermodynamic equilibrium*, but rather a nonequilibrium steady state, and that the law of mass action is invalid [151, 152], see also [153–156].

1.3.4.4 **Bone Cell Types**

The living substance of the bone, the bone cells (or bone corpuscles, German: Knochenkörperchen), account for only 1–5% of the bone volume in the adult skeleton. There are five types of bone cells:

- Osteoprogenitors – immature cells which differentiate to form osteoblasts. Only at this stage, bone cells may divide. Mesenchymal stem cells (MSCs) residing in bone marrow are the progenitors for osteoblasts and for several other cell types, [157]. Osteoprogenitors are induced to differentiate under the influence of growth factors, in particular, the bone morphogenetic proteins (BMPs).

- Osteoblasts are the bone-forming cells. They secrete osteoid, which forms the bone matrix. They also begin mineralization. Osteoblasts arise from osteoprogenitor cells located in the periosteum and the bone marrow. Osteoblasts, when entombed within the osteoid, become osteocytes, with cytoplasmic processes that communicate with each other.
- Osteocytes are the mature osteoblasts that no longer secrete matrix, but being surrounded by it maintain metabolism, and participate in nutrient/waste exchange *via* blood. Osteoblasts and osteocytes develop in the mesenchyme. (Mesenchyme is the meshwork of embryonic CT in the mesoderm from which the CTs of the body and the blood and lymphatic vessels are formed.)
- Osteoclasts function in the resorption and degradation of the existing bone; in this role, they are the opposite of osteoblasts. Monocytes (white blood cells) fuse together to create these huge cells, which are concentrated in the endosteum. Osteoclasts play a key role in bone remodeling: they destroy bone cells and reabsorb calcium.
- Bone lining cells are essentially inactive osteoblasts; they cover all of the available bone surface and function as a barrier for certain ions [157, 158].

Bone is a dynamic tissue. It is constantly being reshaped by osteoblasts, which build bone, and osteoclasts, which resorb bone. An osteoblast (Gr. bone and germ) is a mononucleate cell, responsible for bone formation. Osteoblasts produce osteoid, which is composed mainly of Type I collagen, and osteoblasts are responsible for mineralization of the osteoid matrix. Osteoblast cells tend to decrease as individuals become older, thus decreasing the natural renovation of the bone tissue, cf. [158, 159].

Osteocytes are networked to each other *via* long cytoplasmic extensions that occupy tiny canals called *canaliculi*, which are used for exchange of nutrients and waste. Hence, osteocytes *in vivo* possess a distinctive morphology – that of dendricity – connecting osteocyte to osteocyte creating the osteocyte syncytium and also connecting osteocytes with cells on the bone surface. It is thought that bone fluid surrounding the dendrite within the canaliculi is responsible for the transmission of mechanical strain through fluid flow shear stress. Dendrites may be essential for osteocyte function, viability, and response to load [160–162]. Osteocytes, dendritic or star-shaped cells, are the most abundant cells found in a compact bone, cf. Figures 1.22 and 1.23. There are about 10 000 cells per cubic millimeter and 50 processes per cell.

Cell contains a nucleus and a thin ring of cytoplasm. When osteoblasts get trapped in the matrix they secrete, they become osteocytes. The space that an osteocyte occupies is called a *lacuna*. Although osteocytes have reduced synthetic activity and, like osteoblasts, are not capable of mitotic division, they are actively involved in the routine turnover of bony matrix, through various sensory mechanisms. They destroy the bone through a rapid, transient (relative to osteoclasts) mechanism called *osteocytic osteolysis*. Hydroxyapatite, calcium carbonate, and calcium phosphate are deposited around the cell.

Interesting images of osteocyte lacuno-canalicular network have been obtained. For example, in [160] one can see morphology of osteocytes, osteoblasts, and

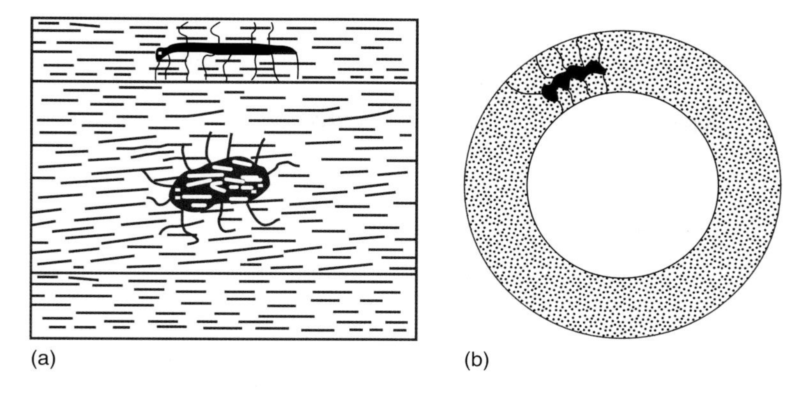

Figure 1.22 Osteocytes under a polarized light microscope (parallel polaroids, i.e., white background), in the cross sections parallel (a) and perpendicular (b) to the axis of osteon. After [104].

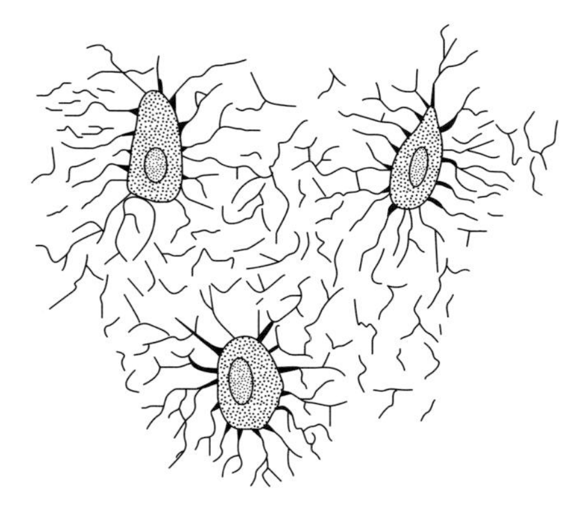

Figure 1.23 Osteocytes (bone cells) and their processes (long extensions), contained in the bone lacunae and their canaliculi, respectively, from a section through the vertebra of an adult mouse. After [63].

periosteal fibroblasts isolated from embryonic chicken calvariae and observed by fluorescence or phase contrast microscopy after 2×24 h of culture.

The osteocytes function to keep the bone alive. Since osteon is lamellar, the lacunae are located between adjacent lamellae; hence, they are arranged in rows, which coincide with the division between one lamella of the osteon and the next. Osteocytes have long extensions (processes): the processes of adjacent osteocytes remain in contact by passing through the minute canaliculi. The canaliculi radiate outward from each lacuna, each one filled by an osteocyte process which contacts that of another osteocyte. At the points of contact, gap junctions allow for a cell-to-cell pathway for nutrients and waste products through the otherwise impermeable hard substance of the bone [161].

Each canaliculus carries a fine osteocyte process; at the points of contact of these processes, gap junctions between the adjacent plasma membranes are the sites of interchange of materials and information between the cells. This system is essential to the survival of compact osteonal bone. Canaliculi are not extensively developed in the flat plates of cancellous (spongy) bone, because the plates are thin and diffusion transfer of products can occur between the osteocyte and the nearby marrow space, cf. Figure 1.11. Osteocytes are thought to function as a network of sensor cells in bones, which can mediate the effects of mechanical loading through their extensive communication network of Kendall's type, cf. [23, 24], also [26, 27].

It was found by Tanaka *et al.* that osteocytes from chick embryonic calvariae stimulated the osteoclast formation and function while maintaining osteocytic features *in vitro*. Isolated chick osteocytes stimulate formation and bone-resorbing activity of osteoclast-like cells. These results suggested that osteocytes may play a role in osteoclast recruitment [163].

The problem was also studied by Rubinacci *et al.*, who have proposed a model system for integrated osseous responses to mechanical, pharmacological, and endocrine signals [164].

The distribution of the osteocyte processes through the bulk of bone differs from their distribution on the bone surface. In [161] and [165], Bonewald presented observations made under a magnification of 3000× in scanning electron: micrographs of resin-embedded acid-etched mouse bone samples. It is visible as to how osteocyte cell processes pass through the bone in thin canals (canaliculi), connecting osteocytes with each other and with cells on the bone surface. In [165], Bonewald gives an analysis indicating that the osteocytes (not only osteoclasts) have both matrix forming and matrix destroying activities and that the osteocytes can remodel bone's local environment including lacunae and canaliculi, cf. also [101].

1.3.4.5 **Osteoclasts**

Osteoclasts are specialized cells responsible for bone resorption. They are derived from the monocyte/macrophage hematopoietic lineage. They develop and adhere to bone matrix and then secrete acid and lytic enzymes that degrade the bone matrix in a specialized, extracellular compartment.

Osteoclasts are large multinucleated cells (Figure 1.24). The nuclei resemble the nuclei of the osteoblasts and osteocytes. The cytoplasm has often a foamy appearance due to a high concentration of vesicles and vacuoles. An osteoclast cell frequently has branching processes.

Osteoclasts may arise from stromal cells of the bone marrow, being related to monocyte/macrophage cells, derived from granulocyte/macrophage-forming colony units (CFU-GM). They may represent fused osteoblasts or may include fused osteocytes liberated from resorbing bone. Osteoclasts lie in shallow cavities (depressions, pits, or irregular grooves) called *Howship's lacunae* (or resorption lacuna), formed in the bone that is being resorbed by osteoclasts [166–169].

At the site of active bone resorption, the osteoclast forms a specialized cell membrane, *the ruffled border*, that touches the surface of the bone tissue. The ruffled border increases the surface area interface for bone resorption, facilitates removal of the bone matrix, and is a morphologic characteristic of an osteoclast that actively resorbs the bone. The mineral portion of the matrix (hydroxyapatite) includes calcium and phosphate ions. These ions are absorbed into small vesicles, which move across the cell and eventually are released into the extracellular fluid, thus increasing the levels of the ions in the blood.

Osteoclasts possess an efficient pathway for dissolving crystalline hydroxyapatite and degrading organic bone matrix rich in collagen fibers. When initiating bone resorption, osteoclasts become polarized, and three distinct membrane domains appear: a ruffled border, a sealing zone, and a functional secretory domain. Simultaneously, the cytoskeleton undergoes extensive reorganization. During this process, the actin cytoskeleton forms an attachment ring at the sealing zone, the membrane domain that anchors the resorbing cell to the bone matrix.

The ruffled border appears inside the sealing zone, and has several characteristics of late endosomal membrane. Extensive vesicle transport to the ruffled border delivers hydrochloric acid and proteases to an area between the ruffled border and the bone surface called the *resorption lacuna*. In this extracellular compartment, crystalline hydroxyapatite is dissolved by acid, and a mixture of proteases degrades the organic matrix. The degradation products of collagen and other matrix components are endocytosed, transported through the cell, and exocytosed through a secretory domain. This transcytotic route allows osteoclasts to remove large amounts of matrix-degradation products without losing their attachment to the underlying bone. It also facilitates further processing of the degradation products intracellularly during the passage through the cell.

1.3.5 Cellular Image – OPG/RANK/RANKL Signaling System

Recently, it has become clear that osteoclasts are not simply trench digging cells, but that they have important regulatory functions as immunomodulators in pathologic states and that they may also regulate osteoblast function. Proper growth and functioning of osteoclasts is controlled by a pathway in which three factors, osteoprotegerin (OPG), receptor activator of nuclear factor-kappaB (RANK), and receptor activator for nuclear factor-kappa B ligand (RANKL), play the main role. RANK and its ligand (RANKL) are important members of the tumor necrosis factor receptor (TNFR) and tumor necrosis factor (TNF) superfamilies, respectively.

OPG is secreted by osteoblasts and osteogenic stromal stem cells and protects the skeleton from excessive bone resorption by binding to RANKL and preventing it from interacting with RANK. The RANKL/OPG ratio in bone marrow is an important determinant of bone mass in normal and disease states. RANKL/RANK signaling also regulates the lymph node formation and mammary gland lactational hyperplasia in mice, and OPG protects large arteries from medial calcification. OPG and RANKL proteins are mainly located in Golgi areas.

RANK, also known as tumor necrosis factor-related activation-induced cytokine (TRANCE) receptor, is a type I membrane protein, which is expressed on the surface of osteoclasts and is involved in their activation upon ligand binding. RANK is also

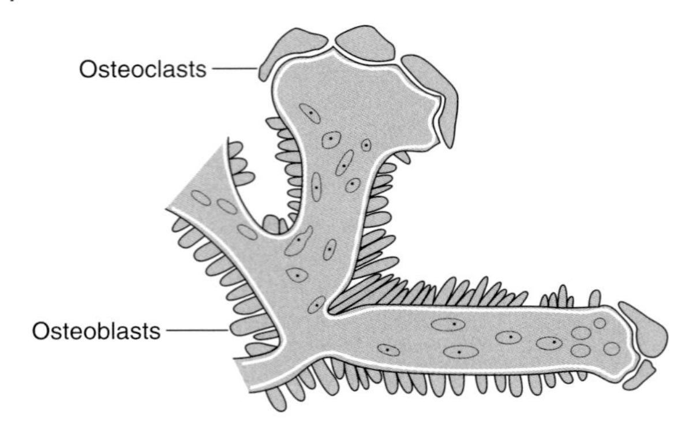

Figure 1.24 Osteoblasts and osteoclasts on trabecula of lower jaw of calf embryo. After [63].

expressed on dendritic cells and facilitates immune signaling. RANKL is found on the surface of stromal cells, osteoblasts, and T cells. RANK is expressed on osteoclasts, T lymphocytes, and dendritic cells, and its ligation with RANKL leads to cellular activation. However, another member of the TNFR family, OPG, acts as a decoy receptor, binding to RANKL and preventing its interaction with RANK.

RANKL is a member of the TNF, and is essential in osteoclastogenesis. RANKL, also known as *tumor necrosis factor-related activation-induced cytokine*, osteoprotegerin ligand (OPGL), and osteoclast differentiation factor (ODF), is a molecule participating in bone metabolism. This surface-bound molecule activates osteoclasts, cells involved in bone resorption.

RANKL knockout mice exhibit a phenotype of osteopetrosis and defects of tooth eruption, along with an absence or deficiency of osteoclasts. RANKL activates NF-$\kappa\beta$ (nuclear factor-kappa B) and NFATc1 (nuclear factor of activated T-cells, cytoplasmic, calcineurin-dependent 1) through RANK. NF-$\kappa\beta$ activation is stimulated almost immediately after the occurrence of RANKL–RANK interaction, and is not upregulated. Overproduction of RANKL is implicated in a variety of degenerative bone diseases, such as rheumatoid arthritis and psoriatic arthritis.

1.3.5.1 **Osteoprotegerin**

Cytokines (Greek *cyto*: cell; and *kinos*: movement) are substances (proteins, peptides, or glycoproteins) that are secreted by specific cells of the immune system, which carry signals between cells, and thus have an effect on other cells. Thus, they belong to a category of signaling molecules that are used in cellular communication. The term *cytokine* encompasses a large family of polypeptide regulators that are widely produced throughout the body by cells of diverse embryological origin.

Researchers in the field of bone biology have, for a long time, sought to understand the mechanisms responsible for the cross-talk between osteoblasts and osteoclasts. A major step toward answering this question was provided by the discovery of OPG. OPG was identified in 1997 by three different groups working in different areas [170–173].

It was isolated as a secreted glycoprotein that blocked osteoclast differentiation from precursor cells, prevented osteoporosis (decreased bone mass) when administered to ovariectomized rats, and resulted in osteopetrosis (increased bone mass) when overexpressed in transgenic mice. It was shown that OPG inhibits the differentiation of osteoclast precursors into osteoclasts and also regulates the resorption of osteoclasts *in vitro* and *in vivo*.

Studies in mutant mice have validated the idea that OPG is identical to the osteoblast-derived osteoclast inhibitory factor. Transgenic mice overexpressing OPG exhibit increased bone density and increased mineralization due to a decrease in osteoclasts terminal differentiation.

OPG is a member of the TNF receptor superfamily. OPG is a protein that plays a central role in regulating bone mass: it is a cytokine, which can inhibit the production of osteoclasts. OPG is also known as osteoclastogenesis inhibitory factor (OCIF). It is a basic glycoprotein comprising 401 amino-acid residues arranged into seven structural domains, cf. Figure 1.25. It is found as either a 60 kDa monomer or a 120 kDa dimer linked by disulfide bonds.

Studies have shown that OPG inhibits osteoclastogenesis by binding ODF (RANKL/OPGL/TRANCE) and blocking its interaction with its receptor, RANK, on osteoclasts (TRANCE receptor or TRANCE-R).

OPG contains a cysteine-rich amino-terminal domain, a putative death domain, and a COOH-terminal heparin-binding domain, but unlike other members of the TNF receptor family, it does not contain a transmembrane domain (Figure 1.25). Therefore, it is thought to act as a soluble receptor. OPG has been detected in the bone, heart, lung, liver, stomach, placenta, calvaria, dendritic cells, and blood vessels [175].

In addition, a role for OPG in the development of germinal centers in secondary lymphoid tissues has been postulated. OPG has also been implicated as a cell survival factor: OPG protects endothelial cells from apoptosis induced by serum withdrawal and NF-κB inactivation.

1.3.5.2 **RANK/RANKL**

RANKL is produced by a variety of cell types and its expression is regulated by many physiologic and pathologic factors. Preclinical studies in mice and studies of human tissues have revealed the functions of RANKL/RANK signaling in normal

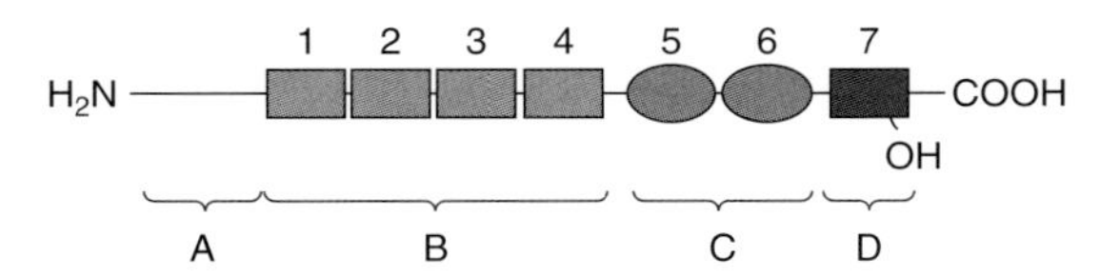

Figure 1.25 Schematic structure of osteoprotegerin OPG polypeptide: A, signaling peptide; B, cysteine-rich amino-terminal domains; C, putative death domains; and D, COOH-terminal heparin-binding domain. After [174].

and pathologic states. The role of the RANKL/RANK system is important not only in bone but also in other tissues.

OPG is a RANK homolog, and works by binding to RANKL on the osteoblast/stromal cells, thus blocking the RANKL–RANK ligand interaction between osteoblast/stromal cells and osteoclast precursors. This has the effect of inhibiting the differentiation of the osteoclast precursor into a mature osteoclast. Recombinant human OPG acts on the bone, increasing bone mineral density and bone volume [176–181]. OPG can bind to RANKL and prevent its interaction with RANK to inhibit osteoclast formation, but its effects on other cellular functions of RANKL have yet to be determined.

Discovery of the RANK signaling pathway in the osteoclast has provided insight into the mechanisms of osteoclastogenesis and activation of bone resorption, and how hormonal signals impact bone structure and mass. Further study of this pathway has provided the molecular basis for developing therapeutics to treat osteoporosis and other diseases of bone loss [180].

Osteoclast formation requires the presence of RANK osteoblasts (RANK) and macrophage colony-stimulating factor (M-CSF). These membrane-bound proteins are produced by neighboring stromal cells and osteoblasts. Thus, a direct contact between these cells and osteoclast precursors is required [168, 169, 178–183].

1.3.5.3 TACE

Tumor necrosis factor-alpha converting enzyme (TACE) is a kind of metalloprotease disintegrins, also known as *ADAM17*. It is a modular transmembrane protein with a zinc-dependent catalytic domain. TACE can cleave or shed the ectodomain of several membrane-bound proteins. In particular, TACE can shed several cytokines from the cell membrane, including RANKL [184, 185].

1.3.5.4 Bone Modeling and Remodeling

Bone modeling develops during the organism's youth and deals with the bone growth in length and width. In this process, a new bone is added (a subprocess called *ossification* or *bone formation*) to a side of the periosteal surface, and the old bone is removed from the skeleton (a subprocess called *bone resorption*) on the side of the endosteal surface. The modeling differs from remodeling in that processes of bone formation and bone resorption are realized at different surfaces of the bone Figure 1.26.

A bone is constantly renewed. The old bone is removed and the new bone is laid down. This process is called *bone remodeling*. Thus, bone remodeling is a lifelong process, where an old bone is removed from the skeleton and a new bone is added. These processes control not only the reshaping or replacement of bone during growth and following injuries like fractures but also microdamage, which occurs during normal activity. Remodeling responds also to functional demands of the mechanical loading. As a result bone is added where needed and removed where it is not required.

In the first year of life, almost 100% of the skeleton is replaced. In adult compact bone, remodeling proceeds at about 10% per year, and in spongy bone, it proceeds

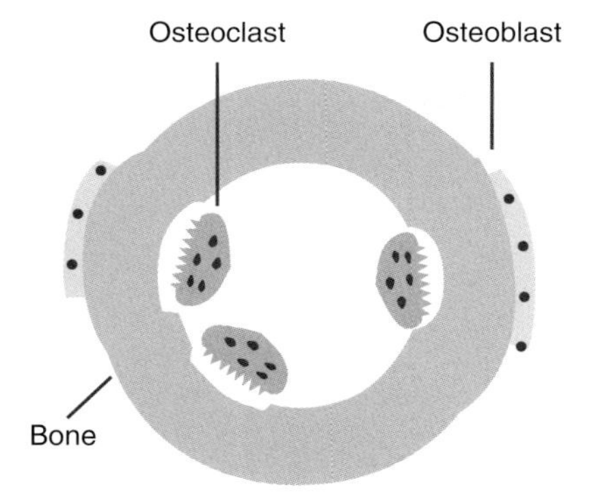

Figure 1.26 Bone modeling. Osteoclasts remove the bone from the endosteal surface. The new bone is added as a result of the osteoblast action from the side of periosteal surface. After [174].

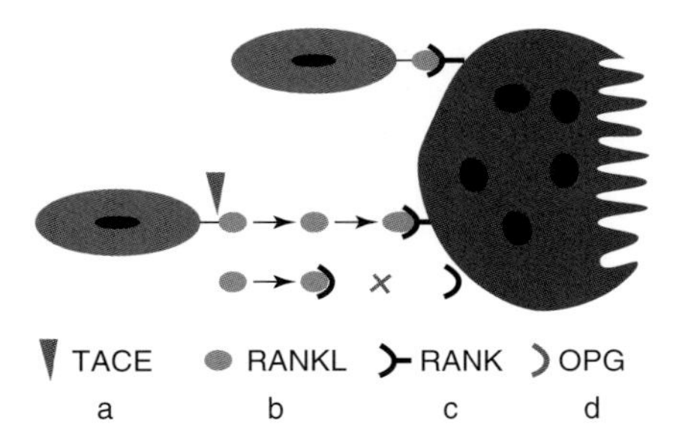

Figure 1.27 The ways of communication between an osteoblast and a maturing osteoclast. Inhibiting action of OPG; TACE (tumor necrosis factor-α converting enzyme): a, ectodomain RANKL after cleavage by the enzyme TACE; b, ectodomain RANKL; c, RANK – membrane receptor of RANKL; and d OPG – solvable receptor-inhibitor of RANKL. After [174].

at about 15–30%. It is so efficient that it is able to exchange constituents of the entire skeleton every 3–10 years [186].

RANKL/RANK/OPG system regulates the differentiation of precursors into multinucleated osteoclasts as well as osteoclast activation and survival, both normally and in most pathologic conditions associated with increased bone resorption (Figure 1.27). Osteoclast differentiation is inhibited by OPG, which binds to RANKL, thereby preventing interaction with RANK, cf. Figure 1.28.

Bisphosphonates exhibit high affinity for hydroxyapatite mineral in the bone and are used to prevent osteoclast-mediated bone loss. The nitrogen-containing bisphosphonate, zoledronic acid (ZOL), influences RANKL expression in human osteoblast-like cells by activating TACE. Bisphosphonates are used to prevent osteoclast-mediated bone loss. ZOL indirectly inhibits osteoclast maturation by increasing OPG protein secretion and decreasing transmembrane RANKL expression in human osteoblasts. The decreased transmembrane RANKL expression seems

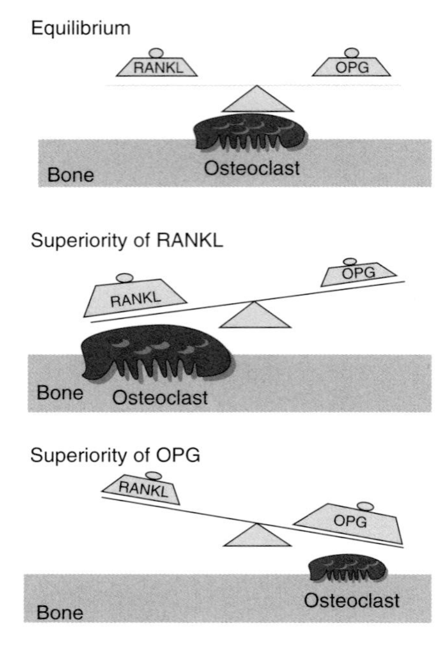

Figure 1.28 Resorption activity of the osteoclast depends on the ratio RANKL/OPG. OPG, osteoprotegerin; RANKL, receptor activator of nuclear factor-κB ligand. After [174].

to be related to the upregulation of the RANKL sheddase, TACE. The reduction in transmembrane RANKL expression was preceded by a marked increase in the expression of the metalloprotease disintegrin, TACE. Studies undertaken by Pan *et al.* indicate that ZOL, in addition to its direct effects on mature osteoclasts, may inhibit the recruitment and differentiation of osteoclasts by cleavage of the transmembrane RANKL in osteoblast-like cells by upregulating the sheddase, TACE [165, 187].

Wedemeyer *et al.* investigated the effect of a single subcutaneous dose of ZOL on particle-induced osteolysis and observed excessive regional new bone formation. They utilized the murine calvarial osteolysis model and polyethylene particles in C57BL/J6 mice. Bone thickness was measured as an indicator of bone growth. Net bone growth was significantly increased in animals with ZOL treatment [166, 188].

Proresorptive molecules that trigger bone loss (hormones and cytokines) induce RANKL expression on osteoblasts. In the inflammatory conditions, activated T-cells also produce RANKL. RANKL binding to RANK on mature osteoclasts and their precursors activates a signal transduction cascade that leads to osteoclast formation and activation. OPG protects bones because of binding to RANKL and inhibiting osteoclastogenesis and osteoclast activation Figure 1.29.

Khosla remarks in his minireview that the identification of the OPG/RANKL/RANK system as the dominant mediator of osteoclastogenesis represents a major advance in bone biology. It ended a long-standing search for the specific factor

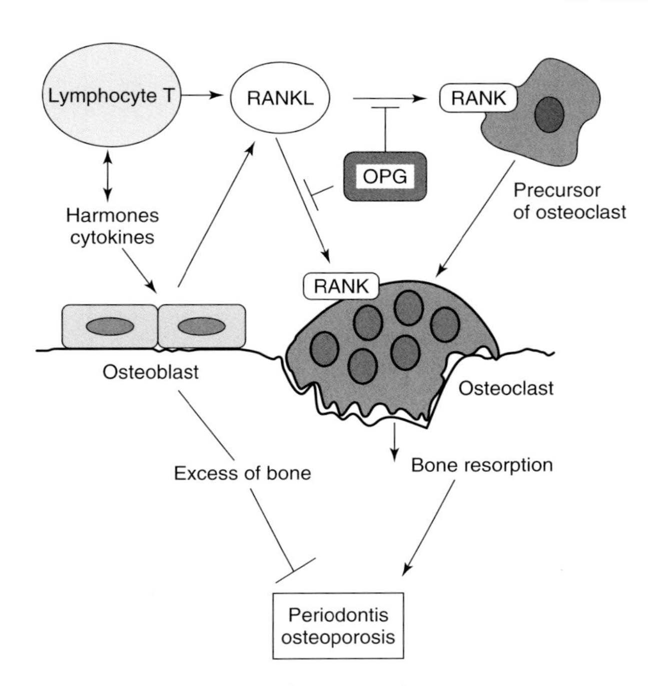

Figure 1.29 The balance of RANKL and osteoprotegerin (OPG) controls osteoclast activity. Most (perhaps all) known inducers of bone resorption and hypercalcemia act indirectly through the production of RANKL; hence OPG can be used in pharmacy to control osteoclast activity, independent of the inducing cytokine. After Stawińska [189].

produced by preosteoblastic/stromal cells that was both necessary and sufficient for osteoclast development. The decisive role played by these factors in regulating bone metabolism was demonstrated by the findings of extremes of skeletal phenotypes (osteoporosis vs osteopetrosis) in mice [168, 169, 190, 191].

When the rate of bone resorption exceeds that of bone formation, destruction of bone tissue occurs, resulting in a fragile skeleton. The clinical consequences, namely, osteoporosis and fragility fractures, are common and costly problems. Treatments that normalize the balance of bone turnover by inhibiting bone resorption preserve bone mass and reduce the risk of fracture. The discovery of RANKL as a pivotal regulator of osteoclast activity provides a new therapeutic target [171, 192].

Thus, the TNF-family molecule OPGL (also known as *TRANCE, RANKL,* and *ODF*) has been identified as a potential ODF and regulator of interactions between T cells and dendritic cells *in vitro*. OPGL is a key regulator not only of osteoclastogenesis but also of lymphocyte development and lymph-node organogenesis [172–174, 193–195].

Mice belong to a commonly used animal research model with hundreds of established inbred, outbred, and transgenic strains. Line MC3T3 is a strain of tissue culture cells derived from *Mus musculus* (house mouse). Various derivatives of this strain have been widely used as model systems in bone biology. The subline MC3T3-E1 is one of the most convenient and physiologically relevant systems for the study of transcriptional control in calvarial osteoplasts.

Kim *et al.* tried to understand the biochemical reaction of RANKL in response to mechanical loading. The MC3T3-E1 cells were biequiaxially stretched. A murine RANKL cDNA with double epitopes, pEF6 HARANKL-V5His, was transfected into MC3T3-E1 cells, which were then stretched. They found that endogenous RANKL protein expression increased in response to mechanical loading. Membrane-bound RANKL (HA-RANKL-V5His) increased in cell lysates while soluble RANKL (RANKL-V5His) decreased in the conditioned media after mechanical loading. This may have resulted from the decreased activity of TACE after mechanical loading. Increased membrane-bound RANKL may be one of the mechanisms through which osteoblasts adapt to mechanical loading by regulating osteoclastogenic activity, [196].

1.3.6
Proteins and Amino Acids

An amino acid is a molecule containing both amine and carboxyl functional groups. There are about 200 amino acids in nature. Protein amino acids or alpha amino acids are the building blocks of proteins. In these amino acids, the amine and carboxyl functional groups are linked to the same atom of carbon, cf. [197–199].

Proteinogenic amino acids are those 22 amino acids that are found in proteins and that are coded for in the standard genetic code. Proteinogenic literally means *protein building*. Proteinogenic amino acids are assembled into a polypeptide (the subunit of a protein) through a process known as *translation* (the second stage of protein biosynthesis, part of the overall process of gene expression).

Peptides (Greek: πεπτίδια, small digestibles) are short polymers formed from the linking, in a defined order, of α-amino acids. The link between one amino-acid residue and the next is called an *amide bond* or a *peptide bond*. Proteins are polypeptide molecules or consist of multiple polypeptide subunits. The distinction is that peptides are short and proteins (polypeptides) are long. Proteins are defined by their sequence of amino-acid residues; this sequence is the primary structure of the protein. It is the genetic code that specifies 20 standard amino acids.

Since the works of Hofmeister (1850–1922) and Fischer (1852–1919), it has been regarded that the proteins are in some fundamental fashion chainlike, that they are constructed from polypeptides of the general formula, given in Figure 1.30, in which $-R'$, $-R''$, $-R'''$, and so on, stand for various univalent groups – 22 different kinds are known – which act as side chains to a common main chain (backbone). Two of these can be specified by the genetic code, but are rare in proteins.

The α-amino acids from which the proteins are formed, and into which they are resolved again on digestion, have the general formula $H_2NCHRCOOH$, where R is an organic substituent, cf. Figure 1.31. In the α-amino acids, the amino and

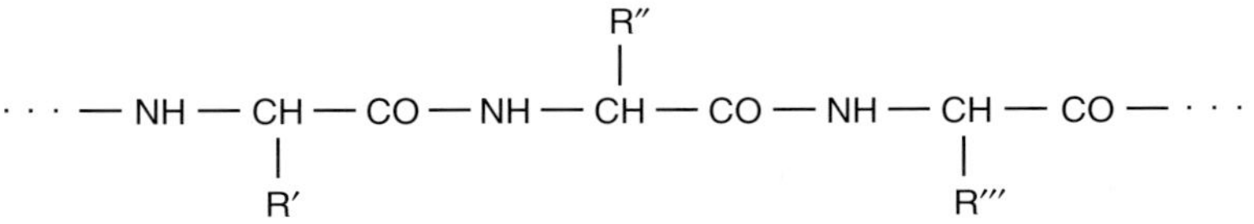

$$\cdots - \mathrm{NH} - \underset{\mathrm{R'}}{\mathrm{CH}} - \mathrm{CO} - \mathrm{NH} - \overset{\mathrm{R''}}{\mathrm{CH}} - \mathrm{CO} - \mathrm{NH} - \underset{\mathrm{R'''}}{\mathrm{CH}} - \mathrm{CO} - \cdots$$

Figure 1.30 "Amino acids in chains are the cause, so the X-ray explains, of the stretching of wool and its strength when you pull, and show why it shrinks when it rains" (A. L. Patterson). After Astbury [200].

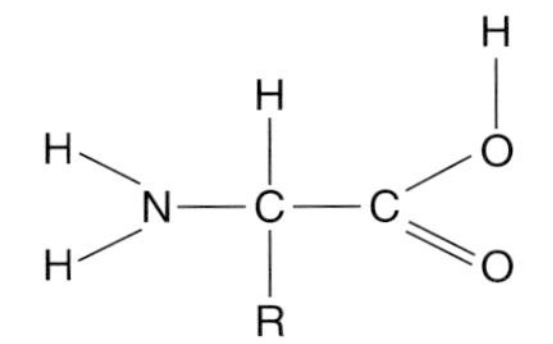

Figure 1.31 An α-amino acid molecule, with the amino group on the left and the carboxyl group on the right. After Pauling Chemistry [9].

carboxyl groups are attached to the same carbon, which is called the *α-carbon*. The various α-amino acids differ in the side chain (R group) that is attached to their α-carbon.

Twenty standard amino acids are used by cells in protein biosynthesis, and these are specified by the general genetic code. These 20 amino acids are biosynthesized from other molecules. Organisms differ in the types of amino acids synthesized by them and the ones that are obtained from food. The amino acids that cannot be synthesized by an organism are called *essential amino acids*. Of the 20 standard amino acids, 8 are essential amino acids.

Glycine and proline, used in building the collagen chain, belong to nonessential amino acids. Hydroxyproline, also appearing in the collagen, is a modification of proline: hydroxyproline differs from proline by the presence of additional hydroxyl (OH) group.

Glycine (abbreviated as *Gly* or G) is the organic compound with the formula NH_2CH_2COOH and is considered a glucogenic amino acid; it is the smallest of the 20 amino acids. Most proteins contain only small quantities of glycine. An exception is collagen, which contains almost one-third of glycine. Glycine is a colorless, sweet-tasting crystalline solid. It is unique among the proteinogenic amino acids in that it is not chiral. It can fit into both hydrophilic and hydrophobic environments, because of its single hydrogen atom side chain. Glycine was discovered in 1820 by Braconnot, who obtained a *gelatin sugar*, named later as glycocolle. It is now called glycine [201].

Proline (abbreviated as *Pro* or P) is an α-amino acid, one of the 20 DNA-encoded amino acids. Its molecular formula is $C_5H_9NO_2$. The distinctive cyclic structure of the proline side chain gives proline an exceptional conformational rigidity as compared to other amino acids. Hydroxyproline (abbreviated as *Hyp*) is an uncommon amino acid and differs from proline by the presence of a hydroxyl (OH) group attached to the C atom.

Because glycine is the smallest amino acid with no side chain, it plays a unique role in fibrous structural proteins. In collagen, *Gly* is required at every third position, because the assembly of the triple helix puts this residue at the interior (axis) of the helix, where there is no space for a larger side group than glycine's single hydrogen atom. For the same reason, the rings of *Pro* and *Hyp* must point outward. These two amino acids, *Pro* and *Hyp*, help stabilize the triple helix. A lower concentration of them is required in animals such as fish, whose body temperatures are lower than those of warm-blooded animals. Proline or hydroxyproline constitute about one-sixth of the total sequence. With glycine accounting for one-third of the sequence, this means that approximately half of the collagen sequence is not glycine, proline, or hydroxyproline.

The element phosphorus is not present in any of the 20 amino acids from which proteins are made (but is present in DNA, cf. the Hershey–Chase experiments [202]).

1.3.7
Collagen and its Properties

Collagen belongs to the long fibrous structural proteins. These are main components of the ECM that supports most tissues and assures cells structure from the outside. Collagen is the main protein of CT in animals and the most abundant protein in mammals, making up about 25% of the total protein content. Thus, collagen is found in large quantities in tendon, bone, skin, cornea, and cartilage. Collagen is also found inside certain cells, cf. [127, 128].

The TC or "collagen macromolecule" is a subunit of larger collagen aggregates such as fibrils. It is approximately 300 nm long and 1.5 nm in diameter, made up of three polypeptide strands (called *α-chains*), each possessing the conformation of a left-handed helix. These three left-handed helices are twisted together into a right-handed triple helix or "super helix" [11]. The triple helix is composed of three polypeptide chains, each with the repeating triplet Gly-X-Y, where X and Y are frequently proline and hydroxyproline, respectively.

The TC macromolecules are synthesized within fibroblast cells, pass into the intercellular tissue spaces, and in particular, aggregate at the appropriate places and time to form fibers. Possibly in all fibrillar collagens if not in all collagens, each TC triple helix associates into a right-handed super-super-coil, which is referred to as the *collagen microfibril*.

The molecular conformation of collagen has been determined primarily from an interpretation of its high-angle X-ray diffraction pattern. Following the pioneering work [203] of Herzog and Jancke, a number of investigators have attempted to find the structure of collagen (and of gelatin, which gives similar X-ray photographs). Astbury showed that there were drastic changes in the diffraction of moist wool or hair fibers as they were stretched significantly (100%). The data suggested that the unstretched fibers had a coiled molecular structure with a characteristic repeat of 5.1 Å (= 0.51 nm). Astbury proposed that (i) the unstretched protein molecules formed a helix (which he called the *α-form*) and (ii) the stretching caused the helix to

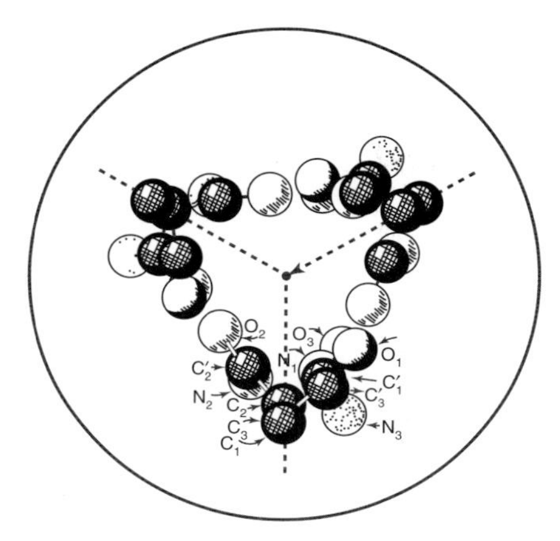

Figure 1.32 The three-chain configuration proposed for collagen and gelatin by Pauling and Corey in 1951 [204].

uncoil, forming an extended state (which he called the *β-form*) [200]. The α-helix has now been recognized as a feature of the majority of protein structures. Astbury's models were correct in essence and correspond to modern elements of secondary structure, the α-helix and the β-strand (Astbury's nomenclature was used), which were developed later by Pauling and Corey. These authors have attempted to account for the positions and intensities of the X-ray diffraction maxima, and pointed out that the equatorial reflections correspond to a hexagonal packing of circular cylinders [204]. They proposed a structure (now withdrawn) having three chains helically intertwined, see Figure 1.32.

The X-ray pattern of an unstretched collagen fiber is rather diffuse and poor in detail. Its principal features are a strong meridional axe at about 2.9 Å and weaker arcs at 4.0 and 9.5 Å. There are also strong equatorial reflections with spacings corresponding to approximately 6 and 12 Å (which depend on the humidity), and a diffuse distribution of intensity around 4.5 Å, mainly near the equator. This pattern is improved in orientation and detail if the fiber is kept stretched during the X-ray exposure [205].

A structural model was built based on various stereochemical properties of the polypeptide chains, and the intensity distribution expected of it was calculated using the helix diffraction theory proposed in 1952 by Cochran *et al.* [206]. Next, the calculated intensity distribution was compared with the observed one [207–211]. The effect of introducing the various side chain atoms on the calculated intensity distribution has been studied in detail by the Madras group of Ramachandran [207].

An interpretation of the X-ray pattern led Rich and Crick in 1955 to the conclusion that the collagen helix has a unit height of approximately 3 Å, the number of units per turn being close to 10/3, which corresponds to a unit twist of 108° [208, 209].

However, the determination of the molecular structure of collagen from X-ray diffraction data has proven extremely difficult, despite the progresses of fiber

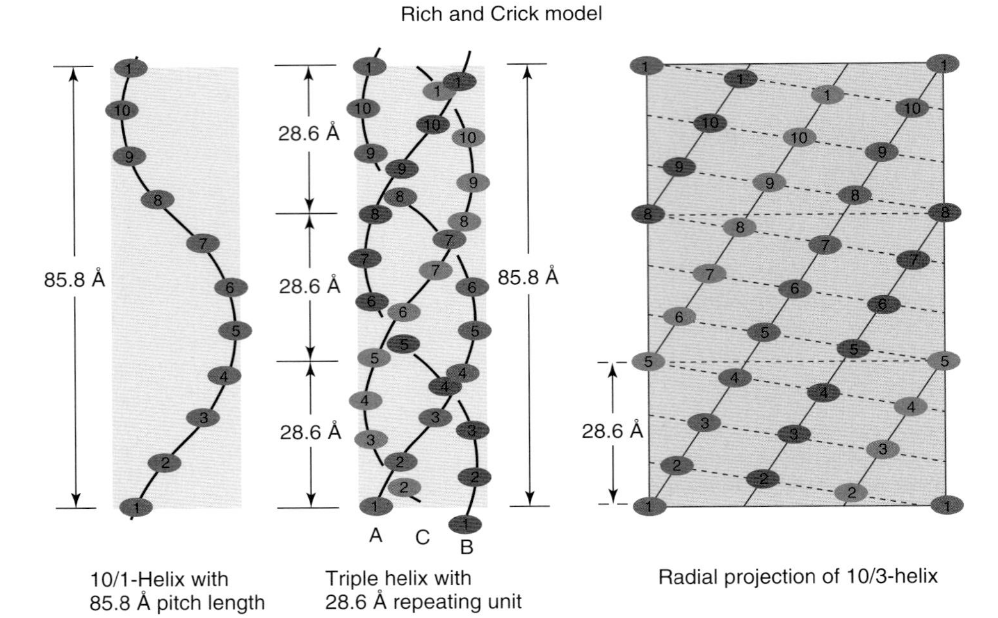

Figure 1.33 The three-chain configuration proposed for collagen by Rich and Crick [208, 209]. For comparison, see Figure 1.5, in which phyllotaxis disposition is given, as depicted by Bravais brothers in 1837. The analogy indicates the screw type development appearing at different levels of living structures and growth of molecules and organs. Courtesy of Kenji Okuyama.

diffraction techniques over the last eight decades. Because of a deficiency of diffraction spots on the layer lines in the wide-angle region (about 1–30 Å resolution), it could not even be determined whether the average helical symmetry of the collagen superhelix was 7/2 (seven tripeptide units for every two turns) or 10/3, [212]. In an article published in 2006, *Microfibrillar structure of type I collagen in situ*, Orgel *et al.* reported the three-dimensional molecular and packing structure of type I collagen determined by X-ray fiber diffraction analysis, which was based on 414 reflections with a completeness of 5% in the range of 5–113 Å resolution. The collagen molecule is made of three chains of more than 1000 residues each.

However, as remarked by Okuyama, it is difficult to determine the three-dimensional molecular conformation based on such a small number of reflections at low resolution [213]. In particular, there are significant differences between the triple helical parameters from the Rich and Crick model. These differences, which led to a 7/2 (with fiber period ~20 Å) as opposed to 10/3 (with fiber period ~29 Å) triple helical symmetry, initiated a debate regarding the actual symmetry of the natural collagen.

In Figures 1.33 and 1.34, two models of the molecular structure of collagen are compared: the first one proposed by Rich and Crick [208, 209] and the second one proposed by Okuyama *et al.* [212–216].

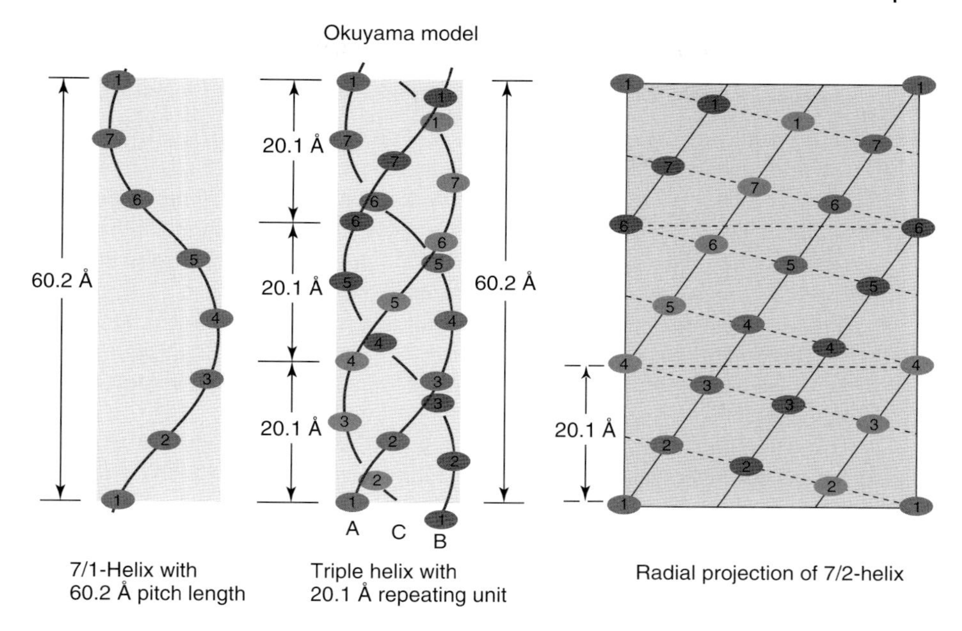

Figure 1.34 The three-chain configuration proposed for collagen by Okuyama [216]. Courtesy of Kenji Okuyama.

The triple-helix motif has now been identified in proteins other than collagens, and it has been established as being important in many specific biological interactions as well as being a structural element. Triple-helix binding domains consist of linear sequences along the helix, making them amenable to description by simple model peptides. Advances, principally through the study of peptide models, have led to an enhanced understanding of the structure and function of the collagen triple helix. In particular, the first crystal structure has clearly shown the highly ordered hydration network that is critical for stabilizing both the molecular conformation and the interactions between triple helices.

Collagen is almost unique among proteins in its use of triple helical secondary structure. Collagen is also unique among animal proteins in its high content of hydroxyproline, which is formed as a post-translational modification of prolines, which are incorporated in the Y position of *Gly*-X-Y triplets. The analysis of collagen structure emphasizes the dominance of enthalpy and hydrogen bonding in the stabilization of the triple helix [217–230].

1.3.7.1 Molecular Structure

There are 29 types of collagens known. Over 90% of the collagens in the body, however, are of types I, II, III, and IV.

- **Collagen I** – skin, tendon, vascular, ligature, organs, bone (main component of bone);
- **Collagen II** – cartilage (main component of cartilage);

- **Collagen III** – reticulate (main component of reticular fibers), commonly found alongside type I;
- **Collagen IV** – forms bases of cell basement membrane.

Type I collagen is the most abundant protein in human body, and it helps to maintain the integrity of many tissues *via* its interactions with cell surfaces, other ECM molecules, and growth and differentiation factors. Nearly 50 molecules have been found to interact with type I collagen, and for about half of them, binding sites on this collagen have been elucidated. In addition, over 300 mutations in type I collagen, associated with human CT disorders, have been described.

The type I collagen is the main component of the bone; it is present in the scar tissue, the end product when the tissue heals by repair. It is found in tendons, skin, artery walls, the endomysium of myofibrils, fibrocartilage, and the organic part of bones and teeth.

The type II collagen is the basis for articular cartilage and hyaline cartilage. It makes up 50% of all proteins in the cartilage and 85–90% of collagen of articular cartilage. The fibrillar network of collagen II allows cartilage to entrap the proteoglycan aggregate as well as provide tensile strength to the tissue. Type II is present in small amounts, with salts, sugars, and vitrosin, in vitreous humor of the eye [222–225].

1.3.8
Geometry of Triple Helix

The triple helix is a unique secondary structural motif that is primarily found within the collagens, and a distinctive feature of collagen is the regular arrangement of amino acids in each of the three chains of collagen subunits.

Coxeter [231, 232] suggested an extension of the concept of a regular polygon. A regular polygon as usually defined is a cycle of vertices ... 1, 2, 3, ... and edges ... 12, 23, ... which is obtained from a single point by repeated action of a rotation. Coxeter's extension replaces "rotation" by the more general "isometry" (distance-preserving transformation). A screw transformation generates a helical polygon (or polygonal helix), an infinite sequence of vertices $\ldots -1, 0, 1, 2, \ldots$, and edges joining consecutive vertices. A Coxeter helix is a polygonal helix such that every set of four consecutive vertices forms a regular tetrahedron. This produces a twisted rod of tetrahedra, the Boerdijk–Coxeter (B–C) helix (Figures 1.35 and 1.36).

Helices and dense packing of spherical objects are two closely related problems. By stacking regular tetrahedral along one direction, one obtains a configuration called the *Bernal* or *B–C helix*. Also, the name tetrahelix given to the chain of tetrahedra by Fuller is used, cf. also [20, 236, 237].

The construction of helix is as follows: to one face of the tetrahedron, the next tetrahedron is glued, and this process of gluing new tetrahedron is continued, with the condition that at one vertex six triangular faces meet (Figure 1.36). The chain of tetrahedra built in such a manner is not periodic because of incommensurability between the distances separating the centers of neighboring tetrahedra and the pitch of the three helices.

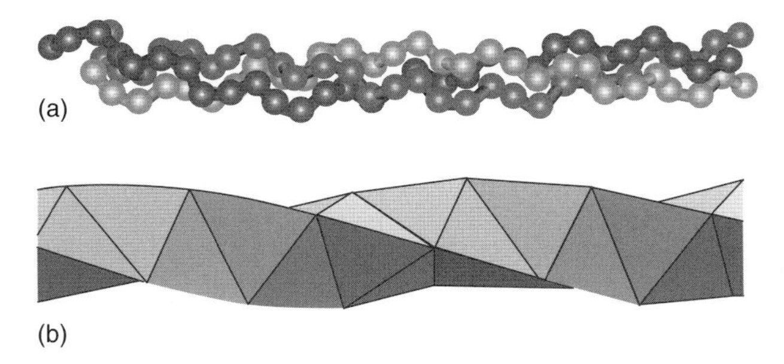

Figure 1.35 Tropocollagen triple helix (a) and the Boerdijk–Coxeter helix (b).

Figure 1.36 The B–C helix as a deltahedron (polyhedron whose faces are all equilateral triangles). Helical structure is generated by the repeated action of a screw transformation acting on a subunit (regular tetrahedron). The tetrahedral helix is called the *Bernal spiral* in association with discussions of liquid structure [233, 234]. Fuller [235] named this helical structure *the tetrahelix* [233].

An infinite strip of a tiling of the Euclidean plane by equilateral triangles, bounded by two parallel lines, can be wrapped around a circular cylinder so that the two strip edges meet. The resulting structure is referred to as a *cylindrical hexagonal lattice.* Alternatively, instead of rolling the strip around a cylinder, corresponding points on the edges may be brought into coincidence by folding along the fundamental lattice lines, keeping the triangular facets flat. The resulting structure is referred to as a *triangulated helical polyhedron* (THP). A THP is an "almost regular" polyhedron, in that its symmetry group, a rod group, acts transitively on the vertices and faces, although not on the edges. The rodlike sphere packings investigated by Boerdijk are derived from the Coxeter helix, which is the simplest THP. In a nanotube, the atomic positions correspond to a subset of the vertices of a THP.

The geometrical properties of the THPs are of relevance in structural chemistry for several reasons. As Sadoc and Rivier [238, 239] have shown, the helical structures commonly occurring in proteins are metrically quite close to polygonal helices consisting of edges of THPs. Sadoc and Rivier proposed a B–C helix with the collagen sequence *Gly*-X-Y.

The value of angle ψ is found from the symmetry: the edges AB, BC, and CD of a tetrahedron projected on the cross section of the circumscribed cylinder are all equal to a certain value c.

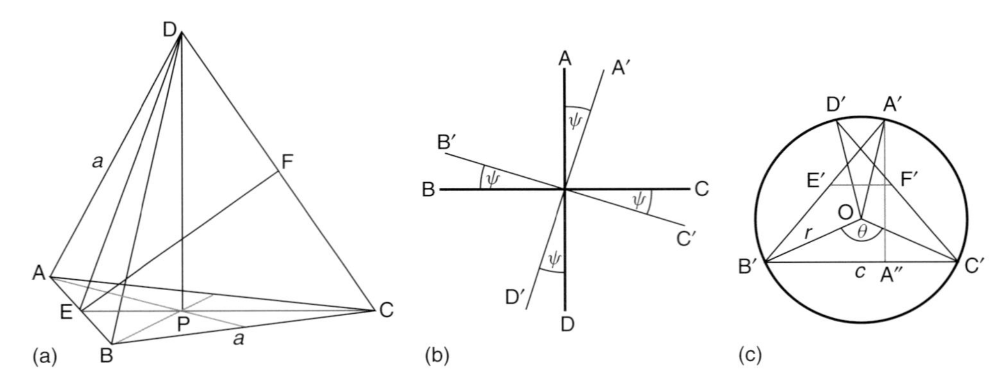

Figure 1.37 (a) A tetrahedron has four faces, four vertices, and six edges. A regular tetrahedron is composed of four equilateral triangular faces. All edges are equal: AB = BC = AC = AD = BD = CD = a. The height of the triangular face, for example, EC = $\sqrt{3}\,a/2$, the height of tetrahedron: PD = $\sqrt{2/3}\,a$, the distance between perpendicular edges, for example, AB and CD is equal to EF = $a/\sqrt{2}$. (b) Top view of two perpendicular tetrahedron edges: AD and BC. In order to inscribe the tetrahedron in a cylinder of radius r, the edges should be turned with respect to the axis of the cylinder through an angle of ψ. The distance $a/\sqrt{2}$ between the perpendicular edges AD and BC remains unchanged. (c) Projection of tetrahedron edges on the cross section of the cylinder circumscribed on the tetrahedron. The segment A′A″ $= a/\sqrt{2}$ (the distance between perpendicular edges AD and BC). The length scale between figures (a), (b), and (c) is not preserved.

According to Figure 1.37, we have projections $\mathrm{A'D'} = a\sin\psi$ and $\mathrm{B'C'} = a\cos\psi$. Hence, $\mathrm{B'A''} = (a/2)(\sin\psi + \cos\psi)$. Further $(\mathrm{A'A''})^2 + (\mathrm{B'A''})^2 = (\mathrm{A'B'})^2$. Subsequently, $\cos\psi = 3/\sqrt{10}$ and $\mathrm{A'B'} = \mathrm{B'C'} = \mathrm{C'D'} = c = 3a/\sqrt{10} \cong 0.9486\,a$. Moreover, $\mathrm{A'D'} = a/\sqrt{10}$. Because $\mathrm{A'D'} = 2r\sin((3\theta - 2\pi)/2)$ and $\mathrm{B'C'} = 2r\sin(\theta/2)$, we get the angle θ and radius r of the cylinder. We get $\sin(\theta/2) = \sqrt{5/6}$, $\cos\theta = -2/3$. Thus $\angle\mathrm{A'OB'} = \angle\mathrm{B'OC'} = \angle\mathrm{C'OD'} = \theta$ and $r = 3\sqrt{3}a/10 = \sqrt{3/10}c$.

The spiral moves along the tetrahelix by an angle $\theta = \arccos(-2/3) \approx 131.81^\circ$. The angle θ is an irrational number, and the tetrahelix has no period. It belongs to aperiodic crystals (according to Schrödinger [3]) or quasicrystals (according to the contemporary notion).

Let the edge of the tetrahedron be equal to 1. The whole structure of a B–C helix is generated by a screw transformation $\mathbf{x} \to R\mathbf{x} + \mathbf{a}$ where

$$R = \frac{1}{3}\begin{bmatrix} 221 \\ 2-1-2 \\ -12-2 \end{bmatrix}, \quad a = \begin{bmatrix} 1/3\sqrt{2} \\ 1/3\sqrt{2} \\ 1/3\sqrt{2} \end{bmatrix}$$

Sadoc and Charvolin have indicated that the idea of three-sphere fibrations may be a tool for analyzing twisted materials in condensed matter. They indicated that chiral molecules, when densely packed in soft condensed matter or biological materials, build organizations that are most often spontaneously twisted. The formation of these organizations is driven by the fact that compactness, which tends to align the molecules, enters into conflict with torsion, which tends to disrupt this alignment. This conflict of topological nature, or frustration, arises because of the flatness of the Euclidean space, but does not exist in the curved space of the three-sphere where particular lines, its fibers, can be drawn which are parallel and nevertheless twisted. As these fibrations conciliate compactness and torsion, they can be used as geometrical templates for the analysis of organizations in the Euclidean space [240].

1.3.9 Polymer Thermodynamics

The first experiments suggested that TC molecules are rigid, rod-shaped structures. However, on the basis of hydrodynamic methods and transmission electron microscopy (TEM), it was found that TC molecules demonstrate some flexibility, which can be measured by the persistence length [241].

1.3.9.1 Thermodynamics

From the first law of thermodynamics, the increase in internal energy $\mathrm{d}U$ during any change in a system is equal to the sum of the elementary amount of heat $\bar{\mathrm{d}}Q$ added to it and the elementary amount of work $\bar{\mathrm{d}}L$ performed on it.

$$\mathrm{d}U = \bar{\mathrm{d}}Q + \bar{\mathrm{d}}L$$

The second law states that the increase in heat $\bar{\mathrm{d}}Q$ is expressed in any reversible process by the relation

$$\bar{\mathrm{d}}Q = T\,\mathrm{d}S$$

where $\mathrm{d}S$ denotes the entropy differential and T denotes the (absolute) temperature. Hence, for a reversible process

$$\mathrm{d}U = T\,\mathrm{d}S + \bar{\mathrm{d}}L$$

The Helmholtz free energy is defined by

$$A = U - TS$$

Its change at constant temperature is given by

$$\mathrm{d}A = T\,\mathrm{d}U + T\,\mathrm{d}S$$

which means that

$$\mathrm{d}A = \bar{\mathrm{d}}L$$

The work done on the system by a tensile force F in a displacement $\mathrm{d}x$ is

$$\bar{\mathrm{d}}L = F\,\mathrm{d}x$$

Then,

$$F = \left(\frac{\partial A}{\partial x}\right)_{\tau}$$

This means that the force of tension is equal to the change in Helmholtz free energy per unit extension. In the normal unstressed state, the free energy has a minimum $(\partial A/\partial x)_T = 0$ at certain $x = x_0$, and for small strains we get

$$F = \left(\frac{\partial^2 A}{\partial x^2}\right)_T (x - x_0)$$

where x denotes the length of the system under the force F. In this approximation, the force is a linear function of the deformation $(x - x_0)$. The tension force, like free energy, may be expressed as the sum of two terms

$$F = \left(\frac{\partial A}{\partial x}\right)_T = \left(\frac{\partial U}{\partial x}\right)_T - T\left(\frac{\partial S}{\partial x}\right)_T$$

In general, the force results in both the internal energy and entropy changes.

1.3.9.2 **Ideal Chain**

Ideal chain model (Gaussian chain) or the freely joined chain assumes that there are no interactions between chain monomers. The ideal chain is the simplest model of a polymer. In this model, fixed length polymer segments are linearly connected, and all bond and torsion angles are equiprobable: the polymer can be described by the random walk statistics [242–244].

Monomers are regarded as rigid rods (segments) of a fixed length l, and their orientation is independent of the orientations and positions of neighboring monomers. This means that no interactions between monomers are considered, the energy of the polymer is taken to be independent of its shape, and all of its configurations are equally probable. If N monomers form the polymer chain, its total length is $L = Nl$.

Let $\mathbf{b}_0, \ldots, \mathbf{b}_{N-1}$ be the vectors corresponding to individual monomers, and let $\boldsymbol{r}$ denote the end-to-end vector of a chain, that is, the distance between the ends of the chain, Figure 1.38. The monomer vectors have randomly distributed components in the three directions of space. With the assumption that the number of monomers N is large, and the central limit theorem applies, the length r is described by the following Gaussian probability density function:

$$p(r) = \left(\frac{3}{2\pi Nl^2}\right)^{3/2} e^{-\frac{3r^2}{2\pi Nl^2}}$$

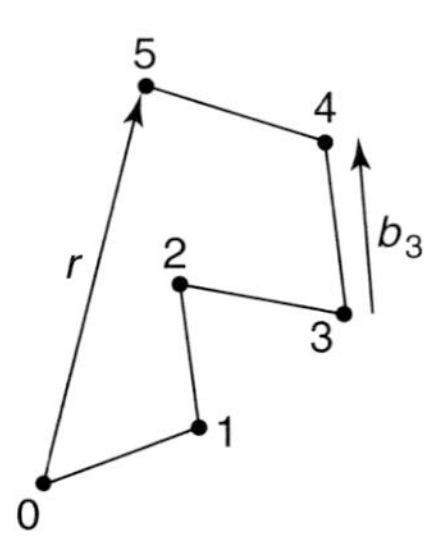

Figure 1.38 Chain of rigid, randomly oriented rods. In this example, we have five rods only, with directions described by vectors $\boldsymbol{b}_0, \ldots, \boldsymbol{b}_4$, and with end-to-end vector $\boldsymbol{r}$.

The entropy of the ideal chain is given by

$$S(r) = k_B \ln p(r) + C$$

where k_B is Boltzmann's constant and C is a constant. As ideal chain has no internal energy, the force appearing as a result of stretching the ideal chain is given by

$$F_{IC} = -T\left(\frac{\partial S}{\partial x}\right)_T$$

This force has purely entropic origin. If the stretching of polymer ideal chain is made in direction of the end-to-end vector $\boldsymbol{r}$, the force exerted on the chain is given by

$$F_{IC} = -T\frac{\mathrm{d}S}{\mathrm{d}r} = k_B T\frac{3r}{2Nl^2}$$

The above result gives the equation of state of the ideal chain: the force is linearly proportional to the temperature. Since the result was derived using the central limit theorem, it is exact for polymers containing a large number of monomers (known as the *thermodynamic limit*).

1.3.9.3 **Wormlike Chain**

The ideal chain model provides a starting point for the investigation of more complex systems.

An important model for polymers in solution is the Kratky–Porod model, sometimes referred to as the wormlike chain (WLC). The WLC model is used to describe the behavior of semiflexible polymers. The model replaces a polymer macromolecule by an isotropic rod that is continuously flexible, in contrast to the freely joined chain model that is flexible only between discrete segments [245–247].

For a polymer chain of length L, parametrize the path along the polymer chain with variable s, $0 \leq s \leq L$. Consider $\mathbf{t}(s)$ being the unit tangent vector to the chain at s. It can be shown that the orientation correlation function for a WLC follows an exponential decay law

$$\langle \mathbf{t}(s)\mathbf{t}(0)\rangle \equiv \langle \cos\theta(s)\rangle = \mathrm{e}^{-s/\xi}$$

In this expression, the quantity ξ is a characteristic constant for a given polymer, known as *persistence length*, regarded as a mechanical property describing the stiffness of a macromolecule.

The stretching force F_{WLC} acting on a WLC with contour length L and persistence length ξ is described by the following interpolation formula:

$$F_{WLC}(r) = \frac{k_B T}{\xi}\left(\frac{1}{4\left(1-\frac{r}{L}\right)^2} - \frac{1}{4} + \frac{r}{L}\right)$$

For small values of fraction r/L, one obtains

$$F_{WLC}(r) \approx \frac{k_B T}{\xi}\left[\frac{1}{4}\left(1+\frac{2r}{L}\right) - \frac{1}{4} + \frac{r}{L}\right] = k_B T\frac{3r}{2\xi L}$$

The force F_{WLC} becomes divergent for the straightened chain, when $r \rightarrow L$ [245–247].

Stretching individual biomolecules are now achieved by a variety of techniques including flow stress, microneedles, optical tweezers, and magnetic tweezers that allow measurement of forces from 10 fN to hundreds of piconewtons.

Different biological molecules have now been analyzed and the accuracy of these techniques has sufficiently improved so that the theoretical models used to analyze force-extension curves must be refined. In particular, Bustamante *et al.* have shown that the force-extension diagram of a DNA molecule is well described by a WLC model [248, 249], and DNA has persistence length $\xi \approx 50$ nm.

The TC molecules have lengths of $L \approx 300$ nm, and are roughly 1.5 nm in diameter. Hydrodynamic methods and TEM permitted us to show that TC molecules exhibit some flexibility. Hydrodynamic methods suggested a persistence length of 130–180 nm; and TEM-based methods estimated the range between 40 and 60 nm. Experiments using optical tweezers suggested a much lower persistence length, between 11 and 15 nm.

Buehler and Wong reported molecular modeling of stretching single molecule of TC, the building block of collagen fibrils and fibers that provide mechanical support in CTs. For small deformation, they observed a dominance of entropic elasticity. At larger deformation, they have found a transition to energetic elasticity, which is characterized by first stretching and breaking of hydrogen bonds, followed by deformation of covalent bonds in the protein backbone, eventually leading to molecular fracture. Their force–displacement curves obtained at small forces show excellent quantitative agreement with optical tweezer experiments. Their model predicts a persistence length $\xi \approx 16$ nm, confirming experimental results suggesting the flexible elastic nature of TC molecules [241].

1.3.9.4 **Architecture of Biological Fibers**

Most biological tissues are built with polymeric fibers. Two of the most important and abundant fibers found in nature are cellulose and collagen. Cellulose is the structural component of the cell walls of green plants, many forms of algae, and the oomycetes. Some species of bacteria secrete it to form biofilms. As a constituent of the plant cell wall, it is responsible for the rigidity of the plant stems. One of the questions deals with the relation between the arrangement of cellulose fibrils inside the cell wall and its mechanical properties, and the way in which the cellulose architecture is assembled and controlled by the cell.

Similar problems arise with the structure of collagen fibrils. The TC molecules are arranged in microfibrils, which are seen under an electron microscope. Microfibrils are arranged into fibrils visible under a light microscope, and fibrils built connective tissues (CTs). The TC molecules and microfibrils are also cross-linked. Examples of fibrous system that achieves high tensile strength by a lateral bonding of macromolecular polymers are the materials such as skin, tendon, and other forms of CTs containing the protein collagen. The tensile strength is about 700 MPa; it is of the order of greatness observed for silk and stainless steel. The tensile strength of a bone is about 100 MPa [17, 250].

1.3.9.5 Architecture of Collagen Fibers in Human Osteon

Gross [15] proposed that collagen develops in seven steps:

1) the starting materials are free amino acids;
2) hydroxyproline is produced from proline after molecular chain has been formed;
3) the chain twists itself into a left-handed helix;
4) three chains intertwine to form a right-handed superhelix, the TC molecule;
5) many molecules line up in a staggered fashion;
6) the molecules overlap by one-quarter of their length to form a fibril;
7) fibrils in CTs are often stacked in layers with fibrils aligned at right angles.

Collagen is synthesized within fibroblast cells as a precursor, procollagen, which also consists of three chains. Molecular mass of procollagen molecules is $\approx$140 000. The polypeptides synthesized on the ribosomes do not contain hydroxyproline or hydroxylysine, which are generated as post-translational modifications before the procollagen is extruded. The TC chains that pass into the intercellular space spontaneously form microfibrils. Then, a variety of cross-links are formed that contribute to the strength of collagen.

This monomeric and microfibrillar structure of the collagen fibers that was discovered by Fraser, Miller, Wess (amongst others) was closest to the observed structure, although their description was oversimplified in topological progression of neighboring collagen molecules and hence did not predict the correct conformation of the discontinuous *D*-periodic pentameric arrangement [251–254].

The properties of collagen and associated polymers are important to understand the structure and functional mechanisms of biocomposites such as bone and cartilage on the microscopic level. The TC subunits spontaneously self-assemble, with regularly staggered ends, into even larger arrays in the extracellular spaces of tissues. In the fibrillar collagens, the TC molecules are staggered from each other by about 67 nm (a distance that is referred to as "*D*" and changes depending upon the hydration state of the aggregate). Each *D*-period (or *D*-spacing) contains approximately 4.4 of TC molecules. This is because 300 nm divided by 67 nm does not give an integer (the length of the collagen molecule divided by the stagger distance *D*). Therefore, in each *D*-period repeat of the microfibril, there is a part containing five molecules in cross section – called the *overlap* and a part containing only four molecules, Figure 1.39.

Studies are carried out on the assembly properties of concentrated solutions of type I collagen molecules. They are, for example, compared before and after sonication, breaking the 300 nm triple helices into short segments of about 20 nm, with a strong polydispersity. Whereas the nonsonicated solutions remain isotropic, the sonicated solutions transform after a few hours into a twisted liquid crystalline phase, well recognizable in polarizing microscopy. The evidence of a twisted assembly of collagen triple helices *in vitro* is relevant in a biological context since it was reported in various collagen matrices.

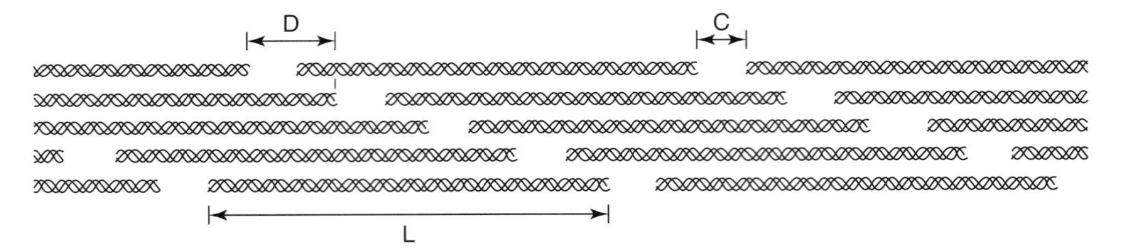

Figure 1.39 Scheme of the structure of collagen microfibril. Microfibrils have a characteristic striation with a repeated distance of $D = 668$ Å because of the end-to-end alignment of the TC molecule. The *D*-period distance corresponds to one hole zone and one region of short overlap: *C* – an overlap region; *L* – length of the tropocollagen molecule, the length of TC molecule $L = 4.4 \times D$; *C* – a hole zone.

The triple helices are also arranged in a hexagonal or quasi-hexagonal array in cross section, in both the gap and overlap regions.

Owing to the noninteger relationship between molecular length and the *D*-period, the projection of the axial structure repeat for a type I collagen fibril consists of a gap region and an overlap region in the ratio of 0.54 : 0.46 *D*, respectively. The ends of the triple helical region are defined by telopeptides, which do not conform to the repeating *Gly*-X-Y pattern of the helical region. The telopeptide region is known to be responsible for the formation of intermolecular cross-links.

Progress has been made in describing the relation between structure and deformation mechanisms of collagen-rich tissues. The principles for the self-assembly of collagen fibrils into larger scale structures still remain a mystery [250].

Cameron, Cairns, and Wess observed the Bragg peaks of diffracted rat tail tendon with sample-to-detector distances of 2000 mm, and obtained an X-ray diffraction image of the meridional region from fibrillar type I collagen. The meridional Bragg peaks were formed from the interactions of the X-rays with the regular repeating structure of the axial packing scheme of the collagen molecules. Next, they investigated the modeling of the nonuniform axial translation of amino acids in the collagen helix through the utilization of a genetic algorithm, and tried to evolve a sequence-based structure that obtains the best fit to the observed meridional X-ray diffraction data. Three structural elements of the model were investigated – the telopeptide inter-residue spacing, folding of the telopeptide, and the helical inter-residue spacing. They found that the variability in amino acid axial rise *per* residue of the collagen helix is an important parameter in structural models of fibrillar collagen [255].

1.3.9.6 **Collagen Elasticity**

Collagen has great tensile strength, and is the main component of fascia, cartilage, ligaments, tendons, bone, and skin. It strengthens blood vessels and plays a role in tissue development. It is present in the cornea and lens of the eye in crystalline form.

Collagen has well-defined mechanical properties (strength, reversible extensibility through only a small range) that make it suited to the special goals for which it is present in the different parts of animal body. Collagen characteristics, in particular, the specific shape of stress–strain curve, are important while considering the properties of a number of tissues, like bones, tendons, or arteries. The experimental trials to determine the mechanical properties of collagen molecule were performed by several groups of researchers, using the X-ray diffraction technique and Brillouin light scattering, beginning from Cowan *et al.* in 1955 [256, 257]. Despite its importance to the biological function of collagen, there is still a lack of understanding of the correlation between the specific shape of the stress–strain curve and the deformations of collagen at the molecular level.

Contemporary efforts to determine the shape of the stress–strain curve of collagen widely exploit synchrotron X-ray diffraction. Synchrotron radiation studies of fibril behavior in tissues have been critical to the investigation of structural details and changes, since they allow transient structural features to be monitored in realistic timescale.

Three experiments using this radiation, by Misof *et al.*, Puxkandl *et al.*, and Gupta *et al.*, should be mentioned here [258–261].

Misof *et al.* have performed *in situ* synchrotron X-ray scattering experiments, which show that the amount of lateral molecular order increases upon stretching of collagen fibers. In strain cycling experiments, the relation between strain and diffuse equatorial scattering was found to be linear in the "heel" region of the stress–strain curve [258].

The stress–strain curve of collagen is characterized by a region of relatively low elastic modulus at small strains ("toe region") followed by an upward bend of the curve ("heel region") and, finally, a linear region with high elastic modulus at large strains. The toe region of the stress–strain curve had been linked to a macroscopic crimp with a period of about 100 μm, found in unstretched collagen fibers by polarized light microscopy, cf. [262, 263].

In tendons, this crimp disappears upon stretching at extensions of the order of 4% – a value that may depend on the age of the animal. At larger strains, X-ray diffraction measurements of the axial molecular packing were interpreted as a side-by-side gliding of the molecules, accompanied by a stretching of the cross-linked telopeptide terminals and a stretching of the triple helices themselves, as was shown by Mosler *et al.* [264].

Although the axial packing of collagen fibrils is regular, as was shown in 1963 by Hodge and Petruska [265], there is a disorder in the structure, in the lateral packing of the molecules.

The following form of the stress–strain relation ($\sigma = \sigma(\varepsilon)$) was proposed in [258] by Misof *et al.*

$$\sigma = \frac{K\varepsilon}{1 - \varepsilon/\varepsilon_0}$$

with

$$K = k_B T \rho \frac{\nu}{\varepsilon_0^2 D} \quad \text{and} \quad \varepsilon_0 = \nu l \frac{\delta_0}{D}$$

Here $k_B = 1.38 \times 10^{-23}$ J K^{-1} is the Boltzmann constant, T is the absolute temperature, ρ is the number of molecules per unit surface in the equatorial plane, D is the length of the axial collagen period after the removal of all kinks, and ν is the total number of kinks of a molecule *per* D-period. Moreover, δ_0 denotes the average value $<\delta>$, obtained in the absence of external stress. Thus, the entropic regime may account for the initial regime of stress–strain behavior for fibril strains up to 8%, as is shown in figure 6 in the Misof *et al.* paper, cf. also [244].

Hence, when ε is small compared to ε_0, the prediction is a linear elastic behavior with the elastic modulus K. When ε approaches ε_0 (that is, when almost all of the kinks are removed), the tension σ required to get a further extension by the removal of the remaining kinks tends to infinity, which means that other mechanisms must become dominant for the sample behavior. The stress–strain curve corresponding to the Misof relation is bent upward. When tension σ becomes too large, other mechanisms of collagen elasticity, like a direct stretching or a side-by-side gliding of the molecules, come into play and lead to a linear stress–strain relation instead of the Misof relation.

When all the kinks are straightened, another mechanism of deformation must prevail and explain the linear dependence of stress and strain in this region of the force–elongation curve. The most probable processes are thought to be the stretching of the collagen triple helices and the cross-links between the helices, implying a side-by-side gliding of neighboring molecules, leading to structural changes at the level of the collagen fibrils.

A hierarchical structure of a collagen tendon for the model of strain-rate-dependent effects was proposed by Puxkandl *et al.* [259]. The tendon was considered as a composite material with collagen fibrils embedded in a proteoglycan-rich matrix. This matrix is mostly loaded under shear. Because the spacing between fibrils is much smaller than their length, the shear stress τ effectively applied to the matrix is much smaller than the tensile stress σ on the tendon. In reality, $\tau \approx \sigma H/L$, where L is the length of the fibrils and H is their spacing, [266]. The aspect ratio L/H is of the order of 100–1000. One may suppose that the elastic response of the matrix is mostly due to the entanglement of molecules attached to the collagen fibrils, such as proteoglycans. In addition, there is a considerable viscosity in the matrix due to the many hydrogen bonds that can form in this glassy structure.

Gupta *et al.* in 2004 carried out tensile measurements by using synchrotron X-ray diffraction [260]. A load cell was mounted on one grip while the other was moved with a motorized translation. To maintain the tendon in a native state, the sample was partially immersed in a physiological solution of phosphate-buffered saline during the test.

The most striking result was a peak splitting of the diffraction spectrum for large macroscopic tendon strains, $\geq$2–3%. The splitting implies an inhomogeneous fibril elongation, with part of the fibrils relaxed back to their unstressed length and the remainder elongated by more than 4–5%. The peak splitting occurs shortly after the mechanical transition from approximately linear elastic to inelastic behavior in the stress–strain curve.

1.4 Remarks and Conclusions

The structure of a bone is discussed in this chapter; this discussion applies to a wide range of species, families, and orders, since, in a rough treatment, the bones of all mammals could be described in a similar manner. Figures and photographs of human bones and those of different animals are presented here without differentiation, although one should be aware of the fact that every species has its own characteristic features.

The bone components described above are essential for development and maintenance of a whole bone. For example, clinical data indicate that the nervous system influences skeletal development. Sciatic nerve injury in infants is sometimes followed by subnormal foot growth [267]. Moreover, Janet McCredie reported that the limbs of children with thalidomide malformations show changes analogous to those that can occur in the adult as a consequence of pathological alterations to peripheral nerves. The original animal tests did not show indications of this unexpected and serious side effect [268, 269].

Scurvy is a disease resulting from a deficiency of vitamin C, which is required for the synthesis of collagen in humans. Collagen is an important part of the bone, so bone formation is also affected. Most plant and animal species synthesize vitamin C. Notable exceptions in the mammalian group include most or all of the order chiroptera (bats), and one of the two major primate suborders, including human beings.

Vitamin D deficiency results in impaired bone mineralization and leads to bone-softening diseases including, rickets, a childhood disease characterized by impeded growth and deformity of the long bones [270].

Astronauts lose significant bone mass during spaceflight, in conditions of microgravity, despite rigorous musculoskeletal conditioning exercises. Femoral neck-bending strength index decreased 2.55%/month for spaceflights lasting 4–6 months.

Similarly, in a study on the influence of long-term immobilization in dogs, 32 weeks of disuse resulted in significant bone loss in both young and old dogs. After 28 weeks of remobilization, young dogs (respectively, old dogs) recovered only 70% (respectively, 40%) of the cortical bone lost during disuse.

Bears and humans have similar lower limb skeletal morphology and bears are plantigrade like humans.

In bears, during hibernation disuse, the bone mass and strength losses are small, if any. It is hypothesized that bears maintain bone cross-sectional properties and strength during annual periods of disuse because they maintain bone formation.

It would mean that bears have biological mechanisms to prevent disuse osteoporosis. This may be accomplished by targeting genes and circulating hormones that are differentially expressed in bears and humans during disuse [271].

1.5 Comments

Both the structure and function of a bone are complex characteristics that have not been understood well by researchers in the fields of medical sciences and mechanics. Since a bone could be treated as an organ, a tissue, or a hard tissue, different methods of describing behaviour of the bone and its physiology could be used, including mathematical topology, partial differential equations, elasticity, strength of materials, mechanics of porous media, diffusion and electrophoretic theories, chemistry, and optical and X-ray experimental methods. In such a description, different models of a bone could be presented including those resembling living and nonliving matter. In the present paper a number of the above methods of describing a bone, are presented.

V. I. Arnold, in his address at the Conference on Teaching of Mathematics in Palais de Découverte in Paris, on 7 March 1997, said, "Mathematics is a part of physics. Physics as an experimental science is a part of natural science. Mathematics is a part of physics where experiments are cheap. ... The mathematical technique of modeling consists of ignoring some experiments and providing a deductive model in such a way as if it coincided with reality. The fact that this path, which is obviously incorrect from the point of view of natural science, often leads to useful results in physics is called 'the inconceivable effectiveness of mathematics in natural sciences' (or 'the Wigner principle'). Here one can add a remark by I. M. Gel'fand: there exists yet another phenomenon which is comparable in its inconceivability with the inconceivable effectiveness of mathematics in physics noted by Wigner – this is the equally inconceivable ineffectiveness of mathematics in biology" [272].

A mathematical structure is hidden in the world around us. A moderate opinion by Andrzej Lasota expresses belief that mathematics is the very structure of the world. "Not a description of the structure, but structure itself. There is no doubt that a mathematician can create strange objects and he may think that he has strayed far from reality. But this is a resemblance of truth only. If his creation is good mathematics, it will sooner or later prove to be a component of reality" [273].

A number of examples where mathematics is efficient in biology can be found in the literature. In particular, considerable progress has been made in recent years in the understanding of the molecular and microstructural properties of bone components, especially on molecular level, which is of general interest in biology, physics, and medicine. Different approaches have resulted, owing to both the difficulty and importance of the problem. As a recent example, Jakob Bohr and Kasper Olsen suggested that the close packing forms the underlying principle behind the structure of collagen. It is shown that the unique zero-twist structure with no strain–twist coupling is practically identical to the close-packed triple helix. Further, the proposed new geometrical structure for collagen is better packed than both the 10/3 and the 7/2 structures [274].

It is characteristic in the field of medicine that, often, indirect explorations lead to proper results for the problems that are resistant to direct investigation. There is a reason to hope that by studying the bone and cartilage tissues on different levels of their organization and from different points of view, one can arrive at knowledge that can be of useful for patients [275].

1.6 Acknowledgments

I wish to thank Professors Mariusz Gajda and Jan A. Litwin from the Institute of Histology of Jagellonian University for a permission to use their unpublished microphotohraphs of a bone in the present chapter; Professor Andrzej Lissowski (Society of Polish Free University) for permitting me to insert his unpublished figures on a two-dimensional crystallization; and Professor Kenji Okuyama (Department of Macromolecular Science, Graduate School of Science, Osaka University) who let me know of his paper before publication and permitted me to present his figures on collagen structure.

I sincerely thank Professor Brian J. Ford from Cambridge University who kindly provided me with clarifications in respect of the first Leeuwenhoek paper. Professor Józef Ignaczak discussed a number of problems appearing in this chapter with me. Professor Andreas Öchsner (Technical University of Malaysia) encouraged me to complete this chapter.

I am also indebted to the State Committee for Scientific Research (KBN, Polska) for support through the project grant No 4 T07A 003 27.

References

1. Aristotle (2007) On the Generation of Animals, translated by A. Platt, Book II; eBooks@Adelaide.
2. Lowenstam, H.A. and Weiner, S. (1989) *On Biomineralization*, Oxford University Press, New York.
3. Schrödinger, E. (1944, 1945) *What is Life? The Physical Aspects of the Living Cell*, Cambridge University Press, London and Macmillan, Bentley House, New York, Toronto, Bombay, Calcutta, Madras.
4. Pasteur, L. (1922) *Oeuvres de Pasteur*, Tome Premier Dissymétrie Moléculaire, Masson et Cie, Paris.
5. Pasteur, L. (1897) *Researches on the Molecular Asymmetry of Natural Organic Products*, William F. Clay, Edinburgh and Simpkin, Marshall, Hamilton, Kent & Company, London.
6. Thomson, W. (1904) Baltimore Lectures on molecular dynamics and the wave theory of light, founded on Mr. A. S. Hathaway's stenographic report of twenty lectures delivered in Johns Hopkins University, Baltimore, in October 1884 by Lord Kelvin: followed by twelve appendices on allied subjects; C.J. Clay and Sons, Cambridge University Press, London – Glasgow; Publication Agency

of the Johns Hopkins University, Baltimore; Macmillan and Co., Bombay and Calcutta; F. A. Brockhaus, Leipzig.

7. Barron, L.D. (1997) *QJM*, **90**(12), 793–800.
8. Cintas, P. (2007) *Angew. Chem. Int. Ed.*, **46**(22), 4016–4024.
9. Pauling, L. (1970) *General Chemistry*, Dover Publications, New York. 1988; reprint, originally published: 3rd edn, W. H. Freeman, San Francisco.
10. Maton, A., Hopkins, J.J., LaHart, S., Quon Warner, D., Wright, M., and Jill, D. (1997) *Cells: Building Blocks of Life*, Prentice Hall, New Jersey.
11. Schmitt, F.O. (1959) *Rev. Mod. Phys.*, **31**(2), 349–358.
12. Schmitt, F.O., Gross, J., and Highberger, J.H. (1953) *Proc. Natl. Acad. Sci. U.S.A.*, **39**, 459–470.
13. Gross, J., Highberger, J.H., and Schmitt, F.O. (1954) *Proc. Natl. Acad. Sci. U.S.A.*, **40**(8), 679–688.
14. Gross, J. (1956) *J. Biophys. Biochem. Cytol.*, **2**(4), 261–274.
15. Gross, J. (1961) *Sci. Am.*, **204**(5), 121–130.
16. Gustavson, K.H. (1956) *The Chemistry and Reactivity of Collagen*, Academic Press, New York.
17. Revised and expanded by Scott, T. and Eagleson, M. (eds) (1988) *Concise Encyclopedia Biochemistry*, 2nd edn, Walter de Gruyter, Berlin, New York.
18. Gregory, S.G., Barlow, K.F., McLay, K.E., Kaul, R., Swarbreck, D., Dunham, A., Scott, C.E., Howe, K.L., Woodfine, K., Spencer, C.C.A., Jones, M.C., Gillson, C., Searle, S., Zhou, Y., Kokocinski, F., McDonald, L., Evans, R., Phillips, K., Atkinson, A., Cooper, R., Jones, C., Hall, R.E., Andrews, T.D., Lloyd, C., Ainscough, R., Almeida, J.P., Ambrose, K.D., Anderson, F., Andrew, R.W., Ashwell, R.I.S., Aubin, K., Babbage, A.K., Bagguley, C.L., Bailey, J., Banerjee, R., Beasley, H., Bethel, G., Bird, C.P., Bray-Allen, S., Brown, J.Y., Brown, A.J., Bryant, S.P., Buckley, D., Burford, D.C., Burrill, W.D.H., Burton, J., Bye, J., Carder, C., Chapman, J.C., Clark, S.Y., Clarke, G., Clee, C., Clegg, S.M., Cobley, V., Collier, R.E., Corby, N., Coville, G.J., Davies, J., Deadman, R., Dhami, P., Dovey, O., Dunn, M., Earthrowl, M., Ellington, A.G., Errington, H., Faulkner, L.M., Frankish, A., Frankland, J., French, L., Garner, P., Garnett, J., Gay, L., Ghori, M.R., Gibson, R., Gilby, L.M., Gillett, W., Glithero, R.J., Grafham, D.V., Gribble, S.M., Griffiths, C., Griffiths-Jones, S., Grocock, R., Hammond, S., Harrison, E.S.I., Hart, E., Haugen, E., Heath, P.D., Holmes, S., Holt, K., Howden, P.J., Hunt, A.R., Hunt, S.E., Hunter, G., Isherwood, J., James, R., Johnson, C., Johnson, D., Joy, A., Kay, M., Kershaw, J.K., Kibukawa, M., Kimberley, A.M., King, A., Knights, A.J., Lad, H., Laird, G., Langford, C.F., Lawlor, S., Leongamornlert, D.A., Lloyd, D.M., Loveland, J., Lovell, J., Lush, M.J., Lyne, R., Martin, S., Mashreghi-Mohammadi, M., Matthews, L., Matthews, N.S., McLaren, S., Milne, S., Mistry, S., Moore, M.J., Nickerson, T., O'Dell, C.N., Oliver, K., Palmeiri, A., Palmer, S.A., Pandian, R.D, Parker, A., Patel, D., Pearce, A.V., Peck, A.I., Pelan, S., Phelps, K., Phillimore, B.J., Plumb, R., Porter, K.M., Prigmore, E., Rajan, J., Raymond, C., Rouse, G., Saenphimmachak, C., Sehra, H.K., Sheridan, E., Shownkeen, R., Sims, S., Skuce, C.D., Smith, M., Steward, C., Subramanian, S., Sycamore, N., Tracey, A., Tromans, A., Van Helmond, Z., Wall, M., Wallis, J.M., White, S., Whitehead, S.L., Wilkinson, J.E., Willey, D.L., Williams, H., Wilming, L., Wray, P.W., Wu, Z., Coulson, A., Vaudin, M., Sulston, J.E., Durbin, R., Hubbard, T., Wooster, R., Dunham, I., Carter, N.P., McVean, G., Ross, M.T., Harrow, J., Olson, M.V., Beck, S., Rogers, J. and Bentley, D.R. (2006) *Nature*, **441**(7091), 315–321.
19. Grochulski, W., Kawczyński, A., and Lissowski, A. (1978) Crystallization as rearrangement of grain boundaries and dislocations. Symposium on Computer Films for Research in Physics and Chemistry, University of Colorado, 20-22 March 1978, Boulder.
20. (a) Sadoc, J.F. and Mosseri, R. (1997) *Frustration Géometrique*, Eyroles, Paris;

(b) Sadoc, J.F. and Mosseri, R. (1999) *Geometrical Frustration*, Cambridge University Press, Cambridge.
21. Bragg, W.L. and Nye, J.F. (1947) *Proc. R. Soc. London*, **190**, 474–481, Reproduced in *The Feynman Lectures on Physics*, Vol. II, Part 2.
22. Loewenstein, W.R. (1999) *The Touchstone of Life: Molecular Information, Cell Communication, and the Foundations of Life*, Oxford University Press, New York.
23. Kendall, D.G. (1971) *Nature*, **231**(5299), 158–159.
24. Kendall, D.G. (1975) *Philos. Trans. R. Soc. London, A*, **279**(1291), 547–582.
25. Bloom, H. (2007) Who's Smarter: Chimps, Baboons or Bacteria? The Power of Group IQ. Submitted by Howard Bloom on 7 November 2007 – *www.scientificblogging.com.*
26. Fliedner, T.M., Graessle, D., Paulsen, C., and Reimers, K. (2002) *Cancer Biother. Radiopharm.*, **17** (4), 405–426.
27. Lidke, D.S., Huang, F., Post, J.N., Rieger, B., Wilsbacher, J., Thomas, J.L., Pouysségur, J., Jovin, Th.M., and Lenormand, Ph. (2010) *J. Biol. Chem.*, **285**, 3092–3102. Also *http://www.sciencedaily.com/releases/2010/01/100125112209.htm* (accessed 27 January 2010).
28. Shekhter, A.B. (1986) *Connect. Tissue Res.*, **15**(1-2) 23–31.
29. Marcello Malpighi *http://en.wikipedia.org/wiki/Marcello_Malpighi.*
30. Weiss, L. (1988) *Cell and Tissue Biology: A Textbook of Histology*, 6th edn, Urban and Schwarzenberg, Inc., Baltimore, MD.
31. Bilezikian, J.P., Raisz, L.G., and Rodan, G.A. (eds) (2002) *Principles of Bone Biology*, 2nd edn, Academic Press, San Diego, CA.
32. Franklin, R.E. (1955) *Nature*, **175**(4452), 379–381.
33. Ryzhkov, V.L. (1957) Crystallisation of viruses, in *Rost kristallov (Growth of crystals)* (eds A.V. Shubnikov and N.N. Sheftal), Izdatelstvo Akademii Nauk SSSR, Moskva, pp. 351–358, in Russian.
34. Protozoa (2010) *http://en.wikipedia.org/wiki/Protozoa.*
35. Sarindam7 (2010) Osteomyelitis, *en.wikipedia.org/wiki/Osteomyelitis.*
36. Moore L.L., Jr. Osteomyelitis (1969) Leishmania donovani in bone marrow cell. Smear. Parasite; from the Centers for Disease Control and Prevention's Public Health Image Library (PHIL), with identification number #468. *www.bioportfolio.com/search/osteomyelitis_wikipedia.html.*
37. Peters, R.A. (1930) *Trans. Farad. Soc.*, **26**, 797–807.
38. Wheatley, D.N. (2003) *J. Exp. Biol.*, **206**(Pt 12), 1955–1961.
39. Braun, A. (1831) Vergleichende Untersuchung über die Ordnung der Schuppen an den Tannenzapfen als Einleitung zur Untersuchung der Blattstellungen überhaupt. *Nova Acta Phys.-Med. Acad. Caesar. Leop.-Carol. Nat. Curiosorum*, (Verhandlung der Kaiserlichen Leopoldinsch-Carolinschen Akademie der Naturforschung) **15**, 195–402.
40. Schimper, C.F. (1831) Beschreibung des Symphytum Zeyheri und seiner zwei deutschen Verwandten der S. bulbosum Schimper und S. tuberosum Jacqu, *Mag. Pharm.* (hgb. Ph. L. Geiger) **29**, 1–92.
41. Bravais, L. and Bravais, A. (1837) Essai sur la disposition des feuilles curvisériées. *Ann. Sci. Nat. Bot.*, **7**, 42–110.
42. Bravais, L. and Bravais, A. (1837) Essai sur la disposition symétrique des inflorescences. *Ann. Sci. Nat. Bot.*, **7**, 193–221, 291–348; **8**, 11–42.
43. Schwendener, S. (1878) *Mechanische Theorie der Blattstellungen*, Engelmann, Leipzig.
44. Schwendener, S. (1883) Zur Theorie der Blattstellungen. *Sitzungsber. Königlich Preussischen Akad. Wiss. Berlin* **II. S**, 741–773.
45. Wulff, G.W. (1908) Simmetriia i ee Proiavleniia v Prirode, (Symmetry and its manifestations in the nature, in Russian), Lektsii chitannye v 1907g, Moskovskoe obshchestvo Narodnykh Universitetov, Tip. T-va I. D. Sytina, Moskva.
46. Lewis, F.T. (1931) *Anat. Rec.*, **50**, 235–265.

47. Rutishauser, R. (1998) Plastochrone ratio and leaf arc as parameters of a quantitative phyllotaxis analysis in vascular plants, in *Symmetry in Plants*, WS Series in Mathematical Biology and Medicine, vol. 4 (eds R.V. Jean and D. Barabé), (Series editors P.M. Auger and R.V. Jean), World Scientific Publishers, pp. 171–212.
48. Atela, P. and Golé, C. (2007) Phyllotaxis *http://maven.smith.edu/~phyllo/About/math.html*.
49. Thompson D'Arcy, W. (1917, 1942) *On Growth and Form*, Cambridge University Press.
50. Rivier, N. and Lissowski, A. (1982) *J. Phys. A: Math. Gen.*, **15**(3), L143–L148.
51. Hofmeister, W.F.B. (1868) Allgemeine Morphologie der Gewächse, *Handbuch der Physiologischen Botanik*, vol. 1, Part 2, Wilhelm Engelmann, Leipzig, pp. 405–664.
52. Wojnar, R. (2009) Strains in tissue development: a vortex description, in *More Progresses in Analysis, Proceedings of the 5th International ISAAC Congress, Catania, Italy, 25–30 July 2005* (eds H.G.W. Begehr and F. Nicolosi), World Scientific, New Jersey, London, Singapore, Beijing, Shanghai, Hong Kong, Taipei, Chennai, pp. 1271–1281.
53. Newell, A.C., Shipman, P.D., and Sun, Z. (2008) *Plant Signal Behav.*, **3**(8), 586–589.
54. Gebhardt, W. (1911) Knochenbildung und Colloidchemie. *Arch. Entwickl. Mech.*, **32**, 727–734.
55. Counce, S.J. (1994) *Dev. Genes. Evol.*, **204**(2), 79–92.
56. Giraud-Guille, M.M. (1988) *Calcif. Tissue Int.*, **42**, 167–180.
57. Erickson, R.O. (1998) Phyllotactic symmetry in plant growth, in *Symmetry in Plants* (eds R.V. Jean and D., Barabé), World Scientific, Singapore, pp. xvii–xxvi.
58. Dumais, J. and Kwiatkowska, D. (2002) *Plant J.*, **31**(2), 229–241.
59. Gibson, M.C., Patel, A.B., Nagpal, R., and Perrimon, N. (2006) *Nature*, **442**, 1038–1041.
60. Farhadifar, R., Röper, J.-Ch., Aigouy, B., Eaton, S., and Jülicher, F. (2007) *Curr. Biol.*, **17**(24), 2095–2104.
61. Ross, M.H., Romrell, L.J., and Kaye, G.I. (1995) *Histology: A Text and Atlas*, 3rd edn, Williams & Wilkins, Baltimore.
62. Ross, M.H. and Pawlina, W. (2006) *Histology: A Text and Atlas, with Correlated Cell and Molecular Biology*, Lippincott Williams & Wilkins, Hagerstown, MD.
63. Gray, H. (1918) *Anatomy of the Human Body*, Lea & Febiger, Philadelphia. Also (2010) *http://www.bartleby.com/107/*.
64. Gray, H. (2008) *Gray's Anatomy: The Anatomical Basis of Clinical Practice*, 40th edn, Elsevier, Churchill-Livingstone.
65. Downey, P.A. and Siegel, M.I. (2006) *Phys. Ther.*, **86**(1), 77–91.
66. Bochenek, A. (1924) *Anatomia człowieka*, tom 2, wyd.IV, Nakładem Polskiej Akademii Umiejętności, Kraków.
67. National Center for Biotechnology Information (2004) *http://www.ncbi.nlm.nih.gov/About/primer/genetics_cell.html*.
68. Connective Tissue (2010) *http://training.seer.cancer.gov/module_anatomy/unit2_2_body_tissues2_connective.html*.
69. Marino, Th.A., Lamperti, A.A., and Sodicoff, M. (2000) Connective Tissue Web Book, Temple University School of Medicine, *http://astro.temple.edu/~sodicm/labs/CtWeb/sld004.htm*.
70. Strum, J.M., Gartner, L.P., and Hiatt, J.L. (2007) *Cell Biology and Histology*, Lippincott Williams & Wilkins, Hagerstwon, MD.
71. McLean, F.C. (1955) *Sci. Am.*, **192**(2), 84.
72. Bone Development and Growth *http://training.seer.cancer.gov/module_anatomy/unit3_3_bone_growth*.
73. School of Anatomy and Human Biology The University of Western Australia Blue Histology – Skeletal Tissues – Bone (2009) *http://www.lab.anhb.uwa.edu.au/mb140/CorePages/Bone/Bone.htm*.
74. National Cancer Institute, SEER Training Modules: Classification of Bones

(2000) : *http://training.seer.cancer.gov/anatomy/skeletal/classification.html.*

75. Fruitsmaak, S. (2008) Gallery Medical Images. *http://commons.wikimedia.org/wiki/User:Stevenfruitsmaak/Gallery.*
76. General Osteology (2005) *http://commons.wikimedia.org/wiki/File:Illu_long_bone.jpg.*
77. Bernard Dery (2005) Skeleton Human Body. Squeleton, in *The Visual Dictionary*, vol. 3. *http://www.infovisual.info/03/011_en.html.*
78. Hydroxylapatite (2010) *http://en.wikipedia.org/wiki/Hydroxylapatite.*
79. Le Geros, R.Z. (1991) Calcium phosphates in oral biology and medicine, in *Monographs in Oral Science*, vol. 15 (ed. H.M. Myers), Karger Publishing, Basilea, Freiburg, Paris, London, New York, New Delhi, Bangkok, Singapore, Tokyo, Sydney, pp. 110–118.
80. Ślósarczyk, A., Paszkiewicz, Z., and Paluszkiewicz, Cz. (2005) *J. Mol. Struct.*, **744–747**, 657–661.
81. Beevers, C.A. and McIntyre, D.B. (1946) *Mineral. Mag.*, **27**, 254–257 + 3 plates. Also *www.minersoc.org/pages/Archive-MM/Volume_27/27-194-254.pdf.*
82. Wopenka, B. and Pasteris, J.D. (2005) *Mater. Sci. Eng. C*, **25**(2), 131–143.
83. Elliott, J.C., Wilson, R.M., and Dowker, S.E.P. (2002) Apatite structures. Proceedings of the 50th Annual Denver X-ray Conference, Steamboat Springs, Colorado, 30th July – 3rd August 2001, Advances in X-ray Analysis, vol. 45, 2002. International Centre for Diffraction Data, Newtown Square, Pennsylvania. Invited presentation for R.A. Young Rietveld Analysis Session (ed. Huang, T.C.), pp. 172–181.
84. Ellis, D.I. and Goodacre, R. (2006) *Analyst*, **131**(8) 875–885. Also *www.rsc.org/analyst.*
85. Wilson, R.M., Dowker, S.E.P., and Elliott, J.C. (2006) *Biomaterials*, **27**(27), 4682–4692.
86. Dowker, S.E.P., Elliott, J.C., Davis, G.R., Wilson, R.M., and Cloetens, P. (2006) *Eur. J. Oral Sci.*, **114** (Suppl. 1), 353–359.
87. Cooper, R. and Blystone, R. (1999) SLiBS Synergistic Learning in Biology and Statistics. *http://www.trinity.edu/rblyston/bone/intro2.htm.*
88. Piekarski, K. and Munro, M. (1977) *Nature*, **269**(5623), 80–82.
89. Cooper, R.R., Milgram, J.W., and Robinson, R.A. (1966) Morphology of the osteon. An electron microscopic study. *J. Bone Joint Surg. Am.*, **48**(7), 1239–1271.
90. Leeuwenhoek, A. (1674) Microscopical observations about blood, milk, bones, the brain, spitle, cuticula. Letter published in *Philos. Trans. R. Soc. London, Ser.* **9**, 121.
91. Leeuwenhoek, A. (1693) Several observations on the texture of bone of animals compared with that of wood: on the bark of trees: on the little scales found on the cuticula, etc. *Philos. Trans. R. Soc. London*, **17**, 838–843.
92. Leeuwenhoek, A. (1800) The select works of antony van Leeuwenhoek, containing his microscopical discoveries in many of the works of nature, translated from the Dutch and Latin editions published by the author, by Samuel Hoole, Volume the first –part the first. G. Sidney, London MDCCC.
93. Havers, C. (1691) *Osteologia Nova, or Some New Observations of the Bones, and the Parts Belonging to Them, with the Manner of their Accretion and Nutrition*, Samuel Smith, London.
94. Dobson, J. (1952) *J. Bone Joint Surg.*, **34-B**(4), 702–707.
95. Martin, R.B. and Burr, D.B. (1989) *Structure Function, and Adaptation of Compact Bone*, Raven Press, New York.
96. Ford, B.J. (1992, access on line 2009) From dilettante to diligent experimenter, a reappraisal of Leeuwenhoek as microscopist and investigator, *http://www.brianjford.com/a-avl01.htm.*
97. Frost, H.M. (1963) *Bone Remodelling Dynamics*, Charles C. Thomas, Springfield, Il.
98. Frost, H.M. (1964) *Mathematical Elements of Lamellar Bone Remodelling*, Charles C. Thomas, Springfield, Il.

99. Frost, H.M. (1973) *Bone Remodelling and its Relation to Metabolic Bone Diseases*, C. Thomas, Springfield, IL.
100. Frost, H.M. (2001) *Anat. Rec. A*, **262**(4), 398–419.
101. Seeman, E. (2006) *Osteoporos. Int.*, **17**(10), 1443–1448.
102. Osteon (2010) *http://en.wikipedia.org/wiki/Osteon.*
103. BME/ME 456 Biomechanics. *http://www.engin.umich.edu/class/bme456/bonestructure/bonestructure.htm.*
104. Gebhardt, W. (1901, 1905) Über funktionell wichtige Anordnungsweisen der feineren und gröberen Bauelemente des Wirbelthierknochens.II. Spezieller Teil. *Arch. Entwickl. Mech. Roux's Arch. Dev. Biol.)*, **20**, 187–322.
105. Gebhardt, W. (1901) Über funktionell wichtige Anordnungsweisen der gröberen und feineren Bauelemente des Wirbelthierknochens. I. Allgemeiner Teil. *Arch. Entwickl. Mech. Roux's Arch. Dev. Biol.*, **11**, 383–498; **12**, 1–52, 167–223.
106. Ascenzi, A. and Bonucci, E. (1961) *Acta Anat. (Basel)*, **44**(3), 236–262.
107. Ascenzi, A. and Bonucci, E. (1968) *Anat. Rec.*, **161**(3), 377–391.
108. Ascenzi, A. and Bonucci, E. (1967) *Anat. Rec.*, **158**(4), 375–386.
109. Ascenzi, A. and Benvenuti, A. (1986) *J. Biomech.*, **19**(6), 455–463.
110. Ascenzi, A. and Bonucci, E. (1976) *Clin Orthop.*, **121**, 275–294.
111. Ascenzi, M.-G. and Lomovtsev, A. (2006) *J. Struct. Biol.*, **153**(1), 14–30.
112. Ascenzi, M.-G., Gill, J., and Lomovtsev, A. (2008) *J. Biomech.*, **41**(16), 3426–3435.
113. Figurska, M. (2007) *Russ. J. Biomech.*, **11**(3), 26–35.
114. Frasca, P., Hari Rao, C.V., Harper, R.A., and Katz, J.L. (1976) *J. Den. Res.*, **55**(3), 372–375.
115. Weiner, S., Arad, T., and Traub, W. (1991) *FEBS Lett.*, **285**(1), 49–54.
116. Enlow, D.H. (2005) *Am. J. Anat.*, **110**(3), 269–305.
117. Gros, N. (1846) *C. R. Acad. Sci.*, **23**(24), 1106–1108.
118. Ottolenghi, D. (1901) *Atti R. Accad. Sci. Torino*, **36**(15), 611–618.
119. Kuntz, A. and Richins, C.A. (1945) *J. Comp. Neurol.*, **83**, 213–221.
120. Röhlich, K. (1961) *Z. Mikroskopischanat. Forsch.*, **49**, 425–464.
121. Efferent Nerve Fiber (2009) *http://en.wikipedia.org/wiki/Efferent_nerve_fiber.*
122. Vento, P. and Soinila, S. (1999) *J. Histochem. Cytochem.*, **47**(11), 1405–1415.
123. Carter, M.S., Cremins, J.D., and Krause, J.E. (1990) *J. Neurosci.*, **10**(7), 2203–2214.
124. Goto, T. (2002) *Microsc. Res. Tech.*, **58**(2), 59–60.
125. Goto, T. and Tanake, T. (2002) *Microsc. Res. Tech.*, **58**(2), 91–97.
126. Substance P. (2006) *http://en.wikipedia.org/wiki/Substance_P.*
127. Dudás, B. and Merchenthaler, I. (2002) *J. Clin. Endocrinol. Metab.*, **87**(6), 2946–2953.
128. Gajda, M., Litwin, J.A., Cichocki, T., Timmermans, J.-P., and Adriaensen, D. (2005) *J. Anat.*, **207**(2), 135–144.
129. Enchanted Learning. *http://www.enchantedlearning.com/subjects/.*
130. Cytoskeleton (2010) *en.wikipedia.org/wiki/Cytoskeleton.*
131. Frixione, E. (2000) *Cell Motil. Cytoskeleton.*, **46**(2), 73–94.
132. Smith, D.A. and Geeves, M.A. (1995) *Biophys. J.*, **69**(8), 524–537.
133. Streater, R.F. (1997) *Rep. Math. Phys.*, **40**(3), 557–564.
134. Astumian, R.D. and Hanggi, P. (2002) *Phys. Today*, **55**(11), 33–39.
135. Wojnar, R. (2002) *Rep. Math. Phys.*, **49**(2), 415–426.
136. Evert, R.F. and Eichhorn, S.E. (2006) *Esau's Plant Anatomy, Meristems, Cells, and Tissues of the Plant Body: Their Structure, Function, and Development*, 3rd edn, John Wiley & Sons, Inc.
137. Cell_Membrane (2010) *en.wikipedia.org/wiki/Cell_membrane.*
138. Bederson, B. and Silbermann, B. (2000) Cellupedia library, *www.thinkquest.org/C004535/cell_membranes.html.*
139. Poniewierski, A. and Stecki, J. (1982) *Phys. Rev. A*, **25**(4), 2368–2370.
140. Degiorgio, V. and Corti, M. (eds) (1985) *Physics of Amphiphiles: Micelles, Vesicles*

and Microemulsions, Italian Physical Society, North-Holland Physics Publishing, Amsterdam.
141. Stecki, J. (2008) *J. Phys. Chem. B*, **112**(14), 4246–4252.
142. Enzyme (2009) *http://en.wikipedia.org/wiki/Enzyme.*
143. Juliano, R.L. and Ling, V. (1976) *Biochim. Biophys. Acta*, **455**(1), 152–162.
144. Payen, A. and Persoz, J.-F. (1833) Mémoire sur la diastase, les principaux produits de ses réactions et leurs applications aux arts industriels. *Ann. Chim. Phys., 2e Sér.*, **53**, 73–92.
145. Henri, V. (1902) *C. R. Acad. Sci. Paris*, **135**, 916–919.
146. Henri, V. (1903) *Lois Générales de L'action des Diastases*, Hermann, Paris.
147. Michaelis, L. and Menten, M.L. (1913) *Biochem. Z.*, **49**, 333–369.
148. Membrane Transport Protein (2009) *http://en.wikipedia.org/wiki/Membrane_transport_protein.*
149. Becker, W.M., Kleinsmith, L.J., and Hardin, J. (2006) *The World of the Cell*, 6th edn, Pearson and Benjamin Cummings, San Francisco, CA.
150. Van Winkle, L.J. (1999) *Biomembrane Transport*, Academic Press, New York.
151. Voituriez, R., Moreau, M., and Oshanin, G. (2005) *Europhys. Lett.*, **69**(2), 177–183.
152. Voituriez, R., Moreau, M., and Oshanin, G. (2005) *J. Chem. Phys.*, **122**(8), 084103/13.
153. Runge, S.W., Hill, B.J.F., and Moran, W.M. (2006) *CBE Life Sci. Edu.*, **5**(4), 348–352.
154. Stryer, L. (1995) *Biochemistry*, 4th edn, W. H. Freeman and Company, New York.
155. Brozek, J., Bryl, E., Ploszynska, A., Balcerska, A., and Witkowski, J. (2009) *J. Pediatr. Hematol. Oncol.*, **31**(7), 493–499.
156. Morita, Y., Sobel, M.L., and Poole, K. (2006) *J. Bacteriol.*, **188**(5), 1847–1855.
157. D'ippolito, G., Schiller, P.C., Ricordi, C., Roos, B.A., and Howard, G.A. (1999) *J. Bone Miner. Res.*, **14**(7), 1115–1122. Also *http://en.wikipedia.org/wiki/Osteoblast#cite_note-Dippolito1999-0.*
158. Bone Cell (2009) *http://en.wikipedia.org/wiki/Bone_cell.*
159. Osteoblast (2009) *http://en.wikipedia.org/wiki/Osteoblast.*
160. Klein-Nulend, J., van der Plas, A., Semeins, C.M., Ajubi, N.E., Frangos, J.A., Nijweide, P.J., and Burger, E.H. (1995) *FASEB J.*, **9**(5), 441–445.
161. Bonewald, L.F. (2005) *J. Musculoskelet. Neuronal Interact.*, **5**(4), 321–324.
162. Caceci, Th. (2008) BoneVM8054 Veterinary Histology Exercise 8. *http://education.vetmed.vt.edu/curriculum/vm8054/labs/Lab8/lab8.htm.*
163. Tanaka, K., Yamaguchi, Y., and Hakeda, Y. (1995) *J. Bone Miner. Metab.*, **13**, 61–70.
164. Rubinacci, A., Covini, M., Bisogni, C., Villa, I., Galli, M., Palumbo, C., Ferretti, M., Muglia, M.A., and Marotti, G. (2002) *Am. J. Physiol. Endocrinol. Metab.*, **282**(4), E851–E864.
165. Bonewald, L. (2006) *J. Musculoskelet. Neuronal Interact.*, **6**(4), 331–333.
166. Osteoclast (2008) *en.wikipedia.org/wiki/Osteoclast.*
167. Väänänen, H.K., Zhao, H., Mulari, M., and Halleen, J.M. (2000) *J. Cell Sci.*, **113**(Pt 3), 377–381.
168. RANKL (2007) *http://en.wikipedia.org/wiki/RANKL.*
169. RANK (2010) *http://en.wikipedia.org/wiki/RANK.*
170. Simonet, W.S., Lacey, D.L., Dunstan, C.R., Kelley, M., Chang, M.-S., Lüthy, R., Nguyen, H.Q., Wooden, S., Bennett, L., Boone, T., Shimamoto, G., DeRose, M., Elliott, R., Colombero, A., Tan, H.L., Trail, G., Sullivan, J., Davy, E., Bucay, N., Renshaw-Gegg, L., Hughes, T.M., Hill, D., Pattison, W., Campbell, P., Sander, S., Van, G., Tarpley, J., Derby, P., Lee, R., and Boyle, W.J. (1997) *Cell*, **89**(2), 309–319.
171. Tsuda, E., Goto, M., Mochizuki, S., Yano, K., Kobayashi, F., Morinaga, T., and Higashio, K. (1997) *Biochem. Biophys. Res. Commun.*, **234**, 137–142.
172. Yasuda, H., Shima, N., Nakagawa, N., Yamaguchi, K., Kinosaki, M., Mochizuki, S., Tomoyasu, A., Yano, K., Goto, M., Murakami, A., Tsuda, E., Morinaga, T., Higashio, K., Udagawa,

N., Takahashi, N., and Suda, T. (1998) *Proc. Natl. Acad. Sci. U.S.A.*, **95**, 3597–3602.

173. Yasuda, H., Shima, N., Nakagawa, N., Mochizuki, S.I., Yano, K., Fujise, N., Sato, Y., Goto, M., Yamaguchi, K., Kuriyama, M., Kanno, T., Murakami, A., Tsuda, E., Morinaga, T., and Higashio, K. (1998) *Endocrinology*, **139**(3), 1329–1337.
174. Kryokiewicz, E. and Lorenc, I.R.S. (2006) *TERAPIA*, **3**(177), 58–63.
175. Malyankar, U.M., Scatena, M., Suchland, K.L., Yun, Th.J., Clark, E.A., and Giachelli, C.M. (2000) *J. Biol. Chem.*, **275**(28), 20959–20962.
176. Rogers, A. and Eastell, R. (2005) *J. Clin. Endocrinol. Metab.*, **90**(11), 6323–6331.
177. Boyce, B.F. and Xing, L. (2007) *Arthritis Res. Ther.*, **9** (Suppl 1), 1–7.
178. Bucay, N., Sarosi, I., Dunstan, C.R., Morony, S., Tarpley, J., Capparelli, C., Scully, Sh., Tan, H.L., Xu, W., Lacey, D.L., Boyle, W.J., and Scott Simonet, W. (1998) *Genes Dev.*, **12**, 1260–1268.
179. Ducy, P., Schinke, Th., and Karsenty, G. (2000) *Science*, **289**(5484), 1501–1504.
180. Boyle, W.J., Simonet, W.S., and Lacey, D.L. (2003) *Nature*, **423**(6937), 337–342.
181. Schoppet, M., Preissner, K.T., and HofbauerI, L.C. (2002) *Arterioscler. Thromb. Vasc. Biol.*, **22**(4), 549–553.
182. Hofbauer, L.C. and Schoppet, M. (2004) *J. Am. Med. Assoc.*, **292**, 490–495.
183. Bekker, P.J., Holloway, D.L., Rasmussen, A.S., Murphy, R., Martin, S.W., Leese, Ph.T., Holmes, G.B., Dunstan, C.R., and DePaoli, A.M. (2004) A single-dose placebo-controlled study of AMG 162, a fully human monoclonal antibody to RANKL. *J. Bone Miner. Res.*, **19**(7), 1059–1066.
184. Black, R.A. (2002) Tumor necrosis factor-alpha converting enzyme. *Int. J. Biochem. Cell Biol.*, **34**(1), 1–5.
185. Black, R.A., Rauch, C.T., Kozlosky, C.J., Peschon, J.J., Slack, J.L., Wolfson, M.F., Castner, B.J., Stocking, K.L., Reddy, P., Srinivasan, S., Nelson, N., Boiani, N., Schooley, K.A., Gerhart, M., Davis, R., Fitzner, J.N., Johnson, R.S., Paxton, R.J., March, C.J., and Cerretti, D.P. (1997) A metalloproteinase disintegrin that releases tumour-necrosis factor-alpha from cells. *Nature*, **385**(6618), 729–733.
186. Dziedzic-Goc3awska, A., Tyszkiewicz, J., and Uhrynowska-Tyszkiewicz, I. (2000) *Nowa Klin.*, **7**(7), 704–712.
187. Pan, B., Farrugia, A.N., To, L.B., Findlay, D.M., Green, J., Lynch, K., and Zannettino, A.C. (2004) *J. Bone Miner. Res.*, **19**(1), 147–154.
188. Wedemeyer, Ch., Knoch, Fv., Pingsmann, A., Hilken, G., Sprecher, Ch., Saxler, G., Henschke, F., Löer, F., and Knoch, Mv. (2005) *Biomaterials*, **26**(17), 3719–3725.
189. Stawińska, N., Ziętek, M., and Kochanowska, I. (2005) *Dent. Med. Prob.*, **42**(4), 627–635.
190. Khosla, S. (2001) Minireview: the OPG/RANKL/RANK system. *Endocrinology*, **142**(12), 5050–5055.
191. Kearns, A.E., Khosla, S., and Kostenuik, P.J. (2008) *Endocrinol. Rev.*, **29**(2), 155–192.
192. McClung, M. (2007) *Arthritis Res. Ther.*, **9** (Suppl 1), 1–3.
193. Yeung, Rae S.M. (2004) The Osteoprotegerin/Osteoprotegerin ligand family: role in inflammation and bone loss. *J. Rheumatol.*, **31**(5), 844–846.
194. Kong, Y.Y., Yoshida, H., Sarosi, I., Tan, H.L., Timms, E., Capparelli, C., Morony, S., Oliveira-dos-Santos, A.J., Van, G., Itie, A., Khoo, W., Wakeham, A., Dunstan, C.R., Lacey, D.L., Mak, T.W., Boyle, W.J., and Penninger, J.M. (1999) *Nature*, **397**(6717), 315–323.
195. Silvestrini, G., Ballanti, P., Patacchioli, F., Leopizzi, M., Gualtieri, N., Monnazzi, P., Tremante, E., Sardella, D., and Bonucci, E. (2005) Detection of osteoprote gerin (OPG) and its ligand (RANKL) mRNA and protein infemur and tibia of the rat. *J. Mol. Hist.*, **36**(1-2), 59–67.
196. Kim, D.W., Lee, H.J., Karmin, J.A., Lee, S.E., Chang, S.S., Tolchin, B., Lin, S., Cho, S.K., Kwon, A., Ahn, J.M., and Lee, F.Y.-I. (2006) *Ann. N.Y. Acad. Sci.*, **1068**(1), 568–572.

197. Bhagavan, N.V. (2002) *Medical Biochemistry*, 4th edn, Academic Press, Harcourt.
198. Amino Acid (2007) *http://en.wikipedia.org/wiki/Amino_acid*.
199. Proteinogenic Amino Acid (2009) *http://en.wikipedia.org/wiki/Proteinogenic_amino_acid*.
200. Astbury, W.Th. (1938) *Trans. Faraday Soc.*, **34**, 378–388.
201. Braconnot, H.M. (1820) Sur la conversion des matières animales en nouvelles substances par le moyen de l'acide sulfurique. *Ann. Chim. Phys. Ser 2*, **13**, 113–125.
202. Hershey, A.D. and Chase, M. (1952) *J. Gen. Physiol.*, **36**, 39–56.
203. Herzog, R.O. and Jancke, W. (1920) *Ber. Dtsch. Chem. Ges.*, **53**, 2162–2164.
204. Pauling, L. and Corey, R.B. (1951) *Proc. Natl. Acad. Sci. U.S.A.*, **37**(5), 272–281.
205. Ramachandran, G.N. and Ambady, G.K. (1954) *Curr. Sci.*, **23**, 349.
206. Cochran, W., Crick, F.H.C., and Vand, V. (1952) *Acta Crystallogr.*, **5**, 581–586.
207. Ramachandran, G.N. (1967) Structure of collagen at the molecular level, in *Treatise on Collagen*, vol. 1, (ed. G.N. Ramachandran), Chapter 3, Academic Press, New York, pp. 103–183.
208. Rich, A. and Crick, F.H.C. (1955) *Nature*, **176**(4489), 915–916.
209. Rich, A. and Crick, F.H.C. (1961) *J. Mol. Biol.*, **3**, 483–506.
210. Boedtker, H. and Doty, P. (1956) *J. Am. Chem. Soc.*, **78**, 4267–4280.
211. Yonath, A. and Traub, W. (1969) *J. Mol. Biol.*, **43**, 461–477.
212. Okujama, K., Bächinger, H.P., Mizuno, K., Boudko, S., Engel, J., Berisio, R., and Vitagliano, L. (2006) *Proc. Natl. Acad. Sci.*, **103**, 9001–9005 (Comments on *Microfibrillar structure of type I collagen in situ* by Orgel *et al.* PNAS, submitted for publication).
213. Okuyama, K. (2008) *Connect. Tissue Res.*, **49**(5), 299–310.
214. Orgel, J.P.R.O., Irving, Th.C., Miller, A., and Wess, T.J. (2006) *Proc. Natl. Acad. Sci.*, **103**(24), 9001–9005.
215. Okuyama, K., Takayanagi, M., Ashida, T., and Kakudo, M. (1977) *Polym. J.*, **9**(3), 341–343.
216. Okuyama, K., Nagarajan, V., Kamitori, Sh., and Noguchi, K. (1998) *Chem. Lett.*, **27**(5), 385–386.
217. Royce, P.M. and Steinmann, B.U. (eds) (2002) *Connective Tissue and its Heritable Disorders: Molecular, Genetic, and Medical Aspects*, 2nd edn, Wiley-Liss, New York.
218. Bansal, M. (1977) *Pramana*, **9**(4), 339–347.
219. Piez, K.A. (1976) Primary structure, in *Biochemistry of Collagen*, Chapter 1 (eds G.N. Ramachandran and A.H. Reddi), Plenum Press, New York, pp. 1–44.
220. Millane, R.P. (1991) *Acta Crystallogr.*, **A47**, 449–451.
221. Rainey, J.K. and Goh, M.C. (2002) *Protein Sci.*, **11**(11), 2748–2754.
222. Vincent, J. (1990) *Structural Biomaterials*, Princeton University Press, Princeton, NJ.
223. Kucharz, E.J. (1992) *The Collagens: Biochemistry and Pathophysiology*, Springer-Verlag, Berlin.
224. Sikorski, Z.E. (2001) *Chemical and Functional Properties of Food Proteins*, CRC Press.
225. Collagen (2010) *http://en.wikipedia.org/wiki/Collagen*.
226. Privalov, P.L. (1982) *Adv. Protein Chem.*, **35**, 1–104.
227. Privalov, P.L. (1989) *Annu. Rev. Biophys. Biophys. Chem.*, **18**, 47–69.
228. Makhatadze, G.I. and Privalov, P.L. (1996) *Protein Sci.*, **5**(3), 507–510.
229. Brodsky, B., Eikenberry, E.F., Belbruno, K.C., and Sterling, K. (1982) *Biopolymers*, **21**(5), 935–951.
230. Brodsky, B. and Ramshaw, J.A.M. (1997) *Matrix Biol.*, **15**(8-9), 545–554.
231. Coxeter, H.S.M. (1948) *Regular Polytopes*, Macmillan, New York; (Dover, 1973).
232. Coxeter, H.S.M. (1961) *Introduction to Geometry*, 1st edn, John Wiley & Sons, Inc., New York.
233. Bernal, J.D. (1964) *Proc. R. Soc. London A*, **280**(1382), 299–322.
234. Zheng, Ch., Hoffman, R., and Nelson, D.R. (1990) *J. Am. Chem. Soc.*, **112**, 3784–3791.
235. Buckminster Fuller, R. (1975) *Synergetics*, MacMillan, New York, also: Gray, R.W.

(1997) *http://www.rwgrayprojects.com/synergetics/synergetics.html.*

236. Lord, E.A. and Ranganathan, S. (2001) *Eur. Phys. J. D*, **15**(3), 335–343.

237. Lord, E.A. (2002) *Struct. Chem.*, **13**(3/4), 305–314.

238. Sadoc, J.F. and Rivier, N. (1999) *Eur. Phys. J. B*, **12**(2), 309–319.

239. Sadoc, J.F. and Rivier, N. (2000) *Mater. Sci. Eng. A*, **294–296**, 397–400.

240. Sadoc, J.F. and Charvolin, J. (2009) *J. Phys. A: Math. Theor.*, **42**, 465209 (17 pp).

241. Buehler, M.J. and Wong, S.Y. (2007) *Biophys. J.*, **93**(1), 37–43.

242. Kuhn, W. (1934) *Kolloid-Z.*, **68**(1), 2–15.

243. Flory, P.J. (1953) *Principles of Polymer Chemistry*, Cornell University Press, Ithaca, NY.

244. Treloar, L.R.G. (2005) *The Physics of Rubber Elasticity*, 3rd edn, Oxford University Press.

245. Kratky, O. and Porod, G. (1949) *Rec. Trav. Chim. Pays-Bas*, **68**, 1106–1123.

246. Polymer Physics (2010) *en.wikipedia.org/wiki/Polymer_physics.*

247. Worm-like_chain (2010) *en.wikipedia.org/wiki/Worm-like_chain.*

248. Bustamante, C., Marko, J.F., Siggia, E.D., and Smith, S. (1994) *Science*, **265**(5178), 1599–1600.

249. Bustamante, C., Bryant, Z., and Smith, S.B. (2003) *Nature*, **421**(6921), 423–427.

250. Fratzl, P. (2003) *Curr. Opin. Colloid Interface Sci.*, **8**(1), 32–39.

251. Fraser, R.D., MacRae, T.P., and Miller, A. (1987) *J. Mol. Biol.*, **193**(1), 115–125.

252. Wess, T.J., Hammersley, A.P., Wess, L., and Miller, A. (1998) *J. Mol. Biol.*, **275**(2), 255–267.

253. Sionkowska, A., Wisniewski, M., Skopinska, J., Kennedy, C.J., and Wess, T.J. (2004) *Biomaterials*, **25**(5), 795–801.

254. Wess, T.J. (2008) Collagen fibrillar structure and hierarchies, in *Collagen, Structure and Mechanics*, (ed. P. Fratzl), Springer, pp. 49–80.

255. Cameron, G.J., Cairns, D.E., and Wess, T.J. (2007) *J. Mol. Biol.*, **372**(4), 1097–1107.

256. Cowan, P.M., North, A.C.T., and Randall, J.T. (1955) *Symp. Soc. Exp. Biol.*, **9**, 115–126.

257. Cowan, P.M. and McGavin, S. (1955) *Nature*, **176**(4480), 501–503.

258. Misof, K., Rapp, G., and Fratzl, P. (1997) *Biophys. J.*, **72**(3), 1376–1381.

259. Puxkandl, R., Zizak, I., Paris, O., Keckes, J., Tesch, W., Bernstorff, S., Purslow, P., and Fratzl, P. (2002) *Philos. Trans. R. Soc. London B Biol. Sci.*, **357**(1418), 191–197.

260. Gupta, H.S., Messmer, P., Roschger, P., Bernstorff, S., Klaushofer, K., and Fratzl, P. (2004) *Phys. Rev. Lett.*, **93**(15), 158101(4).

261. Gupta, H.S., Seto, J., Wagermaier, W., Zaslansky, P., Boesecke, P., and Fratzl, P. (2006) *Proc. Natl. Acad. Sci*, **103**(47), 17741–17746. Also *www.pnas.org_cgi_doi_10.1073_pnas.0604237103*

262. Diamant, J., Keller, A., Baer, E., Litt, M., and Arridge, R.G.C. (1972) *Proc. R. Soc. London B*, **180**, 293–315.

263. Silver, F.H., Kato, Y.P., Ohno, M., and Wasserman, A.J. (1992) *J. Long-Term Eff. Med. Implants*, **2**(2-3), 165–195.

264. Mosler, E., Folkhard, W., Knörzer, E., Nemetschek-Gonsler, H., Nemetschek, Th., and Koch, M.H.J. (1985) *J. Mol. Biol.*, **182** (4), 589–596.

265. Hodge, A.J. and Petruska, J.A. (1963) in *Aspects of Protein Structure* (ed. G.N. Ramachandran), Academic Press, New York, pp. 289–300.

266. Hull, D. (1981) *An Introduction to Composite Materials*, Cambridge University Press.

267. Edoff, K., Hellman, J., Persliden, J. and Hildebrand, C. (1997) *Anat. Embryol.*, **195**(6), 531–538.

268. McCredie, J. (1973) *Lancet*, **2**(7837), 1058–1061.

269. McCredie, J. (2009) *J. Med. Imaging Radiat. Oncol.*, **53**(5), 433–441.

270. Adams, J.S. and Hewison, M. (2010) *J. Clin. Endocrinol. Metab.*, **95**(2), 471–478.

271. Donahue, S.W., McGee, M.E., Harvey, K.B., Vaughan, M.R. and Robbins, Ch.T. (2006) *J. Biomech.*, **39**(8), 1480–1488.
272. Arnold, V.I. (1998) *Russ. Math. Surv.*, **53**(1), 229–236; also *http://pauli.uni-muenster.de/~munsteg/arnold.html.*
273. Myjak, J. (2008) *Opuscula Math.*, **28**(4), 343–351.
274. Jakob, B. and Kasper, O. (2010) The close-packed triple helix as a possible new structural motif for collagen, *arXiv:1004.1781v1 [physics.bio-ph]* 11 Apr 2010.
275. Sommerfeldt, D.W. and Rubin, C.T. (2001) *Eur. Spine J.*, **10**(S2), S86–S95.

Further Reading

Bostanci, N., Emingil, C., Afacan, B., Han, B., Ilgenli, T., Atilla, G., Hughes, F.J., and Belibasakis, G.N. (2008) *J. Dent. Res.*, **87**(3), 273–277.

Di LulloDagger, G.A., Sweeney, Sh.M., Körkkö, J., Ala-Kokko, L., and San Antonio, J.D. (2002) *J. Biol. Chem.*, **277**(6), 4223–4231.

McMurry, J. (2008) *Organic Chemistry*, 7th edn., Brooks/Cole-Cengage, Belmont, CA.

Sikorski, Z.E. (2001) *Chemical and Functional Properties of Food Proteins*, Boca Raton: CRC Press.

2
Numerical Simulation of Bone Remodeling Process Considering Interface Tissue Differentiation in Total Hip Replacements

Eduardo A. Fancello, Darlan Dallacosta, and Carlos R. M. Roesler

2.1
Introduction

An artificial joint is generally used to replace a natural one that does not function properly because of degenerative diseases such as osteoarthritis, or because the natural joint has to be replaced, for instance, in the process of removing a bone tumor. The treatment of such problems aims at restoring the joint function and relieving pain. In cases where it is not possible to apply conservative methods of treatment, there arises the need for joint replacement with components that aim to reestablish the functionality of the original joint. The technique used in the functional recovery of the hip joint through the replacement of both sides of the joint is called total hip arthroplasty (THA). This is a standard surgical procedure employed in orthopedics.

During the surgery, the head of the femur is removed and replaced with a metal femoral component inserted inside the medullary cavity of the bone (diaphysis – Figure 2.1). The femoral component has a spherical proximal extremity, which allows rotation on the surface of the acetabular component. This component, normally manufactured using polyethylene, covers the acetabular cavity and through a proper fit with the femoral component allows the execution of the joint movements.

Bone is a live tissue capable of modifying its structure according to the mechanical requirements imposed on it, which makes it different from any other material. This adaptive behavior is known as *adaptive bone remodeling* and consists of deposition and resorption of bone material over time, causing external geometric changes and internal changes in the bone microstructure [1].

After the insertion of the femoral component, the equilibrium and the load distribution in the femur are modified, initiating the remodeling and adaptation to the new distribution of stresses resulting from the placement of the prosthesis. One clinically reported consequence of this change is the loss of bone mass (resorption) in the femur, which can lead to a fracture and/or reduce the quantity of bone mass available for a future surgical revision.

In order to predict this mass exchange, different bone adaptation models have been proposed in the literature. Pioneering studies postulated the existence of a

Biomechanics of Hard Tissues: Modeling, Testing, and Materials.
Edited by Andreas Öchsner and Waqar Ahmed

ISBN: 978-3-527-32431-6

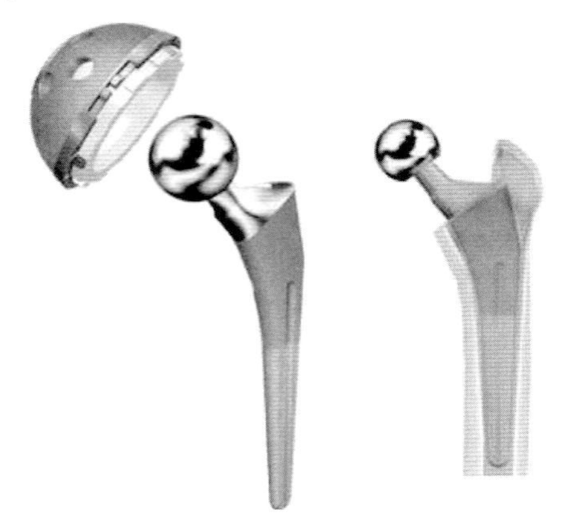

Figure 2.1 Total hip arthroplasty (THA).

homeostatic equilibrium state relating a stress–strain state to a specific bone density distribution. Any local perturbation in the mechanical state would promote a driving force for local changes in bone density in such a way that the equilibrium is recovered [2–7]. Owing to their nature, these models are named *error-driven* models. On the basis of a similar equilibrium concept and introducing continuum damage as a driving force, a group of studies treated remodeling as a continuous process of simultaneous damage and healing, the rates of which depend on local stress and strain levels. Owing to the possibility of including different phenomenological effects within a thermodynamically consistent framework, this approach has been increasingly investigated, for example, in [8–15] among others.

On the basis of the observations of Meyer, Culmann, and Wolf in the second half of the nineteenth century, a set of models described adaptation as a path toward a stable minimum (or maximum) condition, in which the bone behaves as an optimal structure. Later, several studies derived adaptation laws based on optimality conditions of a minimization problem, among them being [16–20]. It is worth mentioning that strong interconnection exists among all these approaches and thus the short classification presented is not strict.

Most of these computational techniques allow an approximate representation of the long-term structural behavior of the bone-implant system. In fact, it has been shown that these models are able to reproduce the normal architecture of bone. The studies [4, 5, 9, 10, 21, 22] are examples of bidimensional simulations, while in [16, 18, 23, 24] *3D* results are found. These models have also been used to predict the osteoporosis process [25] as well as effects related to the resorption phenomenon in host bone due to orthopedic implants [3, 6, 8, 11, 26–29]. The objectives of all these efforts are clearly a better understanding of the mechanisms that control the adaptive bone response, as well as the selection of the most appropriate prosthesis for a given bone state (obtained from personalized digital medical images – computational tomography).

Nevertheless, the problem of long-term stability of orthopedic implants has not been satisfactorily solved yet [30]. Besides the problems related to the resorption of proximal bone, the success of THA is influenced by problems of interface instability as the formation of fibrous tissue around the prosthesis and the inclusion of appropriate interface bone/implant conditions still present open problems. In fact, the majority of the bone adaptation models found in the literature do not distinguish between the interfacial and the periprosthetic adaptation.

This chapter discusses the effects of interface and periprosthetic adaptation and the derivation of bone adaptation models based on the hypothesis of an optimal bone topology. The final purpose of this work is to present an adaptation model for the bone-prosthesis interface, capable of being used in conjunction with formulations of the periprosthetic adaptation. To account for the sparse nature of bone ingrowth that is known to occur *in vivo*, the interfacial model is based on the application of a mixture rule, in which different kinds of tissues may be present in the same region of the interface. Thus, at each interfacial point, two kinds of interfacial conditions could coexist in different proportions: ingrowth (bonded bone) and fibrous tissue (encapsulated bone). Constitutive laws were derived for each one of the interfacial conditions, and appropriate evolution equations were defined for the relative quantities. These evolution equations are dependent on biomechanical criteria related to local micromovements based on experimental results reported in the literature. For the periprosthetic bone volume, an optimization-based remodeling model is employed for the bone density evolution. The bases of these propositions were developed in [31–33].

In the next section, a brief discussion on the biomechanics of bone adaptation is presented. Section 2.3 deals with constitutive models for bone and the interfacial region. The adaptation model is presented in Section 2.4. Numerical experiments are shown in Section 2.5, leaving Section 2.6 for final remarks.

2.2
Mechanical Adaptation of Bone

The process of bone tissue adaptation, also known as *remodeling*, has typically been separated into external and internal remodeling. In reality, they both occur simultaneously [34]. In external remodeling, the material properties are assumed to be fixed, while the geometry changes as a function of time. In internal remodeling, the geometry is assumed to be fixed, whereas the material properties are a function of time. The internal remodeling is currently more studied, since it represents mechanisms relevant to the evaluation of the durability of the prosthetic replacement [35].

The bone adaptation in the interfacial region has frequently been related to theories of bone repair, since during the preparation of the cavity for the prosthesis the bone tissue at this location is damaged. The periimplant repair is a complex process that involves many cellular and extracellular events. The repair is influenced by a variety of factors including the type of bone (cortical or trabecular), blood supply, location of the implant, severity of trauma at the implantation site, degree of fixation during the repair, and sex and age of the patient.

Davies [36] considers the periimplant repair process as a sequence of overlapping stages: hemostasis, formation of granulation tissue, and bone formation. Hemostasis is characterized by the formation of a hematoma (blood coagulation) and lasts a few days. After a few weeks, there begins the formation of a granular tissue, comprising neo-capillaries, fibroblasts, and mesenchymal cells, at the interface. Under favorable conditions, these are able to differentiate into osteoblasts, which deposit bone matrix, directly originating bone. In the formation of bone, the potentially osteogenic cellular population reaches the surface of the implant, inside the area to be repaired, and deposits bone material. The treatment of tissue differentiation in the interfacial region began with Pauwels in 1960, who presented the hypothesis in which the shear deformations are considered the stimulus required for the formation of cartilaginous fibers, while the hydrostatic pressures are the stimulus required for the formation of cartilage [37]. According to Pauwels, bone is not formed directly, but firstly soft tissue formation is necessary, which would allow the stabilization of the mechanical environment. Only after this phenomenon does the formation of bone material occur. Carter and Carter & Giori [38, 39] developed the concepts that relate the tissue differentiation to the mechanical environment. In these studies, the importance of the cyclic loading is discussed in detail and a method to calculate the historical stresses and deformations is proposed. In this formulation, it is considered that regions exposed to a history of low distortional deformation and low hydrostatic stresses are more capable of direct formation of bone tissue, so long as there is an adequate blood supply. Following the same line, Claes & Heigele [40] proposed that when local deformation levels are below 5% there is intermembrane ossification, while compressive hydrostatic stresses greater than 0.15 MPa and local deformation below 15% lead to endochondral ossification. All other environments favor the formation of fibrous tissue and fibrocartilage. Prendergast [41] proposed a mechano-regulation model based on two biophysical stimuli: shear deformations and the flow of interstitial fluids. The basis of this theory is that the fluid flow is related to an increase in stresses and deformations. Using a poroelastic model, Kuiper [42] simulated the process of differentiation in the repair of fractures.

2.3 Constitutive Models

2.3.1 Bone Constitutive Model

As previously stated, bone is a quite complex, live, and changing material, in which progressive scales (macro, meso, micro, etc.) reveal new details that compose a general picture of a nonhomogeneous, anisotropic, nonlinear rate-dependent material. It is thus necessary then to choose the bounds of the modeling, that is, the appropriate level of detail in order to capture the studied phenomenon and keep the model numerically tractable. A common assumption in the literature has been to consider bone as a periodic microstructure from which homogenized macroscale constitutive parameters are obtained as a function of the microstructure design. Simple isotropic material with penalization (SIMP) is an approach taken from topology optimization that was largely used as a first approximation of an artificial isotropic elastic material with behavior dependent on a single parameter: an artificial relative density $\rho \in [0:1]$. This model simply reads [43]:

$$\mathbb{C}^{\rho} = \mathbb{C}\rho^{n} \tag{2.1}$$

where $\mathbb{C}$ is the isotropic linear elasticity tensor corresponding to the solid bone. The bounds $\rho = 0$ and $\rho = 1$ are used to represent a void or solid (cortical) bone, respectively. A second common approach is the assumption of a geometrically defined periodic microstructure from which homogenized properties can be obtained. Among several proposed alternatives in literature, the present study is carried out using the periodic microstructure shown in Figure 2.2, proposed by Bagge [18, 20]. Using homogenization techniques, an orthotropic elastic material elasticity tensor dependent on a relative density $\rho \in [0:1]$ related with the thickness of the microstructure bars is obtained. The microstructure orientation with respect to global axes is given by the Euler angles $\boldsymbol{\theta} = \{\theta_1, \theta_2, \theta_3\}^T$. No distinction is made between trabecular and cortical bones. The latter case is assumed to be the limit case $\rho = 1$ of the former one, although it is known that this is just an approximation. The "solid material" is considered to be isotropic linear elastic, with an elasticity modulus $E = 20\,\mathrm{GPa}$ and Poisson coefficient $\upsilon = 0.3$ [44].

The considered microstructure has cubic symmetry that provides $E_1 = E_2 = E_3$, $G_{12} = G_{13} = G_{23}$, and $\upsilon_{12} = \upsilon_{13} = \upsilon_{23}$ in the material coordinates. The coefficients of the elasticity tensor are given below [18]:

$$\begin{aligned}
\mathbb{C}^{\rho}_{1111} &= \mathbb{C}^{\rho}_{2222} = \mathbb{C}^{\rho}_{3333} = 5409.96\rho^3 + 8.636\rho^2 \\
\mathbb{C}^{\rho}_{1122} &= \mathbb{C}^{\rho}_{1133} = \mathbb{C}^{\rho}_{2233} = 938.144\rho^5 + 720.29\rho^4 \\
\mathbb{C}^{\rho}_{1212} &= \mathbb{C}^{\rho}_{1313} = \mathbb{C}^{\rho}_{2323} = 1789.34\rho^4 + 118.038\rho^3
\end{aligned} \tag{2.2}$$

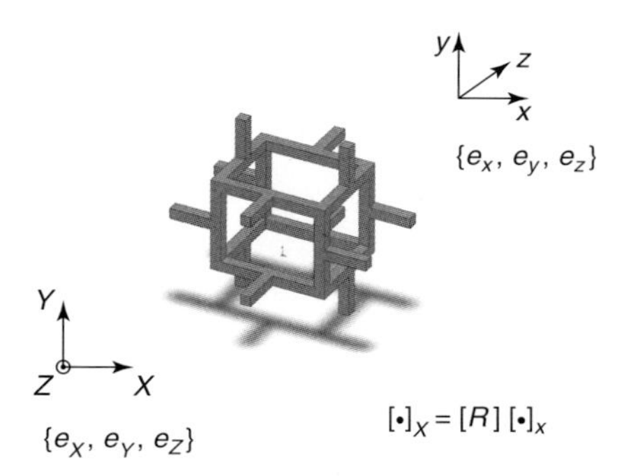

Figure 2.2 Idealized trabecular bone microstructure.

2.3.2 Interface Constitutive Model

In THA, the interface is the region immediately adjacent to the prosthesis, that is, the region through which the forces are transmitted from the prosthesis to the bone tissue (Figure 2.3). This consists of a thin layer of deformable material with particular characteristics. Immediately after the implantation, the shaft is partially stabilized because of its geometric characteristics, since it is inserted into the medullary canal under a moderate pressure, generating a distribution of stresses over the bone and the interface. Experimental studies show that the osseointegration does not occur immediately after the implantation and, even over time, only a small amount of the porous covering is osseointegrated. Some places may also show formation of both the bone and the fibrous tissue, that is, a kind of material mixture at the same place [45–47].

The real mechanisms responsible for the formation of the fibrous tissue, in the interfacial region, are still not completely clear. One of the clearest causes is related to the occurrence of excessive micromovements between the prosthesis and the bone. Some authors state that the mechanical environment is more important than the presence of particles originating from acetabular wear and, therefore, the formation of interfacial tissue is directly related to the historical loadings to which the prosthesis has been subjected [48]. *In vivo* studies have shown that relative micromovements between the prosthesis and the bone of the order of 20 μm do not influence the bone repair process, but for micromovements greater than 150 μm the formation of fibrous tissue around the prosthesis may reach a thickness of 1–2 mm , within six weeks after the implantation [49].

In the structure of the fibrous tissue, the collagen fibers are distributed randomly in several overlapping layers. This tissue has a nonlinear behavior, low stiffness,

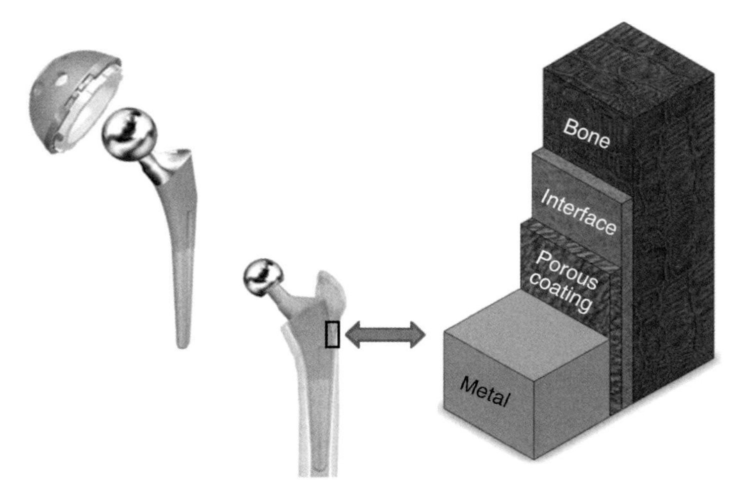

Figure 2.3 Schematic of a hip replacement and interface layer.

and a certain capacity to support compressive loads in the direction normal to the fibrous tissue layers. On the other hand, it fails to support traction and shear forces [50].

The desired condition of structural continuity between the bone and the femoral shaft arises when an effective adherence, condition known as *bone ingrowth*, exists between bone cells and the pores of prosthesis covering. This condition occurs only in the presence of a favorable biomechanical environment that is frequently related to moderate relative displacements between the surfaces in contact. Furthermore, other factors influence the bone growth, for instance, the physiological characteristics of the patient, biocompatibility of the implanted material, and the characteristics of the porous covering, among others. With the aim of defining models that are closer to real physics, different studies have included the modeling of the interface as a new element to be considered [51].

Two main approaches are found in the literature when the interface, in particular, is considered. The first is the direct incorporation of contact-friction-adherence conditions between the surfaces. This approach is used, for example, in the works [16, 32] among others. Another possibility is to assume that the interface is a thin but volumetric region in which mechanical properties follow a different behavior from that of the bodies in contact. This approach may be seen in [51].

In this study, the latter approach is followed. The interface is modeled by a thin volumetric region lying between the bone and the shaft. In this region, volumetric gasket-type elements are defined. Geometrically, these elements are formed by two surfaces, upper and lower, and a local coordinate system that distinguishes (pseudo) normal and tangential directions at each integration point at which the constitutive problem is described (Figure 2.4). The vector $\mathbf{e}_1$ corresponds to the pseudo-normal direction, while $\mathbf{e}_2$ and $\mathbf{e}_3$ are perpendicular to this normal direction. From the kinematic point of view, these elements use the same equations as those

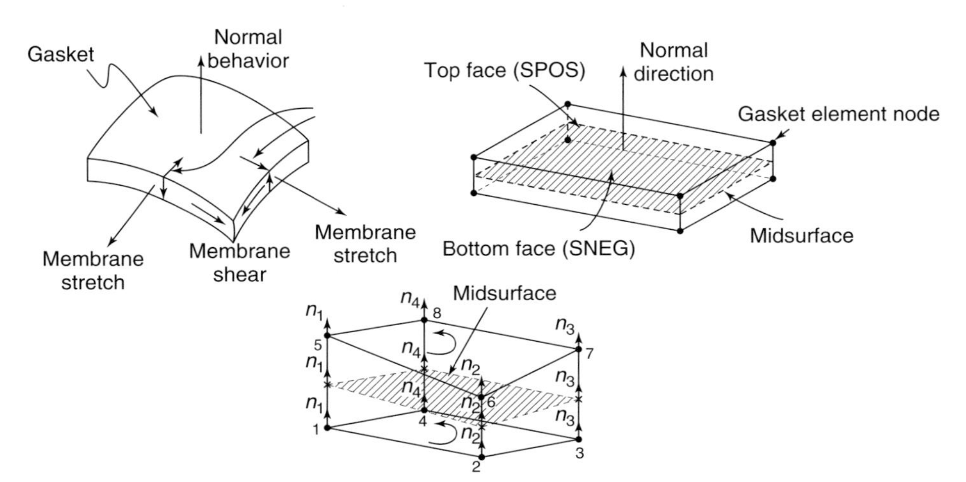

Figure 2.4 Spatial representation of a gasket element.

of conventional volumetric elements to define the strain tensor that, after rotation to local axes, is used to define local stresses. Once in local coordinates, these elements often incorporate a decoupling of normal and tangential effects through the constitutive model. In this study, a bilinear relation is used for the stresses normal $\sigma_{11}(\varepsilon_{11})$ to the reference surface and a linear relation for the shear stresses $\sigma_{13}(\varepsilon_{13})$, $\sigma_{12}(\varepsilon_{12})$. The membrane stresses σ_{22}, σ_{33}, and σ_{23} are considered null. For clarity reasons, in what follow we refer to normal stress σ_n (σ_{11}), normal strain ε_n (ε_{11}), tangential stresses σ_t (σ_{12}, σ_{13}), and tangential strains ε_t (ε_{12}, ε_{13}).

For the normal direction, a bilinear model provides respective stiffness to the bone and fibrous material in the case of tractive effects and a penalization condition for compressive strains (similar to that of contact models, Figure 2.5). For the tangential direction, a linear elastic behavior is assumed for both materials, with a much lower stiffness for a fiber than a bone material.

Thus, the constitutive relationships in local coordinates at an integration point at an element interface are given by

$$\sigma_n^{\mathrm{f}} = \begin{cases} E_n^{\mathrm{f}}\,\varepsilon_n & \text{if } \varepsilon_n \geq 0 \\ E_n^{\mathrm{p}}\,\varepsilon_n & \text{if } \varepsilon_n < 0 \end{cases} \tag{2.3}$$

$$\sigma_t^{\mathrm{f}} = E_t^{\mathrm{f}}\,\varepsilon_t \tag{2.4}$$

$$\sigma_n^{b} = \begin{cases} E_n^{b}(\rho)\,\varepsilon_n & \text{if } \varepsilon_n \geq 0 \\ E_n^{\mathrm{p}}\,\varepsilon_n & \text{if } \varepsilon_n < 0 \end{cases} \tag{2.5}$$

$$\sigma_t^{b} = E_t^{b}(\rho)\,\varepsilon_t \tag{2.6}$$

$$E_n^{b}(\rho) = \rho^{\eta} E_n^{b} \quad E_t^{b}(\rho) = \rho^{\eta} E_t^{b} \tag{2.7}$$

In Eq. (2.3), a bilinear constitutive equation separates positive and negative strains. The superscripts b, f and p in expressions (2.3–2.7) refer to "bone", "fiber" and "penalization", respectively. While positive strains are related to a fiber elasticity coefficient E_n^{f}, negative strains are related to stress by a penalization parameter E_n^{p} in order to prevent penetration. For the tangential behavior (Eq. 2.4), the relationship is linear. Equivalent model is used for the bone. The only difference

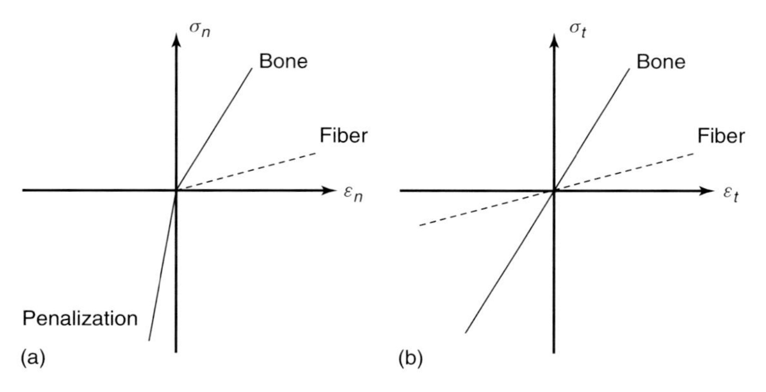

Figure 2.5 (a) Constitutive model for normal direction; (b) constitutive model for tangential direction.

with the fibrous case is that the elastic bone coefficients $E_t^b(\rho)$ and $E_n^b(\rho)$ depend on the relative density of the bone attached to the considered material point of the interface (Eq. 2.7). The parameter η is equivalent to the penalization parameter of the SIMP model.

In what follows, the outlines of the periprosthetic remodeling model is stated. As previously said, a principle of global optimum behavior was chosen as a remodeling engine.

2.3.3 Model for Periprosthetic Adaptation

Consider three bodies in the tridimensional Euclidian space: the femur, occupying the open region Ω_b, the femoral shaft occupying region Ω_s, and the interface body defined in the region denoted by Ω_i (Figure 2.6). The surface Γ_u is the part of the boundary Γ where Dirichlet boundary conditions are applied, while Γ_t is submitted to surface forces $\mathbf{t}$. Both the bone and the shaft are modeled by elastic material. Let us also denote $\boldsymbol{\sigma}$, the Cauchy stress tensor, and $\varepsilon(\cdot) = \nabla^s(\cdot)$, the linearized Green strain tensor. The constitutive behavior of bone follows the equation $\boldsymbol{\sigma} = \mathbb{C}^\rho \varepsilon$ with the homogenized elastic tensor $\mathbb{C}^\rho$ given by Eq. (2.2). This implies a linear dependence on displacements and a nonlinear dependence on the relative density ρ. The interface region Ω_i allows a nonlinear constitutive dependence on strains, depending also on ρ, due to its attachment to Ω_b. The shaft in Ω_s has a linear material behavior and is independent of ρ.

The equilibrium condition is stated by the variational problem: find the displacement field $\mathbf{u} \in \mathcal{U}$ such that

$$\int_{\Omega_b} \boldsymbol{\sigma}(\mathbf{u}) \cdot \varepsilon(\boldsymbol{\lambda}) \mathrm{d}\Omega_b + \int_{\Omega_i} \boldsymbol{\sigma}(\mathbf{u}) \cdot \varepsilon(\boldsymbol{\lambda}) \mathrm{d}\Omega_i + \int_{\Omega_s} \boldsymbol{\sigma}(\mathbf{u}) \cdot \varepsilon(\boldsymbol{\lambda}) \mathrm{d}\Omega_s - l(\boldsymbol{\lambda}) = 0 \quad \forall\, \boldsymbol{\lambda} \in \mathcal{V} \tag{2.8}$$

where $\mathcal{U} := \{\mathbf{u} \in H^1(\Omega) : \mathbf{u}|_{\Gamma_u} = 0\}$ and $\mathcal{V} := \{\boldsymbol{\lambda} \in H^1(\Omega) : \boldsymbol{\lambda}|_{\Gamma_u} = 0\}$ define the set of admissible displacements and admissible variations, respectively. The three first terms of Eq. (2.8) represent the virtual work of the internal forces in each region.

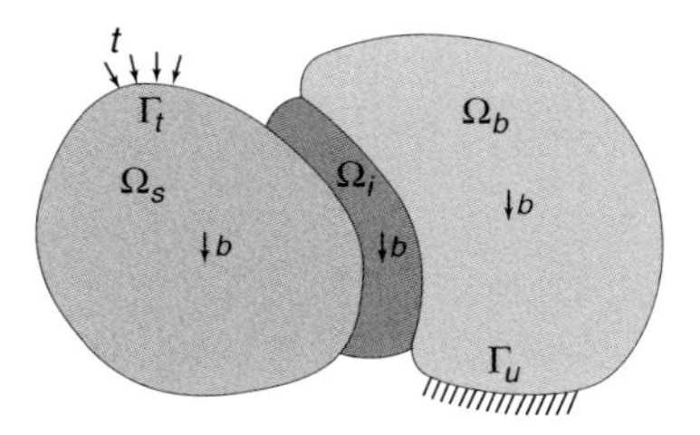

Figure 2.6 Bone, stem, and interface domains.

The last term,

$$l(\boldsymbol{\lambda}) = \int_{\Omega} \mathbf{b} \cdot \boldsymbol{\lambda}\, d\Omega + \int_{\Gamma_t} \mathbf{t} \cdot \boldsymbol{\lambda}\, d\Gamma$$

accounts for the work of the external forces.

Taking the relative densitiy ρ as a design variable, the minimization of the strain energy in Ω_b subject to constraints on the bone mass and satisfaction of equilibrium is then proposed:

$$\min \sum_{j=1}^{\text{nlc}} \frac{1}{2} \int_{\Omega_b} \omega_j \boldsymbol{\sigma}_j(\mathbf{u}) \cdot \varepsilon_j(\mathbf{u}) d\Omega_b \quad \rho \in K \quad \text{subject to} \tag{2.9}$$

$$\int_{\Omega_b} \rho d\Omega = \mathbf{V} \tag{2.10}$$

$$\int_{\Omega_b} \boldsymbol{\sigma}_j(\mathbf{u}) \cdot \varepsilon_j(\boldsymbol{\lambda})\, d\Omega_b + \int_{\Omega_i} \boldsymbol{\sigma}_j(\mathbf{u}) \cdot \varepsilon_j(\boldsymbol{\lambda})\, d\Omega_i + \int_{\Omega_s} \boldsymbol{\sigma}_j(\mathbf{u}) \cdot \varepsilon_j(\boldsymbol{\lambda})\, d\Omega_s - l_j(\boldsymbol{\lambda}) = 0 \quad \forall\, \boldsymbol{\lambda} \in \mathcal{V} \tag{2.11}$$

where nlc is the number of load cases and ω is a weighting factor for each load case. The lateral constraints $\rho(x) \in [0,\ 1]$ are excluded from the equations just for clarity reasons and will be reintroduced at the algorithmic level. Allowing, for simplicity, a single load case ($\text{nlc} = 1,\ \omega_1 = 1$), the Lagrangian of the problem is defined as

$$\begin{aligned} \mathcal{L}(\mathbf{u},\boldsymbol{\lambda},\boldsymbol{\alpha},\boldsymbol{\rho}) = {} & \frac{1}{2} \int_{\Omega_b} \boldsymbol{\sigma}(\mathbf{u}) \cdot \varepsilon(\mathbf{u})\, d\Omega_b + \int_{\Omega_b} \boldsymbol{\sigma}(\mathbf{u}) \cdot \varepsilon(\boldsymbol{\lambda})\, d\Omega_b \\ & + \int_{\Omega_i} \boldsymbol{\sigma}(\mathbf{u}) \cdot \varepsilon(\boldsymbol{\lambda})\, d\Omega_i + \int_{\Omega_s} \boldsymbol{\sigma}(\mathbf{u}) \cdot \varepsilon(\boldsymbol{\lambda})\, d\Omega_s - l(\boldsymbol{\lambda}) \\ & + \alpha \left(\left(\int_{\Omega_b} \rho\, d\Omega - \mathrm{V}_o \right) \right) \end{aligned} \tag{2.12}$$

where $\boldsymbol{\lambda}$ and α play the role of Lagrange multipliers of the equilibrium equation and mass constraint, respectively. The Lagrangian stationarity condition is

$$\left\langle \frac{d\mathcal{L}}{dc}, \mathbf{d} \right\rangle = 0 \tag{2.13}$$

where $\mathbf{c} = (\mathbf{u},\boldsymbol{\lambda},\alpha,\ \rho)$ and $\mathbf{d} = (\boldsymbol{\xi},\ \boldsymbol{\eta},\ \gamma,\ \beta)$.

The stationarity with respect to variables $\boldsymbol{\lambda}$ and α, recovers the nonlinear state equation and mass constraint:

$$\begin{aligned} \left\langle \frac{\partial \mathcal{L}}{\partial \boldsymbol{\lambda}}, \boldsymbol{\eta} \right\rangle = {} & \int_{\Omega_b} \boldsymbol{\sigma}(\mathbf{u}) \cdot \varepsilon(\boldsymbol{\eta})\, d\Omega_b + \int_{\Omega_i} \boldsymbol{\sigma}(\mathbf{u}) \cdot \varepsilon(\boldsymbol{\eta})\, d\Omega_i \\ & + \int_{\Omega_s} \boldsymbol{\sigma}_j(\mathbf{u}) \cdot \varepsilon(\boldsymbol{\eta})\, d\Omega_s - l(\boldsymbol{\eta}) = 0 \quad \forall\, \boldsymbol{\eta} \in \mathcal{V} \end{aligned} \tag{2.14}$$

$$\left\langle \frac{\partial \mathcal{L}}{\partial \alpha}, \gamma \right\rangle = \gamma \int_{\Omega_b} \rho d\Omega_b - \mathbf{V}_b = 0 \quad \forall \gamma \in \mathbb{R} \tag{2.15}$$

Stationarity of $\mathcal{L}$ in relation to the displacement field $\mathbf{u}$ gives

$$\left\langle \frac{\partial \mathcal{L}}{\partial \mathbf{u}}, \boldsymbol{\xi} \right\rangle = \int_{\Omega_b} \frac{\partial \boldsymbol{\sigma}(\mathbf{u})}{\partial \boldsymbol{\varepsilon}} \boldsymbol{\varepsilon}(\mathbf{u}) \cdot \boldsymbol{\varepsilon}(\boldsymbol{\xi})\, d\Omega_b + \int_{\Omega_b} \frac{\partial \boldsymbol{\sigma}(\mathbf{u})}{\partial \boldsymbol{\varepsilon}} \boldsymbol{\varepsilon}(\boldsymbol{\xi}) \cdot \boldsymbol{\varepsilon}(\boldsymbol{\lambda}) d\Omega_b$$
$$+ \int_{\Omega i} \frac{\partial \boldsymbol{\sigma}(\mathbf{u})}{\partial \boldsymbol{\varepsilon}} \boldsymbol{\varepsilon}(\boldsymbol{\xi}) \cdot \boldsymbol{\varepsilon}(\boldsymbol{\lambda})\, d\Omega_i$$
$$+ \int_{\Omega s} \frac{\partial \boldsymbol{\sigma}(\mathbf{u})}{\partial \boldsymbol{\varepsilon}} \boldsymbol{\varepsilon}(\boldsymbol{\xi}) \cdot \boldsymbol{\varepsilon}(\boldsymbol{\lambda})\, d\Omega_s = 0 \quad \forall \boldsymbol{\xi} \in \mathcal{V} \tag{2.16}$$

where the symmetry of the material tensor $\frac{\partial \boldsymbol{\sigma}(\mathbf{u})}{\partial \boldsymbol{\varepsilon}} = \mathbb{C}^\rho$ in Ω_b has been used. Since for both Ω_b and Ω_h the tangent matrix is the linear elasticity tensor, the condition can be rewritten as

$$\int_{\Omega_b} \mathbb{C}^\rho(\rho)\boldsymbol{\varepsilon}(\boldsymbol{\lambda}) \cdot \boldsymbol{\varepsilon}(\boldsymbol{\xi})\, d\Omega_b + \int_{\Omega i} \left(\frac{\partial \boldsymbol{\sigma}(\mathbf{u})}{\partial \boldsymbol{\varepsilon}} \right)^T \boldsymbol{\varepsilon}(\boldsymbol{\lambda}) \cdot \boldsymbol{\varepsilon}(\boldsymbol{\xi})\, d\Omega_i$$
$$+ \int_{\Omega s} \mathbb{C}^s \boldsymbol{\varepsilon}(\boldsymbol{\lambda}) \cdot \boldsymbol{\varepsilon}(\boldsymbol{\xi})\, d\Omega_s = - \int_{\Omega_b} \mathbb{C}^\rho(\rho)\boldsymbol{\varepsilon}(\mathbf{u}) \cdot \boldsymbol{\varepsilon}(\boldsymbol{\xi})\, d\Omega_b \quad \forall\, \boldsymbol{\xi} \in \mathcal{V} \tag{2.17}$$

This equation is called the adjoint problem, and it provides the adjoint solution $\boldsymbol{\lambda}$. Note that, in contrast to the state Eq. (2.18), this problem is linear in $\boldsymbol{\lambda}$, allowing a quite simple calculation.

Finally, the Lagrangian variation in relation to the design variable ρ takes the form

$$\left\langle \frac{\partial \mathcal{L}}{\partial \rho}, \beta \right\rangle = \frac{1}{2} \int_{\Omega_b} \frac{\partial \mathbb{C}^\rho}{\partial \rho} \boldsymbol{\varepsilon}(\mathbf{u}) \cdot \boldsymbol{\varepsilon}(\mathbf{u})\, d\Omega_b + \int_{\Omega_b} \frac{\partial \mathbb{C}^\rho}{\partial \rho} \boldsymbol{\varepsilon}(\mathbf{u}) \cdot \boldsymbol{\varepsilon}(\boldsymbol{\lambda})\, d\Omega_b$$
$$+ \int_{\Omega_i} \frac{\partial \boldsymbol{\sigma}(\mathbf{u})}{\partial \rho} \cdot \boldsymbol{\varepsilon}(\boldsymbol{\lambda})\, d\Omega + \alpha \int_{\Omega_b} d\Omega = 0 \tag{2.18}$$

Using conventional finite elements, Eqs (2.14) and (2.17) are easily discretized leading to the following algebraic equations:

$$\mathbf{F}_{\text{int}} = \mathbf{K}_b\mathbf{U} + \mathbf{F}_i(\mathbf{U}) + \mathbf{K}_s\mathbf{U} = \mathbf{F}_{\text{ext}} \tag{2.19}$$
$$\left[\mathbf{K}_b + (\mathbf{K}_i(\mathbf{U}))^T + \mathbf{K}_s\right] \mathbf{3} = \mathbf{F}_{\text{adj}}(\mathbf{U}) \tag{2.20}$$

where $\mathbf{K}_b$, $\mathbf{K}_s$, and $\mathbf{F}_{\text{ext}}$ denote the conventional stiffness matrices and external forces, $\mathbf{F}_i(\mathbf{U})$ denotes the nonlinear internal forces at the interface, $\mathbf{K}_i(\mathbf{U})$ denotes the tangent matrix at the equilibrium configuration $\mathbf{U}$, and $\mathbf{F}_{\text{adj}}(\mathbf{U})$ denotes the adjoint forces from Eq. (2.17). Given $\mathbf{N}$ and $\mathbf{B}$, the arrays of shape functions and their derivatives, respectively, these matrices are computed as

$$\mathbf{K}_b = \int_{\Omega_b} \mathbf{B}^T \mathbb{C}^\rho \mathbf{B}\, d\Omega \quad \mathbf{K}_s = \int_{\Omega s} \mathbf{B}^T \mathbb{C}^s \mathbf{B}\, d\Omega \tag{2.21}$$
$$\mathbf{F}_i(\mathbf{U}) = \int_{\Omega_b} \mathbf{B}^T \sigma(\mathbf{U})\, d\Omega \quad \mathbf{K}_i = \int_{\Omega i} \mathbf{B}^T \frac{\partial \boldsymbol{\sigma}(\mathbf{U})}{\partial \boldsymbol{\varepsilon}} \mathbf{B}\, d\Omega \tag{2.22}$$
$$\mathbf{F}_{\text{adj}}(\mathbf{U}) = -\int_{\Omega_b} \mathbf{B}^T \mathbb{C}^\rho(\rho)\boldsymbol{\varepsilon}(\mathbf{u})\, d\Omega_b \tag{2.23}$$

Assuming a distribution ρ, constant at each element, condition (2.18) becomes local, allowing one to obtain a stationary condition of the type

$$B_e = \frac{\frac{1}{2}\int_{\Omega_b^e} \frac{\partial \mathbb{C}^\rho}{\partial \rho} \boldsymbol{\varepsilon}(\mathbf{u}) \cdot \boldsymbol{\varepsilon}(\mathbf{u})\, d\Omega_b + \int_{\Omega_b^e} \frac{\partial \mathbb{C}^\rho}{\partial \rho} \boldsymbol{\varepsilon}(\mathbf{u}) \cdot \boldsymbol{\varepsilon}(\boldsymbol{\lambda})\, d\Omega_b + \int_{\Omega_i^e} \frac{\partial \boldsymbol{\sigma}(\mathbf{u})}{\partial \rho} \cdot \boldsymbol{\varepsilon}(\boldsymbol{\lambda}) d\Omega}{\alpha V_b^e} = 1 \tag{2.24}$$

where $V_b^e = \int_{\Omega_b^e} d\Omega$ is the volume of element e of the bone and Ω_i^e is the element (or group of elements) of the interface depending on the density of element e. In practice, this contribution provided by the integral over Ω_i^e is only computed for those elements of the bone that are in contact with a (generally single) element of the interface. The satisfaction of this optimality condition is sought by a fixed point scheme that updates the density values according to the following evolution expression:

$$\rho_e^{n+1} = \left\{ \begin{array}{l} \min\left\{(1-\xi)\,\rho_e^n, \rho_{\min}\right\} \text{ if } B_e^\varsigma \rho_e^n \leq \max\left\{(1-\xi)\,\rho_e^n, \rho_{\min}\right\} \\ B_e^\varsigma \rho_e^n \text{ if } \max\left\{(1-\xi)\,\rho_e^n, \rho_{\min}\right\} \leq B_e^\varsigma \rho_e^n \leq \min\left\{(1+\xi)\,\rho_e^n, \rho_{\max}\right\} \\ \max\left\{(1+\xi)\,\rho_e^n, \rho_{\max}\right\} \text{ if } \min\left\{(1+\xi)\,\rho_e^n, \rho_{\max}\right\} \leq B_e^\varsigma \rho_e^n \end{array} \right\} \tag{2.25}$$

where n refers to the iteration number, ξ is a movable limit, and ς is a numerical stabilization parameter. It can be observed that in Eq. (2.25) lateral constraints overρ ($\rho \in [\rho_{\min}, \rho_{\max}]$) were included. In order to obtain a stable sequence of iterations, the ξ and ς parameters must be appropriately selected. The variable ϕ defining the orientation of the microstructure is not present in the sensitivity expressions. A heuristic reorientation following a weighted sum of the principal strain directions computed at the centroid of each element is used instead [52]:

$$\boldsymbol{\varepsilon}_i^e = \sum_{j=1}^{\mathrm{Nlc}} \omega_j \left(\boldsymbol{\varepsilon}_i^e\right)_j \tag{2.26}$$

If the mass constraint must be satisfied, the α parameter must be updated in Eq. (2.25) at each step. This procedure is used in this study to define the bone distribution at the beginning of the remodeling process, that is, the initial condition of the femur before surgical intervention. With this aim, the bone alone is treated, excluding shaft and interface conditions. The iterations seek a stationary point satisfying Eq. (2.18) plus lateral constraints. Clearly, this step may be replaced by a direct experimental definition of ρ and ϕ by, for example, image processing.

Once the initial ρ and ϕ are defined, the femoral head is removed, and shaft and interface regions are included in the model. The remodeling process starts as a consequence of the stress/strain changes in the system following the evolution law provided by Eq. (2.18). It must be emphasized, however, that the mass is allowed to change during this process, which means that the Lagrange multiplier α remains fixed. In the present study, the value of α is that obtained at the end of the first stage in which stationarity conditions for the bone are satisfied.

It should be noted that it is necessary to solve two finite element problems at each load case of each iteration: the first is to find the solution to the nonlinear equilibrium problem (Eq. 2.27) and the second is the solution to the adjoint linear problem (Eq. 2.17).

2.3.4
Model for Interfacial Adaptation

It is considered here that the mechanical environment is the main aspect regulating the tissue differentiation and that biological factors are directly related to a mechanical stimulus. The one selected is related to the relative micromovements between the prosthesis and the bone, which in this case correspond to the shear strains of the gasket elements. This model is based on a simple mixture rule that allows the coexistence of the bone and fibrous tissue at the same interfacial location. Two relative quantities α_b and α_f quantify the occurrence of bone ingrowth and fibrous tissue respectively, satisfying the condition:

$$\alpha_b + \alpha_f = 1 \tag{2.27}$$

The internal forces F^e_{int} of each gasket element are computed from the contribution of each material, as follows:

$$\alpha_b F^e_b + \alpha_f F^e_f = F^e_{\text{int}} \tag{2.28}$$

where F^e_b and F^e_f are the corresponding internal forces of each material to the same strain. The evolution of quantity α_b (α_f is related to α_b by Eq. (2.27)) is proportional to the difference between the current state and the state that corresponds to the current strain (displacements):

$$\frac{\mathrm{d}\alpha_b}{\mathrm{d}t} = \upsilon[M(\mathbf{u}) - \alpha_b] \tag{2.29}$$

where M is denominated the stimulus for the osteogenesis calculated from the current normal and shear strain of the interfacial element. The parameter υ is related to the evolution rate and was introduced in such a way as to limit the changes between each time-step. The stimulus M for the osteogenesis is based on the tissue differentiation theory from Carter *et al.* [38]. The simplified graphical representation of this theory can be seen in Figure 2.7, where the value of the stimulus M depending of the normal and tangential relative displacements u_n and u_t is shown. Interfacial elements that undergo relative displacements within the limits $u_t \in [-u_{t\,\max}, u_{t\,\max}]$ and $u_n \in [-u_{n\,\max}, 0]$ will contribute to the formation of interfacial bone, whereas the elements submitted to displacements outside these limits tend toward the formation of the fibrous tissue. A transition zone is also incorporated in order to allow for a smooth change between the two extreme states. The values assumed for these limits are $u_{t\,\max} \simeq 150\ \mu\text{m}$ and $u_{n\,\max} \simeq 6\ \mu\text{m}$.

The updating of the relative coefficients follows a simple explicit time integration scheme:

$$\alpha_b^{k+1} = \alpha_b^k + \alpha_b \Delta t, \quad \alpha_f^{k+1} = 1 - \alpha_b^{k+1} \tag{2.30}$$

The evolution law presented allows that, during the entire course of the remodeling simulation, the interfacial elements can evolve toward the formation of fibrous

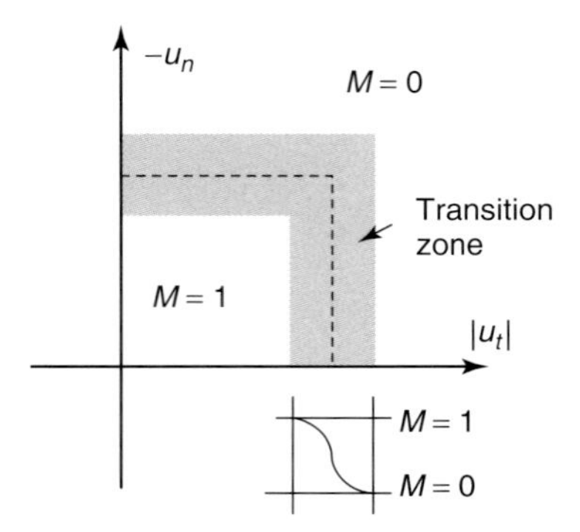

Figure 2.7 Graphic scheme of osteogenesis criterium.

or bone tissue so as to fulfill the adaptation criterion. Therefore, the constitutive behavior of the interfacial tissue is dependent on the relative quantities of each tissue present at the location. In parallel, the interfacial adaptation found in the course of the periprosthetic remodeling process and, therefore, the density of the interfacial elements, must vary according to the density of the bone attached to the interface material point, according to Eq. (2.7).

2.4
Numerical Examples

The examples shown in this section were obtained using the code ABAQUS plus user subroutines in FORTRAN and Python in order to include the proposed models for bone, interface, and adaptation routines. Also, the geometric model of a standard femur from Cheung *et al.* [53] was used.

A simulation of the femur alone is run first. The aim of this run is to obtain a valid initial density distribution of bone that emulates a real one. Three load cases are applied. The loadings due to the body weight are applied only in certain regions by surface forces parabolically distributed along and normal to the surface of the femoral head. The location of the loading surfaces is determined from the angles referring to the gait cycle. According to Allard [54] they vary from -20° to 25° for the hip joint. The application of load to a region greater than the surface is justified by the fact that over time other activities, such as climbing steps and even sitting on and getting up from a chair, must be taken into account during the simulations. In the present simulations, the loadings due to the body weight are divided into three load cases in order to represent, approximately, the forces to which the femur is subjected. Each load case has a body mass (BM) value of 700 N.

The stresses due to the action of the gluteus maximus, medius, and minimum muscles, applied to the greater trochanter (Figure 2.8), are also considered, assuming that these muscles with an insertion point close to the femoral head are sufficient to obtain results that approximate real conditions [55]. The loads referring to the action of each muscle are applied on the anterior, medial, and

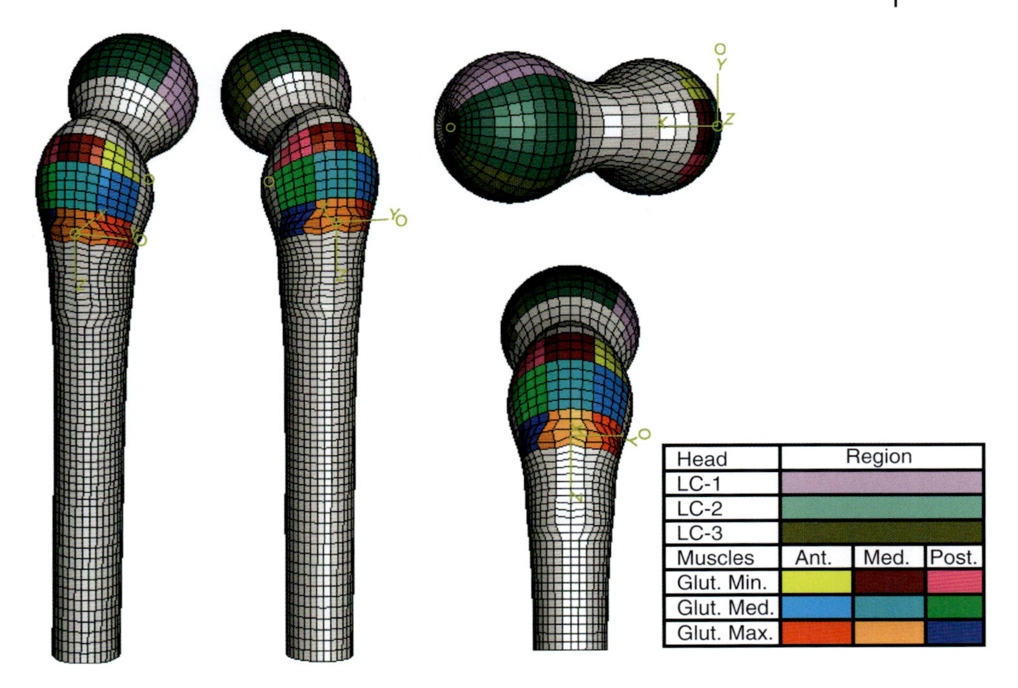

Figure 2.8 Spatial distribution of load cases.

posterior regions. The insertion points and the directions of the application of loadings follow the study presented by Dostal and Andrews [56], whereas the load values follow the observations of Pedersen *et al.* [55] in which 25% of the BM is considered for the gluteus maximus and medius and 12.5% of the BM for the gluteus minimus.

In contrast with the model presented by Bagge [18], which used 10 load cases, this model considers only three load cases obtained from the average of the 10 cases mentioned above (this simplification provides a significative reduction in the computational cost of the next step in which interfacial nonlinear conditions are included). The weighting factors related to each load case (1, 2, and 3) are 20%, 60%, and 20%, respectively.

The loads applied (Table 2.1) must be equilibrated by the reaction forces of the knee joint, which in this model are imposed through a clamped condition on the external ring of the femoral diaphysis. The coordinate system is oriented so that the x axis is in the lateral–medial direction, the y axis is in the anterior–posterior direction, and the z axis in the upper–lower direction.

The solution of the optimization problem with the volume constraint of 50% is given in Figure 2.9. In this case, it is possible to see the formation of the layers of cortical bone with high density along the diaphysis. Also, the formation of a low-density medullary canal, the formation of metaphyseal cortical plates, low density in the Ward's triangular region, and the complex distribution of density in

Table 2.1 Directions of loadings due to the action of muscles.

	Gluteus minimus								
Load case	Anterior			Medial			Posterior		
	x	y	z	x	y	z	x	y	z
1	0.11	−0.43	−0.90	0.16	−0.03	−0.90	0.20	0.10	−0.76
2	0.24	−0.25	−0.90	0.32	0.13	−0.95	0.37	0.26	−0.77
3	0.10	−0.15	−1.05	0.18	0.25	−1.10	0.22	0.42	−0.92
	Gluteus medium								
Load case	Anterior			Medial			Posterior		
	x	y	z	x	y	z	x	y	z
1	0.05	−0.36	−0.88	0.24	−0.94	−0.92	0.32	0.32	−0.80
2	0.08	−0.25	−0.92	0.31	−0.05	−0.91	0.40	0.32	−0.84
3	−0.06	−0.13	−1.07	0.25	0.13	−1.20	0.35	0.48	−0.92
	Gluteus maximus								
Load case	Anterior			Medial			Posterior		
	x	y	z	x	y	z	x	y	z
1	1.67	−3.16	−3.92	−1.33	0.50	−1.50	1.67	3.16	−3.92
2	2.88	−2.50	4.25	−0.62	0.44	−1.87	2.88	2.50	4.25
3	1.67	−1.08	−4.83	−1.50	0.75	−2.17	1.67	1.08	−4.83

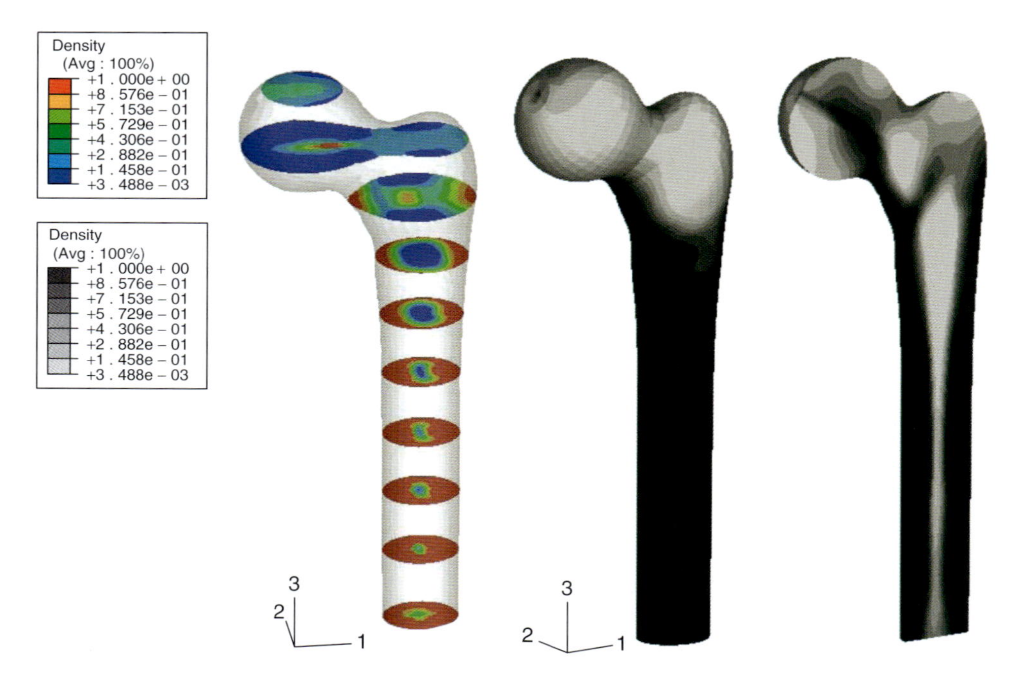

Figure 2.9 Bone density distribution for a volume constraint of 50%.

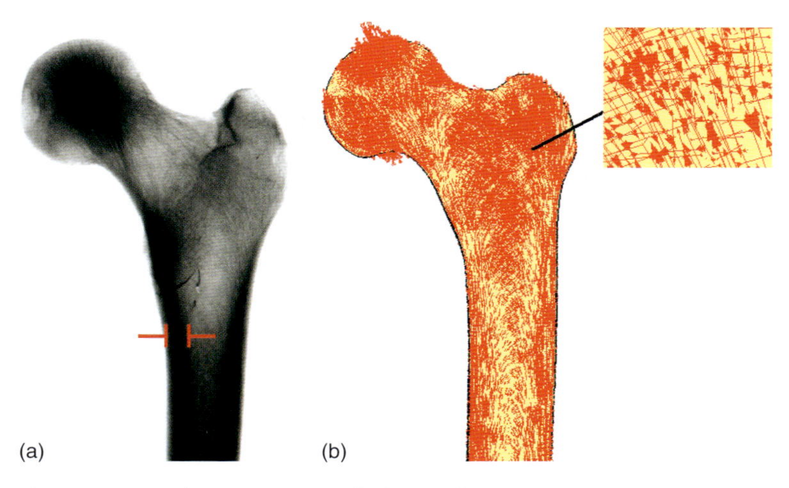

Figure 2.10 (a) The anterior view of a human femur (www.doitpoms.ac.uk/tlplib/bones/images/intact.pdf). (b) Material directions aligned withthe principal strain directions.

the epiphyseal region can be observed. High density values are also present in the region of the femoral neck, following the line of compressive load.

It is important to highlight that the model shows a high sensitivity in relation to the loading conditions. However, the results obtained confirm the use of the optimum structure hypothesis as a possible approach for the determination of an admissible distribution of densities.

The result in Figure 2.10 shows a microstructure orientation following a pattern similar to that of a real femur. In this case, it is possible to note the vertical alignment of the material, both in the medial cortex and the lateral cortex. In the metaphyseal region of the femur it is not possible to identify preferential orientations, consistent with a certain degree of isotropy of this region. In the femoral neck, a certain alignment is recovered because of bending stresses.

This density distribution is now used as the initial condition for the simulation of the adaptation of both the bone and the interface around the femoral shaft. The test is performed starting from two extreme interface initial conditions: a total encapsulation by fibrous tissue and a totally consolidated bone tissue. Figure 2.12 shows the final distribution of the product $\alpha_b \rho^n$ (note that bone stiffness at the interface is proportional to this product). It is possible to see small differences in the final distribution for each case. The total percentage of interfacial bone during time-steps starting from both initial conditions is shown in Figure 2.11. The similarity of results for the two conditions points out that the low quality of the bone tissue in the proximal part of the femur resulting from resorption does not allow a biomechanically favorable environment for bone ingrowth in this region.

An important issue frequently related to bone ingrowth is the stiffness of the prosthesis. There are evidences that the more rigid the prosthesis, the more proximal

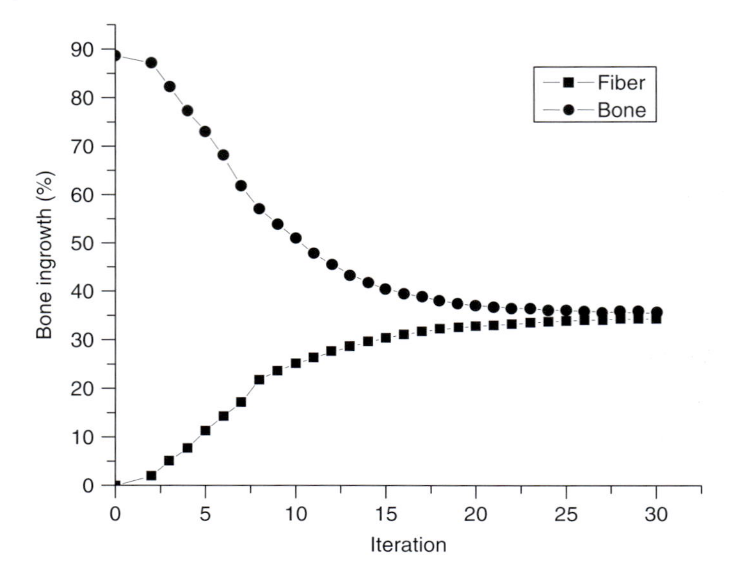

Figure 2.11 Bone ingrowth evolution.

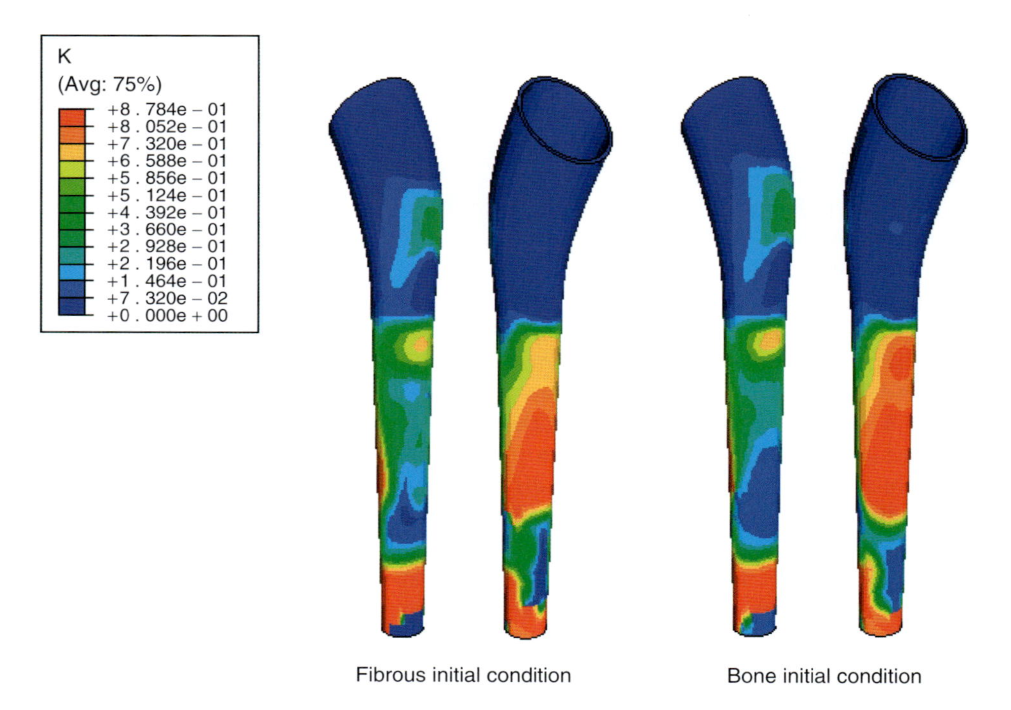

Figure 2.12 Bone ingrowth evolution for each initial condition.

the bone resorption. Conversely, the more flexible the prosthesis, the higher the shear strain in the interface/prosthesis region, hindering bone formation.

To evaluate the influence of this parameter on the model response, two simulations were carried out: one around a Ti alloy shaft ($E = 120$ GPa) and another around a CrCo shaft ($E = 200$ GPa).

In Figures 2.13 and 2.14, it can be observed that the Ti preserves a greater quantity of bone mass than the CrCo shaft. An hypertrophy was noted in the distal part of the femur, with a tendency toward a filling of the medullary canal and more pronounced bone resorption in the proximal part of the femur in the case of the CrCo prosthesis. A loss in bone mass of 9.2% for the CrCo prosthesis and 5.79% for the Ti prosthesis was noted.

For the interfacial region, the bone growth is sparse, in both cases. This situation is consistent with many clinical outcomes showing that bone ingrowth occurs only in a small portion of porous surface. In both cases, bone resorption in the proximal part produces a weak mechanical support and, consequently, conditions nonfavorable to bone ingrowth at these points (Figure 2.15).

Comparing the results of both cases, it is verified that with the Ti prosthesis the bone ingrowth occurs in a smaller region of the porous coating, as expected for a more flexible stem.

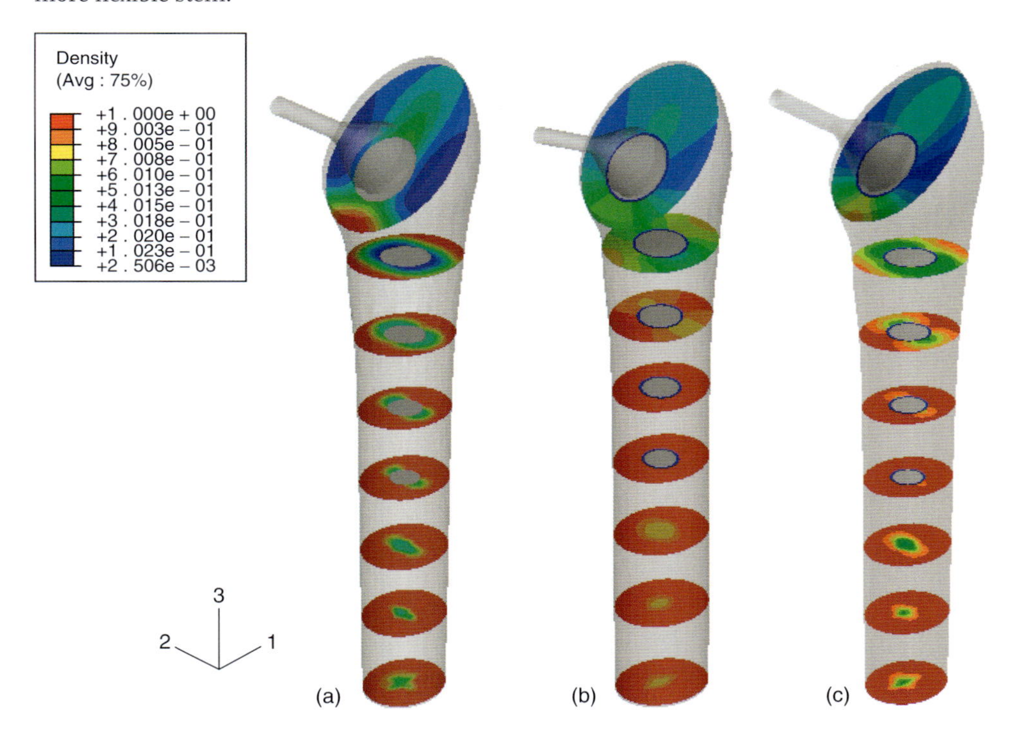

Figure 2.13 (a) Bone density distribution postoperatively. (b) Bone remodeling for CrCo prosthesis. (c) Bone remodeling for Ti prosthesis.

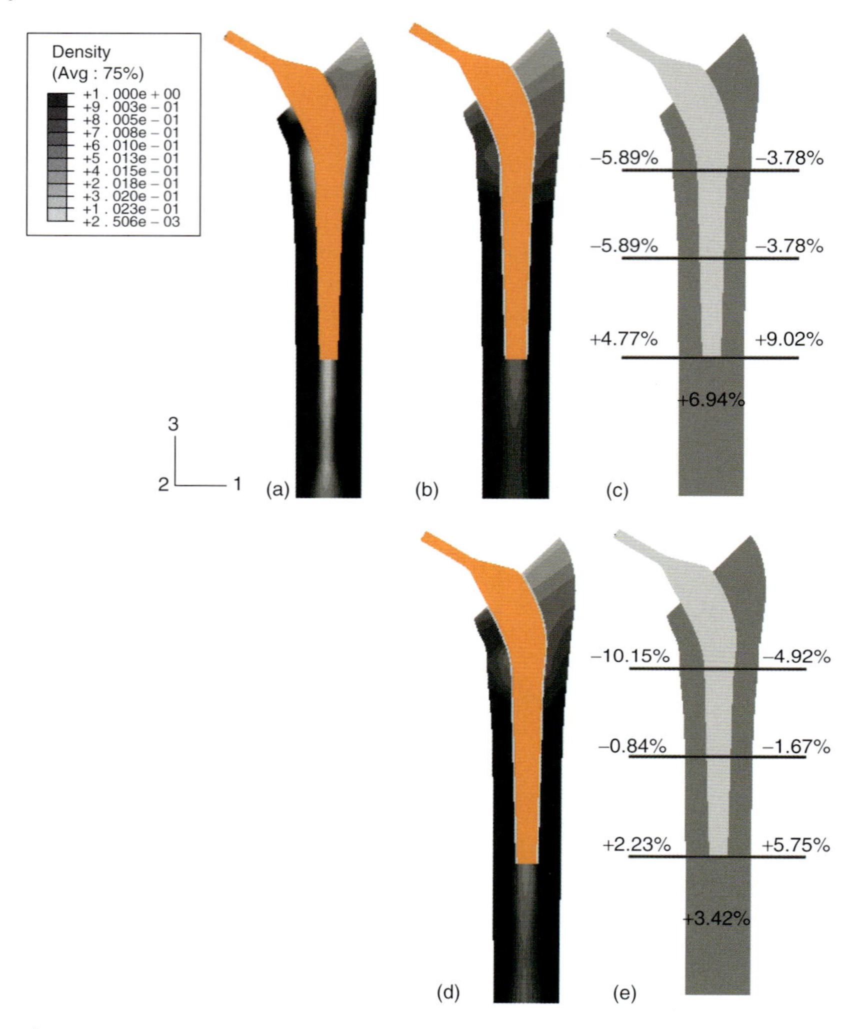

Figure 2.14 Femur anterior view. Bone density distribution postoperatively (a), after bone remodeling (b). Changes in bone mass for CrCo prosthesis (c). Bone density distribution after bone remodeling (d). Changes in bone mass for Ti prosthesis (e).

Owing to the trabecular reorientation, it is possible to see in Figure 2.16(a) (diaphysis) a vertical alignment in the lateral cortex, as expected, and an alignment of 45° in the region next to the prosthesis, in order to support shear forces from the bonded interface. Figure 2.16(b) shows an orientation that is not well-defined in the metaphysis because of the complex and changing stress–strain distribution of the trabeculae.

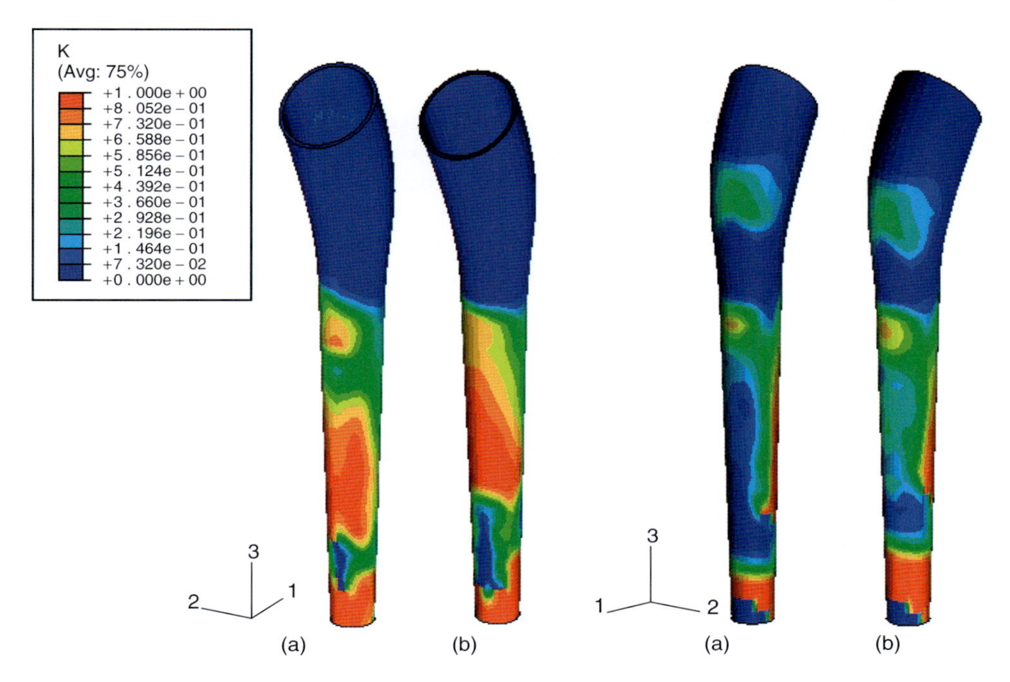

Figure 2.15 Tissue differentiation around the prosthesis. Regions in red show bone ingrowth and blue show fibrous tissue. (a) Ti prosthesis; (b) CrCo prosthesis.

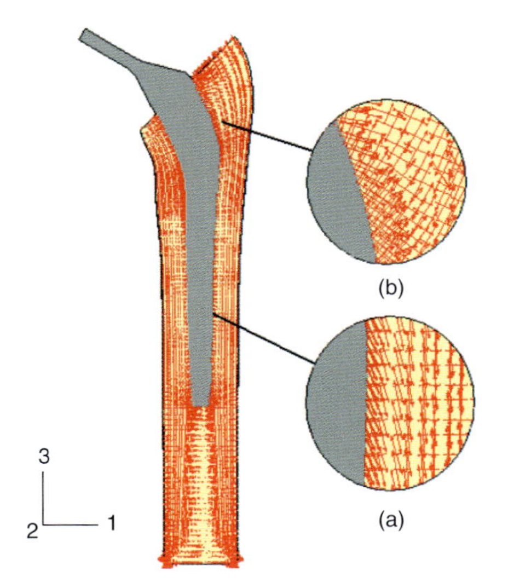

Figure 2.16 Microstruture orientation after bone remodeling. (a) Orientation at distal region, (b) orientation at proximal region.

2.5 Final Remarks

Few studies have considered the coupling effects of the periprosthetic bone remodeling process with interfacial tissue evolution. In order to achieve a better understanding of this subject, a model that simultaneously accounts for both phenomena is investigated here. With this aim, the interface is characterized by a thin volumetric region that allows the introduction of constitutive behaviors and adaptive laws in a similar way as it is done for the periprosthetic region.

A classic homogenized material based on a microstructure model was taken from literature to account for the bone representation. Its material properties change continuously with respect to a relative density parameter. Following an identical concept, an interfacial region is proposed with mechanical properties that continuously change in time and space according to the relative quantity of bone and fibrous material and mechanical environment.

An optimization-based remodeling law was chosen to be the driving force for the bone adaptation in which nonlinear boundary conditions are included. For the interface adaptation, a model based on a relative displacement is proposed, as the information most frequently reported in experimental observations.

Periprosthetic and interface evolutions run simultaneously. Besides the natural coupling of both effects due to global equilibrium, a local dependence of the interface stiffness on the density of bone configurates a kind of local coupling between the two models.

The numerical experiments, although strongly dependent on a set parameters whose values are quite difficult to define precisely, show encouraging results. Both the interface and periprosthetic results differ from classical results under static interface conditions, achieving a pattern of ingrowth along the interface surface that is consistent with clinical observations: low connection in the proximal regions and a kind of two-point support in the distal and medial regions of the shaft. The distribution of the bone ingrowth was also shown to be sensitive to changes in the environment, like the stiffness of the prosthesis.

Different geometries, load conditions, and intensive sensitivity to parameters should be analyzed in further studies, as well as the incorporation of more accurate and consistent models for both interfacial and periprosthetic remodeling phenomena. However, the present results show the relative influence that changing boundary conditions exert on the final stability of the prosthesis. The inclusion of a volumetric interfacial region with changing material properties has been found to provide an adequate and flexible approach.

2.6 Acknowledgments

The authors would like to thank the Brazilian CAPES and *CNPq* who provided partial financial support for this research. We also thank Prof. Tomasz Lekszycki from the *Institute of Fundamental Technological Research – Polish Academy of Sciences* who, through helpful and illustrative technical discussions, suggested the present research topic.

References

1. Hart, R.T., Davy, D.T., and Heiple, K.G. (1984) *Calcif. Tissue Int.*, (36), 104–109.
2. Cowin, S.C. and Hegedus, D.H. (1976) *J. Elast.*, **6**, 313–326.
3. Huiskes, R., Weinans, H., Grootenboer, H.J., Dalstra, M., Fubala, B., and Sloof, T.J. (1987) *J. Biomech.*, **20**(11/12), 1135–1150.
4. Carter, D.R., ORR, T.E., and Fyhrie, D.P. (1989) *J. Biomech.*, **22**(3), 231–244.
5. Beaupré, G.S., Orr, T.E., and Carter, D.R. (1990) *J.Orthop. Res.*, **8**(5), 651–661.
6. Weinans, H., Huiskes, H., and Grootenboer, H.J. (1992) *J. Biomech.*, **25**, 1425–1441.
7. Weinans, H., Huiskes, H., and Grootenboer, H.J. (1994) *J. Biomech. Eng.*, **116**, 393–400.
8. Prendergast, P.J. and Taylor, D. (1992) *J. Biomed. Eng.*, **14**, 499–506.
9. Jacobs, C. (1994) Numerical simulation of bone adaptation to mechanical loading, PhD. Thesis – Standford University.
10. Jacobs, C.R., Simo, J.C., Beaupré, G.S., and Carter, D.R. (1997) *J. Biomech.*, **30**, 603–613.
11. Doblaré, M. and García, J.M. (2001) *J. Biomech.*, **34**, 1157–1170.
12. Doblaré, M. and García, J.M. (2002) *J. Biomech.*, **35**, 1–17.
13. Doblaré, M., García, J.M., and Gómez, M.J. (2004) *Eng. Fract. Mech.*, **71**, 1809–1840.
14. McNamara, M.L. (2004) Biomech. origins of osteoporosis, PhD thesis, University of Dublin.
15. McNamara, M.L. and Prendergast, P.J. (2007) *J. Biomech.*, **40**, 1381–1391.
16. Fernandes, P.R., Rodrigues, H., and Jacobs, C. (1999) *Comput. Methods Biomech. Biomed. Eng.*, **2**, 125–138.
17. Luo, Z.P. and An, K.N. (1998) *J. Math. Biol.*, **36**, 557–568.
18. Bagge, M. (1999) Remodeling of bone structures, Thesis, Technical University of Denmark.
19. Lekszycki, T. (1999) *J. Theor. Appl. Mech.*, **3**(37), 607–623.
20. Bagge, M. (2000) *J. Biomech.*, **33**, 1349–1357.
21. Turner, C.H., Anne, V., and Pidaparti, R.M.V. (1997) *J. Biomech.*, **30**(6), 555–563.
22. Miller, Z., Fuchs, M.B., and Arcan, M. (2002) *J. Biomech.*, **35**, 247–256.
23. Fyhrie, D.P. and Carter, D.R. (1990) *J. Biomech.*, **23**(1), 1–10.
24. Fernandes, P.R., Folgado, J., Jacobs, C., and Pellegrini, V. (2002) *J. Biomech.*, **35**, 167–176.
25. Slawinki, P. and Lekszycki, T. (2000) *Acta Bioeng. Biomech.*, **2**, (Suppl. 1), 488–494.
26. Weinans, H., Huiskes, H., and Grootenboer, H.J. (1993) *J. Biomech.*, **26**(11), 1271–1281.
27. Van Rietbergen, B., Huiskes, R., Weinans, H., Summer, D.R., Turner, T.M., and Galante, J.O. (1993) *J. Biomech.*, **26**(4/5), 369–382.
28. Terrier, A., Racotomanana, R.L., Ramaniraka, R.N., and Leyvras, P.F. (1997) *Comput. Methods Biomech. Biomed. Eng.*, **1**, 17–59.
29. Racotomanana, L.R., Terrier, A., Ramaniraka, N.A., and Leyvraz, P.F. (1999) in *Synthesis in Bio Solid Mechnaics* (eds P. Pedersen and M. Bendsoe), Kluwer Academic Publishers, Dordrecht, pp. 55–66.
30. Cowin, S.C. (2003) Remarks on Optimization and the Prediction of Bone Adaptation to Altered Loading, *http://biopt.ippt.gov.pl.*
31. Fancello, E.A. and Roesler, C.R.M. (2003) Special issues on formulations for bone remodeling around prostheses, Anais do XXIV Iberian Latin-American

Congress on Computational Methods in Engineering, Ouro Preto.

32. Roesler, C.R. (2006) Adaptação mecânica do osso em torno de implantes ortopédicos, Ph.D. Thesis, Federal University of Santa Catarina.
33. Dallacosta, D. (2007) Simulação Tridimensional da Remodelação Óssea em Torno de Prótese de Quadil, M.Eng. Thesis, Federal University of Santa Catarina.
34. Hart, R.T. (2001) in *Bone Mechanics Handbook*, (ed. S.C. Cowin), Chapter 31, CRC Press, pp. 31:1–31:42.
35. Pawlikowski, M., Skalski, K., and Haraburda, M. (2003) *Compos. Struct.*, **81**, 887–893.
36. Davies, J.E. (ed.) (2000) *Bone Engineering*, Squared incorporated, Toronto.
37. Pauwels, F. (1960) *Z. Anat. Entwicklungsgesch.*, **121**, 478–515.
38. Carter, D.R., Beaupre, G.S., Giori, N.J., and Helms, J.A. (1988) *Clin. Orthop.*, **355**, 41–55.
39. Carter, D.R. and Giori, N.J. (1991) Effect of mechanical stress on tissue differentiation in the bony implant bed, *Bone-Biomaterial Interface*, University of Toronto Press, Toronto, 367–379.
40. Claes, L.E. and Heigele, C.A. (1999) *J. Biomech.*, **32**, 255–266.
41. Prendergast, P.J., Huiskes, R., and Soballe, K. (1997) *J. Biomech.*, **30**, 621–630.
42. Kuiper, J.H. (1993) Numerical optimization of artificial hip joint designs, Ph.D. Thesis, Catholic University of Nijmegen, The Netherlands.
43. Bendsoe, M.P. and Sigmund, O. (2003) *Topology Optimization: Theory, Methods and Applications*, 2nd edn, Springer Verlag.
44. Choi, K., Kuhn, J.L., Ciarelli, M.J., and Goldstein, S.A. (1990) *J. Biomech.*, **23**, 1103–1113.
45. Engh, C.A., Glassman, A.H., and Suthers, K.E. (1990) *Clin. Orthop. Relat. Res.*, **261**, 63–81.
46. Kerner, J., Huiskes, R., Van Lente, G.H., Weinans, H., Van Rietbergen, B., Engh, C.A., and Amis, A.A. (1999) *J. Biomech.*, **32**(7), 695–703.
47. Kienapfel, H., Sprey, C., Wilke, A., and Griss, P. (1999) *J. Arthroplasty*, **14**(3), 355–368.
48. Aspenberg, P. and Herbertsson, P. (1996) *J. Bone Joint Surg.*, **78-B**(4), pp. 641–646.
49. Jasty, M., Bragdon, C., Burke, D., O'Connor, D., Lowenstein, J., and Harris, W. (1997) *J. Bone Joint Surg.*, **79-A**(5), 707–714.
50. Hori, R.Y. and Lewis, J.L. (1982) *J. Biomed. Mater. Res.*, **16**, 911–927.
51. Bigoni, D. and Movchan, A.B. (2002) *Int. J. Solids Struct.*, **39**, 4843–4865.
52. Pedersen, P. (1991) *Struct. Opt.*, **3**, 69–78.
53. Cheung, G., Zalzal, Z., Bhandari, M., Spelt, J.K., and Papini, M. (2004) *Med. Eng. Phys.*, **26**, 93–108.
54. Allard, P., Stokes, A.F., and Blanchi, J.P. (1995) *Three-Dimensional Analysis of Human Movement*, Human Kinetics, Champaign.
55. Pedersen, D.R., Brand, R.A., and Davy, D.T. (1997) *J. Biomech.*, **30**(9), 959–965.
56. Dostal, W.F. and Andrews, J.G. (1981) *J. Biomech.*, **14**, 803–812.

3
Bone as a Composite Material

Michelle L. Oyen and Virginia L. Ferguson

3.1
Introduction

Bone is the natural structural material that comprises the bulk of the vertebrate skeleton. Bone tissue not only provides mechanical support and protection but also provides other important biological functions, including acting as a store for calcium and housing the hematopoietic stem cells in the bone marrow. The unusual combination of mechanical properties observed in bone tissue has led to substantial interest in studying natural bone with an aim toward synthesizing bonelike materials, be it for forming replacement bones for medical use [1] or for nonmedical applications [2].

Much of the interest in bone is related to its existence as an organic–inorganic composite material. This combining of phases with grossly different properties and with nanometer-scale feature sizes results in a material with properties that are an improvement on the sum of its parts. Bone is fundamentally a stiff, strong, and tough material and yet relatively lightweight. Table 3.1 illustrates this feature with the example of specific stiffness: bone is compared with PTFE (polytetrafluoroethylene), a polymer of comparable density, and lead, a metal of comparable stiffness (numerical values for the polymer and metal from [3]). There is a stark discrepancy in the specific stiffness (here quantified as the ratio of elastic modulus to physical density) for the composite bone compared with the monolithic materials. It is for this reason that bone is of interest to a broad community outside traditional medicine and there is significant interest in biomimetic materials synthesis techniques to create organic–inorganic composite materials such as bone. In addition to interest in the bonelike materials themselves, there is great interest in the synthesis routes that allow for processing of ceramic materials at room temperature and pressure – problems in technology are typically solved by engineers with the addition of large quantities of energy, in stark contrast to nature's low-energy solutions [4]. Harnessing natural approaches to materials synthesis thus could revolutionize some aspects of ceramic and composite processing, and help to fulfill increasingly important requirements of energy efficient and ecologically sound engineering practices.

Biomechanics of Hard Tissues: Modeling, Testing, and Materials.
Edited by Andreas Öchsner and Waqar Ahmed

ISBN: 978-3-527-32431-6

Table 3.1 Specific stiffness comparison between bone, PTFE, and lead.

	Density, ρ (g cm^{-3})	Elastic modulus, E (GPa)	Specific stiffness, E/ρ (GPa cm^3 g^{-1})
Bone	1.9	15	7.9
PTFE	2.0	0.5	0.25
Lead	11.3	13.5	1.2

PTFE and Lead data from Callister [3].

Given the strong interest in bone and bonelike materials, it is perhaps surprising that little work had been done until recently considering the composite materials aspects of bone. Pioneering work on the subject was performed by Katz in the 1970s [5], but the subject then remained largely dormant until a recent burst of activity. In the time elapsed since the work of Katz, there have been significant and substantial advances in both experimental techniques, including the miniaturization of mechanical testing technology to nanometer length scales, and in computational power, thanks to rapid advances in the microprocessor market. For these reasons, in this work, we take the work of Katz as a starting point, but emphasize the recent developments and outstanding controversies. In this chapter, we first review the components of the composite material bone, including a critical assessment of what is known about the material properties of each constituent phase. We then review basic aspects of composite mechanics, to assess the degree to which bone can be considered in the framework of engineering composite materials. This treatment includes an examination of the macroscopic and microscopic aspects of bone as a composite, highlighting the areas in which substantial further research is yet needed. Finally, following a critical examination of current thinking about bone anisotropy, we conclude with recommendations for future study.

3.2 Bone Phases

Bone is composed of three constituent phases: organic, mineral, and water. Adult compact bone contains approximately, by volume, 30% organic, 50% mineral, and 20% water (Figure 3.1) [6, 7]. The dominant organic phase, type I collagen, forms a scaffold that is mineralized by carbonated hydroxyapatite. Watery interstitial fluid fills pore spaces at scales ranging from nanometers to millimeters. The properties of the composite material bone depend on the properties and molecular structure of each individual constituent phase, the arrangement of each phase in relation to each other, and the organization of bone at the tissue level [8]. Bone is therefore a hierarchical structural composite material.

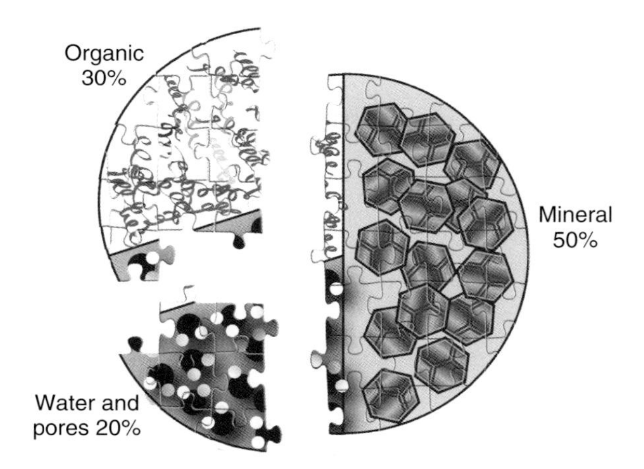

Figure 3.1 Three constituent phases of bone: organic, mineral, and water/pore space collectively contribute to the overall mechanical behavior of the material bone. The volume percent of each component is reported [6].

3.2.1 Organic

The organic phase is dominated by the presence of collagen type I ($\approx$90% by mass). Although noncollagenous proteins and glycoproteins are present, they are likely to have little effect on the overall mechanical behavior of the material. Osteoblasts deposit the collagen matrix in a three-dimensional formation through self-assembly from individual tropocollagen molecules (triple helices). A 67 nm characteristic banding indicates the presence of gaps between the lengths of collagen fibrils and is intrinsically due to the quarter-staggering of 300 nm long triple helices [9]. The fibrils organize and cross-link to form larger fibers [9], which contribute to the viscoelastic response and toughness of bone [10]. The noncollagenous organic phase in bone is mostly composed of glycoproteins, osteopontin, and proteoglycans. This nonfibrillar organic matrix may "glue" mineralized collagen fibers together and may also serve as a source of sacrificial bonds, thus increasing the fracture energy [11].

3.2.2 Mineral

The mineral phase of bone is itself heterogeneous. Multiple forms of poorly crystalline carbonate apatite mineralize and effectively stiffen the organic matrix. The primary phase of hydroxyapatite $Ca_{10}(PO_4)_6(OH)_2$ [12, 13] exists within a milieu of calcium phosphate materials that physically connect through a continuous and rigid structure after the removal of the organic phase. Carbonate (CO_3^{2-}) is commonly substituted into the apatite lattice for OH^- or PO_4^{3-} and a common

Ca^{2+} ion vacancy in apatite often contains sodium, potassium, magnesium, and zinc substitutions [12].

Composite models of bone material are complicated by what is not known about the exact physical characteristics of the constituent phases. The exact shape and size of the nanometer-sized crystals that mineralize the underlying collagen matrix remain unclear. A variety of techniques used to examine bone mineral have quantified bone crystals over a wide range of shapes and sizes [14–16].

Atomic force microscopy (AFM) measurements of bone mineral revealed plate-shaped crystals of 30–200 nm length and width and 1.5–4 nm thickness [14]. Such measurements are certainly complicated by the source and type of bone used for such analysis. Moreover, a composites model is profoundly affected by assumptions of a platelike and needlelike mineral structure or one that is 50 versus 200 nm in length. A survey of the literature shows that the majority of bone mineral (~98%) exists as small platelike crystals measuring ≈1 nm × 10 nm × 15 nm [16–18]. AFM techniques have determined that a few larger crystals exist that measure ≈40 nm × 60 nm × 90 nm [17] and platelike structures of <10 nm thickness × 30–200 nm length and width [14]. Platelike crystals were also observed via transmission electron microscopy (TEM) to be 6–9 nm thick, 20–60 nm wide, and 30–120 nm long [15]. However, the exact structure of these crystals is brought into question by X-ray diffraction measurements that estimate the average length of crystals along their *c*-axis to possibly extend to lengths as large as several hundreds of nanometers [16]. Individual mineral platelets are physically bonded together with "bridging crystallites" such that the bone that has been completely deproteinated resembles a porous ceramic with substantial compressive strength [15].

3.2.3 Physical Structure of Bone Material

The interactions between collagen and mineral are also of considerable interest and debate. The initial formation of crystals between the collagen fibrils in the hole region or "Hodge–Petruska" gaps, originally proposed by Petruska and Hodge in 1964, may drive the resulting shape and size of bone minerals [19]. Ultimately, the fibrillar collagen structure is interrupted and, disrupted by the presence of mineral forming within the gaps [20]. Crystals exist that are too large to fit within the fibril or hole region and have been observed, via TEM, to exist in the interfibrillar region [17, 20, 21]. TEM observations have shown that the majority of bone crystals lie within and on the surface of the collagen fibers [15, 22] where the long axis of the crystals lies parallel to the long axis of the collagen fibrils.

3.2.4 Water

Interestingly, the water in bone has received relatively little attention from a mechanical perspective, especially in the context considering bone as a composite material. However, fluid-filled pore spaces occupy approximately 20% of a bone's volume. Pores ranging in size include large millimeter-sized trabecular spaces, vascular pores including Haverisan canals ($\approx$20 μm), canaliculi and lacunae ($\approx$0.1 μm), and nanometer-sized pore spaces or "matrix micropores" [23] that exist between and within the collagen and mineral crystals. Bound water may stabilize the mineral crystallites by occupying OH^- and possibly Ca^{2+} vacancy sites [24] but likely plays a minor role in the mechanical behavior of the bone tissue.

Porosity in bone facilitates movement of the interstitial fluid (i.e., unbound water) throughout the tissue that contributes to poroviscoelasticity. Fluid that exists in smaller pore spaces between the mineral crystals and collagen fibrils [25] may add plasticity to the mechanical response of bone. In addition, the proteins and glycoproteins within the organic matrix interact with chemically unbound water through charged interactions. The availability of charged sites on the organic matrix has a profound effect on the stiffness of unmineralized matrix in tissues containing other forms of collagen [26]. Similarly, the interaction between water and the organic matrix of bone is subject to similar hydrostatic interactions, where the bone tissue stiffens with occupation of charged sites by polar solvents [27].

3.3 Bone Phase Material Properties

Having established the three-phase nature of bone, we now examine what is known about the individual phase material properties for each phase.

3.3.1 Organic Matrix

The organic matrix within bone is composed primarily of fibrillar type I collagen that is laid down by osteoblasts. While the resulting longitudinal, hierarchical structure is well designed to undergo tension, as in the case of tendon and ligaments, it also directs the pattern of mineralization along load-bearing directions in bone and contributes to bone's resistance to tension, torsion, compression, and bending.

The material properties of collagen are scale dependent and vary with measurement technique, hydration state, and source of the collagen material. Most modeling for mineralized tissues has incorporated a modulus of 1–1.5 GPa [28–31] consistent with the value of 1.2 GPa proposed by Gosline in a comparative analysis of elastic proteins [32]. Young's modulus values of 3–9 GPa were obtained for single tropocollagen molecules via X-ray diffraction [33]. Cusack and Miller [34] used Brillouin light scattering and obtained an elastic modulus value of 11.9 GPa for dry collagen and 5.1 GPa for wet collagen. The extremely large values are probably due to the ultrahigh frequency (10^{10} Hz) nature of the measurement since collagen is viscoelastic and measured (complex) modulus in a viscoelastic solid depends strongly on the measurement frequency. These values, therefore, would not be appropriate for the quasistatic elastic modeling of mineralized tissues.

Atomistic modeling for stretching single tropocollagen molecules has been used to generate comparable values of 2.4–7 GPa [35–37]. Collagen fibrils yielded modulus values of 1–2.7 GPa via X-ray diffraction [38] and AFM testing [39], respectively. In addition, a microelectromechanical system (MEMS) device used to stretch hydrated individual collagen fibers revealed modulus values of 0.4–0.5 GPa at small strains (more relevant for mineralized tissues) and ≈12 GPa at large strains [40, 41]. Molecular multiscale modeling of comparable individual collagen fibers produced modulus values of 4.36–38 GPa for small and large strains, respectively [35, 42].

The scale dependence of collagen is a current subject of controversy. In some works, collagen shows a reduction of the modulus from the tropocollagen level to the fibril level by ≈40% [33, 42] and from the fibril to the fiber level by ≈80% [42]. However, other authors have come to opposite conclusions concerning both scale dependence and the relative range of numerical values for the collagen modulus. Two studies [43, 44] discussed the advances in measuring fundamental collagen elastic properties, taking advantage of technological developments such as the optical tweezers and sophisticated computational techniques. Both these works report extremely low modulus values for collagen at molecular scales, in the single- to double-digit megapascal range and in contrast to many of the values listed above. Both works also note an increased elastic modulus values for larger structural units of collagenous soft tissues, such as whole tendons, typically in the hundreds of megapascal to single-digit gigapascal range. This increase in modulus for larger structural units is attributed to cross-linking [44]. A comparison of the approximate stress–strain for four different hierarchical levels of collagen from [44] gives approximate values of elastic modulus for tendon $E = 636$ MPa, collagen fascicle $E = 162$ MPa, collagen fiber $E = 43$ MPa, and procollagen molecule $E = 27$ MPa. Modulus values for demineralized bone matrix also have been reported as generally lower than those presented for individual collagen structures: from hundreds of megapascals [45] to ≈1–2 GPa [46]. Clearly, there are fundamental issues outstanding concerning the appropriate value to use for the elastic modulus of collagen, and this will affect the fidelity of any composite models used to simulate the mechanical behavior of bone.

3.3.2 Mineral Phase

The elastic modulus of both single-crystal hydroxyapatite and mineralogical fluoroapatite is reported in the range from 100 to 150 GPa depending on the test method and specimen orientation [47–51]. The Poisson's ratio was reported as $\nu = 0.28$ [48]. Biological apatite is an analog of mineral apatite, and so its mechanical properties are quite well known if the mineral is fully dense – which may not be the case in biological systems [29]. The heterogeneity of bone mineral complicates the measurement of properties. Not only is bone mineral heterogeneous within the bone material itself but also localized differences in bone mineral exist within the regions of varying age and in different bone types.

Because hydroxyapatite mineral is anisotropic, the properties do depend on orientation to some degree. Velocity measurements, using scanning acoustic microscopy (SAM), performed on single-crystal mineralogical hydroxyapatite revealed five independent elastic constants that describe a transversely isotropic material [52]:

$$\begin{bmatrix} C_{11} & C_{12} & C_{13} & 0 & 0 & 0 \\ & C_{11} & C_{13} & 0 & 0 & 0 \\ & & C_{33} & 0 & 0 & 0 \\ & & & C_{44} & 0 & 0 \\ & & & & C_{44} & 0 \\ & & & & & 1/2\,(C_{11} - C_{12}) \end{bmatrix}$$

$$= \begin{bmatrix} 137 & 42.5 & 54.9 & 0 & 0 & 0 \\ & 137 & 54.9 & 0 & 0 & 0 \\ & & 172 & 0 & 0 & 0 \\ & & & 39.6 & 0 & 0 \\ & & & & 39.6 & 0 \\ & & & & & 47.25 \end{bmatrix} \text{GPa}$$

The density was reported as $3200\,\mathrm{kg\,m^{-1}}$ [52]. Using the Voigt–Reuss–Hill approximation, anisotropic single-crystal elastic constants converted into isotropic polycrystalline elastic Reuss and Voigt moduli [53] yield values of 116 and 119 GPa, respectively.

3.3.3 Water

Water is a principal constituent of bone, and thus the hydration state of the sample plays a critical role in the measured mechanical properties. Dehydration causes increased modulus of elasticity and tensile and bending strengths and reduced values of fracture toughness [54–57]. Dehydration increases nanoindentation modulus values by $\approx$15–25%, and up to a maximum of 50% [58–60]. While dehydration increases the magnitude of mechanical properties, the inherent material relationships (e.g., anisotropy) are maintained.

Bone behaves in a time-dependent manner due, in part, to its organic phase [61, 62] and its water content with a corresponding poroelastic flow through the bone material [63]. Viscoelasticity in bone has been shown to correlate with hydration state [27, 61, 64, 65] and mineral content [62]. The time that it takes for fluid to flow through the pore spaces in bone conveys poroelasticity. Bone tissue plasticity and viscoelasticity are highly influenced by interactions between water and charged sites on the collagen and other organic matrix constituents [26, 64, 66, 67].

Having examined the individual components of a composite mineralized tissue, we now move on to examine composite mechanics models for prediction of the elastic modulus based on proportions and properties of the constituent materials.

3.3.4
Elastic Modulus of Composite Materials

Properties of a composite material are weighted combinations of the component properties [68]. The characteristic properties for quantifying each phase (i) are the volume (v_i) and weight (w_i). The combination of these properties, the density ($\rho_i = w_i/v_i$), is also frequently used to characterize the phases. From these fundamental properties one can calculate the weight fraction (W_F) or volume fraction (V_F) for the particle or filler phase (F):

$$W_F = \frac{w_F}{w_M + w_F} \tag{3.1}$$

$$V_F = \frac{v_F}{v_M + v_F} \tag{3.2}$$

where the subscript M is used to indicate the matrix or reinforced phase.

A rule-of-mixtures approach is a first approximation to describe the composite in terms of the component phases. For example, the density of the composite (ρ_c) is related to the density and volume fraction of the component phases by

$$\rho_c = \rho_F V_F + \rho_M V_M \tag{3.3}$$

There is a simple relationship between weight fraction and volume fraction for two-phase composite (assuming no porosity):

$$V_F = \frac{w_F}{w_M\left(\frac{\rho_F}{\rho_M}\right) + w_F} = \frac{1}{\left[1 + \frac{\rho_F}{\rho_M}\left(\frac{1}{W_F} - 1\right)\right]} \tag{3.4}$$

For a two-phase material *with* porosity, there is no change in the weight fraction expression but there is a volume contribution from the weightless pores (where v_P is the pore volume):

$$V_F = \frac{v_F}{v_M + v_F + v_P} \tag{3.5}$$

Many material properties of a composite material, including the elastic modulus, are typically expressed as the volume-fraction-weighted combination of the component phase properties. In the simplest approximation for elastic modulus, the two components are volume-fraction-weighted series or parallel springs. The two combinations (series and parallel) form extreme bounds on the actual composite's behavior, and these simple bounds for two-phase composites are frequently called Voigt–Reuss (V–R) bounds. The upper (iso-strain, parallel springs, Voigt) and lower (iso-stress, series springs, Reuss) bounds are [68]

$$E_U = V_2 E_2 + (1 - V_2)\, E_1 \tag{3.6}$$

$$E_L = \left(\frac{V_2}{E_2} + \frac{(1 - V_2)}{E_1}\right)^{-1} \tag{3.7}$$

where by convention $E_2 > E_1$ and $V_1 + V_2 = 1$. These bounds are frequently used to describe the longitudinal (upper bound) and transverse (lower bound) moduli of oriented fiber-reinforced composites with fibers aligned in one primary direction.

A second common set of bounds for composite materials are the Hashin–Shtrikman (H–S) bounds, which have been found to be more physically realistic for many composites, particularly those in which the reinforcing phase is particulate and not continuous aligned fibers. These bounds are formulated in terms of the shear and bulk moduli G and K, remembering that, for isotropic materials,

$$G = \frac{E}{2(1+\nu)} \text{ and } K = \frac{E}{3(1-2\nu)} \tag{3.8}$$

The lower bounds are [69]

$$\begin{aligned} K_L &= K_1 + V_2 \left(\frac{1}{(K_2 - K_1)} + \frac{3(1-V_2)}{3K_1 + 4G_1} \right)^{-1} \\ G_L &= G_1 + V_2 \left(\frac{1}{(G_2 - G_1)} + \frac{6(K_1 + 2G_1)(1-V_2)}{5G_1(3K_1 + 4G_1)} \right)^{-1} \\ E_L &= \frac{9K_L G_L}{3K_L + G_L} \end{aligned} \tag{3.9}$$

and the upper bounds are

$$\begin{aligned} K_U &= K_2 + (1-V_2) \left(\frac{1}{(K_1 - K_2)} + \frac{3V_2}{3K_2 + 4G_2} \right)^{-1} \\ G_U &= G_2 + (1-V_2) \left(\frac{1}{(G_1 - G_2)} + \frac{6(K_2 + 2G_2)V_2}{5G_2(3K_2 + 4G_2)} \right)^{-1} \\ E_U &= \frac{9K_U G_U}{3K_U + G_U} \end{aligned} \tag{3.10}$$

A comparison of the V–R and H–S bounds is shown in Figure 3.2 for a stiff phase with $E_2 = 100$ GPa and a modulus mismatch factor (E_2/E_1) of 10, 100, and 1000. The difference between the upper and lower bounds (either V–R or H–S) increases with increasing modulus mismatch, making it increasingly difficult to predict the expected behavior of the composite material without *a priori* knowledge of the mechanism of strain transfer between the two different phases. The two variations (V–R or H–S) on the lower bound expression become indistinguishable (on linear coordinates) at large modulus mismatch, but the upper bounds remain distinct.

The Halpin–Tsai relationships for composite materials are approximations to exact elasticity solutions in easy-to-calculate simple analytical forms. There are well-known Halpin–Tsai expressions for the elastic modulus of a composite with oriented short fibers or whiskers, precisely the situation being modeled with the FE analysis presented in this section. The analytical expressions for the longitudinal and transverse elastic moduli (E_L, E_T) relative to the compliant matrix modulus (E_M) are [70]

$$\frac{E_L}{E_M} = \frac{1 + (2AR)\eta_L V_F}{1 - \eta_L V_F} \tag{3.11}$$

$$\frac{E_T}{E_M} = \frac{1 + 2\eta_T V_F}{1 - \eta_T V_F} \tag{3.12}$$

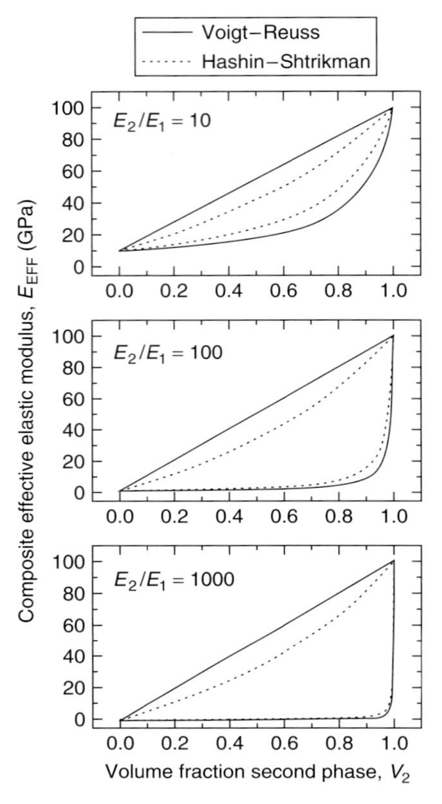

Figure 3.2 Elastic modulus bounds for a material with $E_2 = 100$ GPa and $E_1 = 10$, 1, or 0.1 GPa for modulus mismatch (E_2/E_1) factors of 10, 100, or 1000.

where $AR = l/d$ is the particle aspect ratio, V_F is the volume fraction reinforcing phase, and the coefficients η_L and η_T are related to the filler and matrix phase elastic moduli (E_F, E_M) as

$$\eta_L = \frac{(E_F/E_M) - 1}{(E_F/E_M) + 2AR} \tag{3.13}$$

$$\eta_T = \frac{(E_F/E_M) - 1}{(E_F/E_M) + 2} \tag{3.14}$$

The Halpin–Tsai expressions are empirical and this reasonable but imperfect agreement is to be expected at intermediate volume fractions of reinforcement phase [71].

A more recent model is for a composite with a staggered high aspect-ratio reinforcing phase [72, 73]. The longitudinal modulus is expressed as

$$E_L = V_F^2 \left(\frac{V_F}{E_F} + \frac{4(1 - V_F)}{G_M AR^2} \right)^{-1} + (1 - V_F) E_M \tag{3.15}$$

where G_M is the shear modulus of the protein phase and other terms are as defined above. This expression gives low values (i.e., beneath the other lower bounds) for

Table 3.2 Calculated values GPa for a bonelike material ($V_F = 0.5$) using different composites models (Eqs. 3.6, 3.7, 3.9, 3.10, and 3.15).

	$E_M = 0.1$ GPa, $E_F = 150$ GPa	$E_M = 1$ GPa, $E_F = 150$ GPa	$E_M = 10$ GPa, $E_F = 150$ GPa
Voigt–Reuss lower, upper	0.2, 75.1	2.0, 75.5	18.8, 80
Hashin–Shtrikman lower, upper	0.3, 50	3.0, 50.8	25.7, 58.7
Eq. (3.15), AR = 1	0.05	0.55	5.5
Eq. (3.15), AR = 10	0.52	5.0	34.1
Eq. (3.15), AR = 20	1.9	15.7	58.8
Eq. (3.15), AR = large	75.0	75.5	80

small aspect ratios but approaches the upper bound at very large aspect ratios (Table 3.2).

3.4 Bone as a Composite: Macroscopic Effects

For a first approximation, mineralized tissues are considered as a two-phase composite of mineral and nonmineral phases, building on the work of Katz [5]. The information necessary for examination of mineralized tissues within this simple composite materials framework is twofold: knowledge of the elastic properties of the individual phases, as discussed above, and detailed knowledge of the relative proportions of different phases (e.g., their phase fractions) present in the composite material.

Compositions of mineralized tissues are frequently reported in terms of weight (Table 3.3). From the weight fraction of mineral, a volume fraction for the mineral and nonmineral phases can be calculated (Eq. 3.4). These estimates can then be compared with estimates made on the basis of the material densities (ρ_i) using a rule-of-mixtures approach (Eq. 3.3). In both cases, the density of the apatite mineral phase is taken to be 3.1 g cm^{-3} [74] and the density of the remainder (organic phase and water) is assumed to be unity.

Table 3.3 Mineralized tissue mean composition by weight or by mass density.

% weight [75]	Water	Mineral	Organic	Mass density (g cm^{-3})
Bone	5–10	75	15–20	1.8–2.0
Enamel	2	97	1	2.97
Dentin	10	70	20	2.14

Table 3.4 Mineralized tissue mineral volume fraction (V_F) estimated in two ways and based on raw data in Table 3.3.

	V_F from weight % mineral	V_F from mass density
Bone	0.49	0.38–0.48
Enamel	0.91	0.94
Dentin	0.43	0.54

The calculated mineral volume fractions for each tissue, obtained using the composition data from Table 3.3, are presented in Table 3.4. The calculations differ slightly, particularly in the composition of dentin relative to that of bone, but the mineral volume fractions from either calculation are comparable. The reported volume fraction of mineral in bone is approximately 50% [7], in good agreement with either calculation. For simplicity, many model calculations and finite element simulations of bone or dentin will assume a volume fraction of 50% mineral, in good agreement with either calculation for both materials (Table 3.4).

Katz concluded that the simple examination of elastic modulus bounds based on phase fractions and component moduli resulted in adequate prediction of enamel properties but absolutely did not result in predictive capability for bone. The modulus of bone increases dramatically at nearly constant mineral volume fraction (in the region of mineral volume fraction 0.35–0.5). The result is that bone modulus spans a region between the upper and lower H–S composite bounds, and therefore cannot be predicted on the basis of the mineral volume fraction alone [5].

3.5 Bone as a Composite: Microscale Effects

There have really been few truly successful mechanical models for bone as a composite. Part of the difficulty in modeling bone as a composite material is in the uncertainty at the very finest levels of the ultrastructure. For example, recent work has queried whether proteins or other organic components are directly opposed against the bone mineral [76]. Manipulation of hydration state with polar solvents dramatically alters the hydraulic permeability [77] in a manner which is unclear – likely the organic–mineral interactions are affected due to alterations in hydrogen bonding.

Such varied observations have been translated into a wide range of assumptions about how the organic and mineral phases of bone interrelate. Mineral crystals have been assumed to lie entirely within the collagen fibrils [73, 78], outside of the fibrils to form interpenetrating phases between collagen and mineral [79, 80], both within (~25%) and outside (~75%) of the collagen [81], or predominantly outside of the collagen fibrils [82]. The specific relationship between the collagen fibrils and mineral crystals has a tremendous influence on accurately predicting bone stiffness. Experimentally, mineral has been demonstrated to lie both within and outside of the collagen in dentin [83], mineralized tendon [20], and bone [22], where the distribution of mineral in mature tissues is predominantly in the extrafibrillar compartment. Functionally, the intrafibrillar material in dentin likely plays a dominant role over extrafibrillar material in sustaining load [84].

The most successful attempts to relate bone macroscopic mechanical behavior to the response of individual components at fundamental (ultrastructural) length scales has been via sophisticated multiscale computational modeling [82]. However, it is unclear that this model presents a true solution to the problem, as an extremely large (tens of gigapascals) elastic modulus value is used for the collagen phase. With the collagen essentially as stiff as bone, and a strong influence in overall modulus due to porosity, this model likely represents a first step toward a full description of bone at multiple-length scales based on the nanometer-scale constituent phases.

3.6 Bone as a Composite: Anisotropy Effects

Bone is anisotropic throughout each level of its hierarchical structure. At the macroscopic scale, trabecular (or cancellous) bone possesses a three-dimensional structural organization that is driven by loading patterns. The orientation and spacing of trabeculae determine the degree of anisotropy, which accounts for 72–94% of the variability in the elastic constants for trabecular bone [85–87]. At levels below that of the whole bone, the material behavior is itself directionally dependent [88]. What is known, and not known, about each constituent phase has a profound influence on our ability to understand the true composite nature of bone and also to develop accurate, high-quality composite models. The remainder of this section therefore focuses on anisotropy in osteonal cortical bone, as that is the focus of most composite models of bone.

Bone exhibits directional dependence at multiple scales and within various bone types [89, 90]. Cortical bone, within long bones, demonstrates transverse isotropic behavior in conventional load-deformation testing of macroscopic sections [90–92] and in ultrasound measurements [89]. The elastic modulus and ultimate strength are both greater in the longitudinal than in the transverse direction and show little difference between the radial and transverse directions [89, 90, 92, 93]. Bone loaded in tension also demonstrates a substantially greater strain to failure in the longitudinal versus the transverse direction [90]. Overall, these properties are explained by the longitudinal alignment of osteons and Haversian canals [94] and in that the collagen fibrils [93] and, consequently, the *c*-axis of the mineral crystals [92] generally align with the long axis of the bone. Properties may also vary with the orientation of lamellae, blood vessel networks, and laminae, as well as with anatomical site and age [90, 95]. Stiffness values, measured via ultrasonic measurements, change significantly with bone type and maturity: the elastic coefficients that describe bovine plexiform bone reveal an orthotropic material

Table 3.5 Elastic coefficients (GPa) for single-crystal hydroxyapatite [52] and plexiform and osteonal bone [96]. All data were collected using ultrasonic techniques. Indices 1 and 2 refer to the radial and circumferential directions, respectively, and 3 refers to the long axis of the bone.

	HA single crystal	Plexiform bone	Osteonal bone
C_{11}	137	22.4	21.2
C_{22}	137	25.0	21.0
C_{33}	172	35.0	29.0
C_{44}	39.6	8.2	6.3
C_{55}	39.6	7.1	6.3
C_{66}	47.25	6.1	5.4
C_{12}	42.5	14.8	11.7
C_{23}	54.9	13.6	11.1
C_{13}	54.9	15.8	12.7

that becomes increasingly transversely isotropic as the bone undergoes osteonal remodeling [96] – properties in the transverse direction (C_{11} and C_{22}) are of similar values in single-crystal hydroxyapatite and osteonal bone but vary greatly in plexiform bone (Table 3.5). Reorganization of bone with remodeling likely involves increased organization of bone mineral onto a more highly organized and less randomly deposited collagen structure.

There are clear differences between the overall pattern of anisotropy in the mineral crystal and in the bone samples, as is apparent when the C_{ij} values (from Table 3.5) for each material are normalized by the largest stiffness coefficient, those associated with the longitudinal axis (C_{33}), as shown in Table 3.6. The numerical values for C_{11}, C_{22}, and C_{66} are relatively smaller in bone compared with apatite, while C_{12}, C_{23}, and C_{13} are all relatively greater in bone. Thus, and unsurprisingly, the pattern of anisotropy in bone is not simply the result of anisotropy in the mineral phase itself; other factors must contribute to the anisotropic response.

In contrast to such measurements in healthy bone, metabolic diseases may have a profound influence on anisotropy. Osteoporosis has little effect while osteopetrosis reduces material anisotropy in bone [96]. The degree of anisotropy of this complex structure is affected by changes in the bone due to normal growth [97], type [97–101], and disease [96, 102].

Mineral crystals align with the collagen fibrils [15, 52] to contribute to anisotropy at the level of the material. Acoustic microscopy of whole, demineralized, and deproteinated bone samples showed the organic phase to possess near isotropy while the mineral phase was anisotropic [103]. Knoop microindentation testing revealed directionally dependent behavior on a planar bone surface [104, 105]. Through integration of measured elastic constants over the indented plane, an anisotropy ratio (E) can be determined that accounts for the indentation modulus

Table 3.6 Elastic coefficients (GPa) for single-crystal hydroxyapatite [52] and plexiform and osteonal bone [96] from Table 3.5 normalized to the long axis coefficient (C_{33}) value for comparisons in anisotropy between materials.

	HA single crystal	Plexiform bone	Osteonal bone
C_{11}/C_{33}	0.80	0.64	0.73
C_{22}/C_{33}	0.80	0.71	0.72
C_{33}/C_{33}	1	1	1
C_{44}/C_{33}	0.23	0.23	0.22
C_{55}/C_{33}	0.23	0.20	0.22
C_{66}/C_{33}	0.27	0.17	0.19
C_{12}/C_{33}	0.25	0.42	0.40
C_{23}/C_{33}	0.32	0.39	0.38
C_{13}/C_{33}	0.32	0.45	0.44

ratio (M) in anisotropic materials [97, 106]. For bone, the E ratio of 1.75 (ratio of longitudinal to transverse moduli) corresponds to an M ratio of $\approx$1.4 [106]. This analytical approach was used to demonstrate that bone mineral has greater connectivity in the longitudinal versus the transverse direction [107].

Nanoindentation testing, performed in various directions, reveals properties that vary significantly with direction. In one example, individual lamellae within dry samples were $\approx$55% greater in modulus than in the transverse direction: 25.7 ± 1.0 and 16.6 ± 1.1 GPa, respectively [98]. Equine cortical bone processed to remove the organic phase remains highly anisotropic – with a 34% difference in transverse versus longitudinal indentation modulus – and bone samples with the mineral fraction removed show only a 10% difference [66] (Table 3.7). In comparison to the

Table 3.7 Anisotropy in equine cortical bone from nanoindentation measurements for plane strain modulus (E') in planes that lie transverse and longitudinal to the long axis of the bone. Student's t-test between directions[a].

Condition	E' Longitudinal (GPa)	E' Transverse (GPa)
Wet	11.7 (1.7), 326	11.8 (1.9), 68 (N.S.)
Ethanol dehydrated	15.0 (2.2), 309	24.8 (2.2), 68**
PMMA (polymethylmethacrylate) embedded	19.4 (2.1), 283	25.8 (2.1), 74**
Deproteinated (embedded)	13.4 (1.0), 62	20.4 (2.2), 62**
Decalcified (embedded)	4.9 (0.3), 59	5.5 (0.3), 49*

Data are from Bembey *et al.* [66]

[a]Results reported as not significantly different (N.S.), significant (*), or highly significant (**).

organic phase, the mineral phase is the major contributor to measured anisotropy at the material level although, as discussed above, the pattern of anisotropy does not precisely mimic that of the hydroxyapatite itself.

The porosity and density of bone can be measured either within the material itself or within a region of the bone material that includes organic and mineral constituent phases, pore spaces, and microcracks or other inherent structural flaws. Within the larger structure of cortical bone, large Haversian canals (diameter ≈50 μm) run generally parallel to the long axis of the bone. Smaller pore spaces made of networks of canaliculi (≈1 μm), Volkman's canals (≈5–10 μm), and osteocyte lacunae (≈5 μm) permeate the bone material. Most measurements of bulk samples cannot avoid the influence of porosity and so report an effective measurement of mineralization (i.e., percent mineral) or density. The interpretation of the composite nature of bone depends heavily on the inclusion or exclusion of voids within the tissue. These voids are filled by soft tissue or water *in vivo*.

Further, the exact nature (e.g., shape, size, density) of the pore spaces within bone may highly influence the material anisotropy. The influence of porosity on material properties is directional anisotropy [108, 109]. Young's and shear modulus in human femoral cortical bone samples correlate with porosity in the longitudinal, but not the transverse, direction [109]. In addition, increasing porosity correlates with decreased anisotropy [108, 109]. In general, pore spaces in bone preferentially align with the long axis of the bone. As compared to the bone tissue, pores introduce an additional anisotropy of 16–20% of the effective medium for human cortical bone samples ranging in porosity from 2 to 15% [108].

Simple composites theory shows that no simple relationship exists between the elastic modulus and mineral content for large specimens of bone [5]. Nanoindentation, to examine the tissue level of bone, has helped to better elucidate the relationship between modulus and mineral volume fraction (V_F) [50]. Analysis of micrometer-sized volumes has extended the study of material heterogeneity to the tissue level. Similar to the groundbreaking work of Katz [5], no simple relationship exists between modulus and mineral content at the tissue level (Figure 3.3) [50].

Similar to the work of Katz in larger volumes of bone [5], the modulus–mineral content relationship was examined in bone samples ranging from very poorly mineralized, osteomalacic bone to the exceptionally dense whale rostrum (Figure 3.3). Data for nanoindentation modulus and mineral volume fraction (V_F) fell within H–S composite bounds, which describe the continuity between stiff particles in a continuous compliant phase (lower bound) and compliant particles in a continuous stiff phase (upper bound) [69]. Elastic indentation modulus was shown to exist as a function of mineral volume fraction (as calculated from calibrated quantitative backscattered electron images of each indentation site) [110]. The heterogeneity in the data, as demonstrated by the amorphous relationship between modulus and V_F throughout the range of bone types, resulted from a complex interplay of factors that include the composition of the mineral phase, crystallinity, collagen orientation, and nano- to micrometer-sized pore spaces.

Anisotropy is mainly influenced by the intrinsic factors of tissue-level organization: orientation of collagen fibers and mineral crystals within the bone material

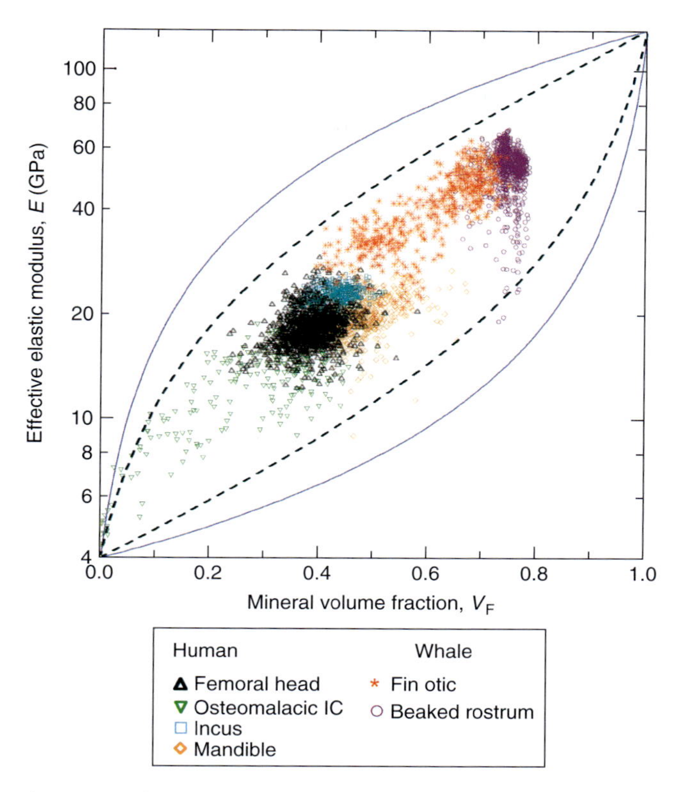

Figure 3.3 Elastic modulus (E) versus mineral volume fraction (V_F) of PMMA-embedded bone samples: human femoral head ("normal"); human osteomalacic iliac crest (OM); human incus; human mandible ("jaw"); fin whale otic bone; and dense beaked whale rostrum. The solid lines are Voigt–Reuss bounds and dashed lines are the Hashin–Shtrikman bounds. From Oyen *et al.* [50] with permission.)

as well as into specifically organized structures such as lamellae, mineral volume fraction, and porosity. In addition to the topics discussed, factors such as the prevalence and directionality of microcracks, fatigue damage, and bony features such as cement lines may influence the anisotropy of bone material.

3.7 Bone as a Composite: Implications

There is significant interest in bone as a composite, and the mechanical properties of bone have been much studied for the potential that biomimicked bonelike materials have in both medical and engineering applications [2]. For exploitation of a bottom-up materials synthesis based on biomimicry, it is necessary to understand natural bone, in terms of the means by which its structure influences its mechanical properties. To a first approximation, detailed understanding of the organic matrix–mineral interactions is limited by intrinsically small length scales, and there exists a clear need for new tools and experimental approaches to examine materials at these scales. The ability to model – and thus predict – the mechanical properties of bonelike composites are currently limited by our lack of knowledge about mineral–matrix interactions at molecular length scales.

We know that the apatite in bone deposits onto the organic matrix by heterogeneous nucleation, but it is unclear as to what is the nature of the nucleation sites – collagen, noncollagenous proteins (including glycoproteins such as biglycan). Recent examination of natural bone by solid-state NMR has indicated hydroxyapatite–sugar interfaces as the fundamental interaction, and not hydroxyapatite–protein interfaces [76]. Obviously, this degree of uncertainty influences our ability to model bone as a composite material.

Further complicating matters is the significant lack of uncertainty in the elastic modulus of collagen. With a large value for $E_{collagen}$, lower-bounds-type models with discontinuous mineral "particles" or "platelets" embedded in a continuous collagen matrix result in reasonable elastic modulus values compared with the known values for bone (Table 3.2). However, these models are inconsistent with the known continuous nature of bone mineral, as can be demonstrated by the existence of the porous solid that remains following removal of the organic component of bone.

Current composites models of bone have primarily emphasized two-phase behavior with the two components, the lumped water plus the organic and mineral phases. A second common approach is to consider a poroelastic model in which the organic plus mineral phases are lumped into a single "solid skeleton" that is porous, where the second phase is the water. What is truly missing is a model that considers individually the three phases – mineral, organic, and water – at fundamental length scales without lumping two of the components together. This provides opportunity for further research into bone as a composite.

In reality, there is no single model likely to predict *the* elastic modulus of bone. Small-scale, local measurements of mineral content (volume fraction) and mechanical properties (elastic modulus) demonstrate substantial point-to-point variability in both values even in the same bone sample (Figure 3.3). A model for bone consistent with the observed local variations in both mineral content and stiffness [50, 111] must include a stochastic element and some sort of averaging process to arrive at the macroscopic bone modulus from the very small-scale, fundamental measurements. It has been speculated that this local randomness is a key mechanism in obtaining the toughness of bone [111] such that any model incorporating a single value of V_F or obtaining a single global value of E cannot be realistic. With computational capabilities increasing all the time, it is entirely possible to increase the complexity of bone composites models, to perhaps someday achieve the aim of fully modeling macroscopic behavior based on the composition and individual phase properties for a material as hierarchical and complicated as bone.

References

1. Ko, C.-C., Oyen, M.L., Fallgatter, A.M., Kim, J.-H., Douglas, W.H., Fricton, J., and Hu, W.-S. (2006) *J. Mater. Res.*, **21**, 3090–3098.
2. Oyen, M.L. (2008) *Mater. Res. Soc. Bull.*, **33**, 49–55.
3. Callister, W.D. (2000) *Materials Science and Engineering: An Introduction*, 5th edn, John Wiley & Sons, Inc., New York.
4. Vincent, J.F., Bogatyreva, O.A., Bogatyrev, N.R., Bowyer, A., and Pahl, A.K. (2006) *J. R. Soc. Interface*, **3**(9), 471–482.
5. Katz, J.L. (1971) *J. Biomech.*, **4**(5), 455–473.
6. Gong, J.K., Arnold, J.S., and Cohn, S.H. (1964) *Anat. Rec.*, **149**(3), 325.
7. Hayes, W.C. (1991) Biomechanics of cortical and trabecular bone: implications for the assessment of fracture risk, in *Basic Orthopaedic Biomechanics* (eds V.C. Mow and W.C. Hayes), Raven Press, New York, pp 93–142.
8. Rho, J.Y., Kuhn-Spearing, L., and Zioupos, P. (1998) *Med. Eng. Phys.*, **20**(2), 92–102.
9. Alberts, B., Bray, D., Lewis, J., Raff, M., Roberts, K., and Watson, J.D. (1994) *Molecular Biology of the Cell*, 3rd edn, Garland Publishing, New York.
10. Bowman, S.M., Zeind, J., Gibson, L.J., Hayes, W.C., and McMahon, T.A. (1996) *J. Biomech.*, **29**(11), 1497–1501.
11. Fantner, G.E., Hassenkam, T., Kindt, J.H., Weaver, J.C., Birkedal, H., Pechenik, L., Cutroni, J.A., Cidade, G.A.G., Stucky, G.D., Morse, D.E., and Hansma, P.K. (2005) *Nat. Mater.*, **4**(8), 612–616.
12. Elliott, J.C. (2002) Calcium Phosphate Biomaterials in Phosphates: Geochemical, Geobiological, and Materials Importance, in *Reviews in Mineralogy* (eds M.J. Kohn, J. Rakovan, and J.M. Hughes), Mineralogical Society of America, Washington, DC, pp. 427–453.
13. Skinner, H.C.W. and Jahren, A.H. (2004) *Treatise on Geochemistry*, Elsevier, vol. 8, pp. 117–184.
14. Hassenkam, T., Fantner, G.E., Cutroni, J.A., Weaver, J.C., Morse, D.E., and Hansma, P.K. (2004) *Bone*, **35**(1), 4–10.
15. Rosen, V.B., Hobbs, L.W., and Spector, M. (2002) *Biomaterials*, **23**(3), 921–928.
16. Ziv, V. and Weiner, S. (1994) *Connect. Tissue Res.*, **30**(3), 165–175.
17. Eppell, S.J., Tong, W.D., Katz, J.L., Kuhn, L., and Glimcher, M.J. (2001) *J. Orthop. Res.*, **19**(6), 1027–1034.
18. Wilson, R.M., Dowker, S.E.P., and Elliott, J.C. (2006) *Biomaterials*, **27**(27), 4682–4692.
19. Petruska, J.A. and Hodge, A.J. (1964) *Proc. Natl. Acad. Sci. U.S.A.*, **51**(5), 871–876.
20. Landis, W.J., Hodgens, K.J., Song, M.J., Arena, J., Kiyonaga, S., Marko, M., Owen, C., and McEwen, B.F. (1996) *J. Struct. Biol.*, **117**(1), 24–35.
21. Katz, E.P. and Li, S. (1973) *J. Mol. Biol.*, **80**(1), 1–15.
22. McKee, M.D., Nanci, A., Landis, W.J., Gotoh, Y., Gerstenfeld, L.C., and Glimcher, M.J. (1991) *J. Bone. Miner. Res.*, **6**(9), 937–945.
23. Knothe-Tate, M.L. and Cowin, S.C. (eds) (2001) Interstitial Fluid Flow, in *Bone Mechanics Handbook*, 2nd edn, CRC Press, New York, pp. 22.21–22.29.
24. Wilson, E.E., Awonusi, A., Morris, M.D., Kohn, D.H., Tecklenburg, M.M.J., and Beck, L.W. (2006) *Biophys. J.*, **90**(10), 3722–3731.
25. Neuman, W.F., Toribara, T.Y., and Mulryan, B.J. (1956) *J. Am. Chem. Soc.*, **78**(17), 4263–4266.
26. Eisenberg, S.R. and Grodzinsky, A.J. (1985) *J. Orthop. Res.*, **3**(2), 148–159.
27. Bembey, A.K., Bushby, A.J., Boyde, A., Ferguson, V.L., and Oyen, M.L. (2006) *J. Mater. Res.*, **21**(8), 1962–1968.
28. Akiva, U., Wagner, H.D., and Weiner, S. (1998) *J. Mater. Sci.*, **33**(6), 1497–1509.
29. Qin, Q.H. and Swain, M.V. (2004) *Biomaterials*, **25**(20), 5081–5090.
30. Kotha, S.P. and Guzelsu, N. (2002) *J. Theor. Biol.*, **219**(2), 269–279.

31. Wagner, H.D. and Weiner, S. (1992) *J. Biomech.*, **25**(11), 1311–1320.
32. Gosline, J., Lillie, M., Carrington, E., Guerette, P., Ortlepp, C., and Savage, K. (2002) *Philos. Trans. R. Soc. Lond.: B Biol. Sci.*, **357**(1418), 121–132.
33. Sasaki, N. and Odajima, S. (1996) *J. Biomech.*, **29**(9), 1131–1136.
34. Cusack, S. and Miller, A. (1979) *J. Mol. Biol.*, **135**(1), 39–51.
35. Buehler, M.J. (2006) *J. Mater. Res.*, **21**(8), 1947–1961.
36. Lorenzo, A.C. and Caffarena, E.R. (2005) *J. Biomech.*, **38**(7), 1527–1533.
37. Vesentini, S., Fitie, C.F., Montevecchi, F.M., and Redaelli, A. (2005) *Biomech. Model. Mechanobiol.*, **3**(4), 224–234.
38. Gupta, H.S., Messmer, S.P., Roschger, P., Bernstorff, S., Klaushofer, K., and Fratzl, P. (2004) *Phys. Rev. Lett.*, **93**(15), 158101/1–158101/4.
39. van der Rijt, J.A., van der Werf, K.O., Bennink, M.L., Dijkstra, P.J., and Feijen, J. (2006) *Macromol. Biosci.*, **6**(9), 697–702.
40. Eppell, S.J., Smith, B.N., Kahn, H., and Ballarini, R. (2006) *J. R. Soc. Interface*, **3**(6), 117–121.
41. Shen, Z.L., Dodge, M.R., Kahn, H., Ballarini, R., and Eppell, S.J. (2008) *Biophys. J.*, **95**(8), 3956–3963.
42. Buehler, M.J. (2008) *J. Mech. Behav. Biomed. Mater.*, **1**(1), 59–67.
43. An, K.N., Sun, Y.L., and Luo, Z.P. (2004) *Biorheology*, **41**(3-4), 239–246.
44. Freeman, J.W. and Silver, F.H. (2004) *J. Theor. Biol.*, **229**(3), 371–381.
45. Catanese, J. III, Iverson, E.P., Ng, R.K., and Keaveny, T.M. (1999) *J. Biomech.*, **32**(12), 1365–1369.
46. Currey, J.D. (1964) *Biorheology*, **2**, 1–10.
47. Broz, M.E., Cook, R.F., and Whitney, D.L. (2006) *Am. Mineral.*, **91**, 135–142.
48. Katz, J.L. and Ukraincik, K. (1971) *J. Biomech.*, **4**(3), 221–227.
49. Oyen, M.L. (2005) Ultrastructural characterization of time-dependent, inhomogeneous materials and tissues. Dissertation. University of Minnesota.
50. Oyen, M.L., Ferguson, V.L., Bembey, A.K., Bushby, A.J., and Boyde, A. (2008) *J. Biomech.*, **41**(11), 2585–2588.
51. Zioupos, P., Currey, J.D., and Hamer, A.J. (1999) *J. Biomed. Mater. Res.*, **45**(2), 108–116.
52. Gardner, T.N., Elliott, J.C., Sklar, Z., and Briggs, G.A. (1992) *J. Biomech.*, **25**(11), 1265–1277.
53. Hill, R. (1952) *Proc. Phys. Soc. A*, **65**, 349–354.
54. Broz, J.J., Simske, S.J., Greenberg, A.R., and Luttges, M.W. (1993) *J. Biomech. Eng.*, **115**(4), 447–449.
55. Evans, F.G. and Lebow, M. (1951) *J. Appl. Physiol.*, **3**(9), 563–572.
56. Hoffler, C.E., Guo, X.E., Zysset, P.K., and Goldstein, S.A. (2005) *J. Biomech. Eng.*, **127**(7), 1046–1053.
57. Nyman, J.S., Roy, A., Shen, X., Acuna, R.L., Tyler, J.H., and Wang, X. (2006) *J. Biomech.*, **39**(5), 931–938.
58. Bushby, A.J., Ferguson, V.L., and Boyde, A. (2004) *J. Mater. Res.*, **19**(1), 249–259.
59. Hengsberger, S., Kulik, A., and Zysset, P. (2002) *Bone*, **30**(1), 178–184.
60. Rho, J.Y. and Pharr, G.M. (1999) *J. Mater. Sci. Mater. Med.*, **10**(8), 485–488.
61. Sasaki, N. and Enyo, A. (1995) *J. Biomech.*, **28**(7), 809–815.
62. Sasaki, N. and Yoshikawa, M. (1993) *J. Biomech.*, **26**(1), 77–83.
63. Cowin, S.C. (1999) *J. Biomech.*, **32**(3), 217–238.
64. Bembey, A.K., Oyen, M.L., Bushby, A.J., and Boyde, A. (2006) *Philos. Mag.*, **86**(33–35), 5691–5703.
65. Yamashita, J., Furman, B.R., Rawls, H.R., Wang, X., and Agrawal, C.M. (2001) *J. Biomed. Mater. Res.*, **58**(1), 47–53.
66. Bembey, A.K., Koonjul, V., Bushby, A.J., Ferguson, V.L., and Boyde, A. (2005) *Mater. Res. Soc. Symp. Proc.*, **844**, R2.7.1–R2.7.6.
67. Tseretely, G.I. and Smirnova, O.I. (1992) *J. Therm. Anal.*, **38**(5), 1189–1201.
68. Chawla, K.K. (1987) *Composite Materials*, Springer-Verlag, New York.
69. Hashin, Z. and Shtrikman, S. (1963) *J. Mech. Phys. Solids*, **11**, 127–140.
70. Agarwal, B.D. and Broutman, L.J. (1990) *Analysis and Performance of*

Fiber Composites, 2nd edn, Wiley Interscience, New York.

71. Hull, D. and Clyne, T.W. (1996) *An Introduction to Composite Materials*, Cambridge University Press, United Kingdom.
72. Gupta, H.S., Schratter, S., Tesch, W., Roschger, P., Berzlanovich, A., Schoeberl, T., Klaushofer, K., and Fratzl, P. (2005) *J. Struct. Biol.*, **149**(2), 138–148.
73. Jager, I. and Fratzl, P. (2000) *Biophys. J.*, **79**(4), 1737–1746.
74. Deer, W.A., Howie, R.A., and Zussman, J. (1966) *An Introduction to the Rock-Forming Minerals*, Longman, London.
75. Currey, J.D. (2002) *Bones: Structure and Mechanics*, Princeton University Press, Princeton.
76. Jaeger, C., Groom, N.S., Bowe, E.A., Horner, A., Davies, M.E., Murray, R.C., and Duer, M.J. (2005) *Chem. Mater.*, **17**(12), 3059–3061.
77. Oyen, M.L., Bembey, A.K., and Bushby, A.J. (2007) *Mater. Res. Soc. Symp. Proc.*, **975**, DD.7.5–DD.7.11.
78. Weiner, S., Traub, W., and Wagner, H.D. (1999) *J. Struct. Biol.*, **126**(3), 241–255.
79. Fritsch, A. and Hellmich, C. (2007) *J. Theor. Biol.*, **244**(4), 597–620.
80. Pidaparti, R.M., Chandran, A., Takano, Y., and Turner, C.H. (1996) *J. Biomech.*, **29**(7), 909–916.
81. Sasaki, N., Tagami, A., Goto, T., Taniguchi, M., Nakata, M., and Hikichi, K. (2002) *J. Mater. Sci. Mater. Med.*, **13**(3), 333–337.
82. Hellmich, C. and Ulm, F.J. (2002) *J. Biomech.*, **35**(9), 1199–1212.
83. Balooch, M., Habelitz, S., Kinney, J.H., Marshall, S.J., and Marshall, G.W. (2008) *J. Struct. Biol.*, **162**(3), 404–410.
84. Kinney, J.H., Habelitz, S., Marshall, S.J., and Marshall, G.W. (2003) *J. Dent. Res.*, **82**(12), 957–961.
85. Cowin, S.C. (2007) *J. Biomech.*, **40** (Suppl. 1), S105–S109.
86. Cowin, S.C. and Mehrabadi, M.M. (1989) *J. Biomech.*, **22**(6-7), 503–515.
87. Cowin, S.C. and Turner, C.H. (1992) *J. Biomech.*, **25**(12), 1493–1494.
88. Martin, R.B. and Ishida, J. (1989) *J. Biomech.*, **22**(5), 419–426.
89. Ashman, R.B., Cowin, S.C., Van Buskirk, W.C., and Rice, J.C. (1984) *J. Biomech.*, **17**(5), 349–361.
90. Reilly, D.T. and Burstein, A.H. (1975) *J. Biomech.*, **8**(6), 393–405.
91. Reilly, D.T., Burstein, A.H., and Frankel, V.H. (1974) *J. Biomech.*, **7**(3), 271.
92. Sasaki, N., Matsushima, N., Ikawa, T., Yamamura, H., and Fukuda, A. (1989) *J. Biomech.*, **22**(2), 157–164.
93. Riggs, C.M., Vaughan, L.C., Evans, G.P., Lanyon, L.E., and Boyde, A. (1993) *Anat. Embryol.*, **187**(3), 239–248.
94. Petrtyl, M., Hert, J., and Fiala, P. (1996) *J. Biomech.*, **29**(2), 161–169.
95. Katz, J.L. and Meunier, A. (1993) *J. Biomech. Eng.*, **115**(4), B543–B548.
96. Katz, J.L., Yoon, H.S., Lipson, S., Maharidge, R., Meunier, A., and Christel, P. (1984) *Calcif. Tissue Int.*, **36** (Suppl. 1), S31–S36.
97. Fan, Z., Swadener, J.G., Rho, J.Y., Roy, M.E., and Pharr, G.M. (2002) *J. Orthop. Res.*, **20**(4), 806–810.
98. Rho, J.Y., Roy, M.E., Tsui, T.Y., and Pharr, G.M. (1999) *J. Biomed. Mater. Res.*, **45**(1), 48–54.
99. Rho, J.Y., Tsui, T.Y., and Pharr, G.M. (1997) *Biomaterials*, **18**(20), 1325–1330.
100. Zysset, P.K., Guo, X.E., Hoffler, C.E., Moore, K.E., and Goldstein, S.A. (1998) *Technol. Health Care*, **6**(5–6), 429–432.
101. Zysset, P.K., Guo, X.E., Hoffler, C.E., Moore, K.E., and Goldstein, S.A. (1999) *J. Biomech.*, **32**(10), 1005–1012.
102. Ferguson, V.L., Bushby, A.J., and Boyde, A. (2003) *J. Anat.*, **203**(2), 191–202.
103. Turner, C.H., Chandran, A., and Pidaparti, R.M. (1995) *Bone*, **17**(1), 85–89.
104. Amprino, R. (1958) *Acta Anat. (Basel)*, **34**(3), 161–186.
105. Riches, P.E., Everitt, N.M., Heggie, A.R., and McNally, D.S. (1997) *J. Biomech.*, **30**(10), 1059–1061.
106. Swadener, J.G., Rho, J.Y., and Pharr, G.M. (2001) *J. Biomed. Mater. Res.*, **57**(1), 108–112.

107. Oyen, M.L., Ko, C.C., Bembey, A.K., Bushby, A.J., and Boyde, A. (2005) *Mater. Res. Soc. Symp. Proc.*, **874**, L1.7.1–L1.7.6.
108. Baron, C., Talmant, M., and Laugier, P. (2007) *J. Acoust. Soc. Am.*, **122**(3), 1810.
109. Dong, X.N. and Guo, X.E. (2004) *J. Biomech.*, **37**(8), 1281–1287.
110. Boyde, A., Travers, R., Glorieux, F.H., and Jones, S.J. (1999) *Calcif. Tissue Int.*, **64**(3), 185–190.
111. Tai, K., Dao, M., Suresh, S., Palazoglu, A., and Ortiz, C. (2007) *Nat. Mater.*, **6**(6), 454–462.

4
Mechanobiological Models for Bone Tissue. Applications to Implant Design

José Manuel García-Aznar, María José Gómez-Benito, María Ángeles Pérez, and Manuel Doblaré

4.1
Introduction

Bone is an evolving tissue that is active throughout the whole life. In spite of being one of the hardest living tissues, it has an amazing capacity to adapt and self-heal. In fact, bone changes its shape and mechanical properties as a response to the mechanical environment. For instance, changes in bone mass could be observed after long stance in bed, or changes in bone shape also appear in fracture healing and after implantation of fixation devices.

Bone is a connective tissue formed by cells and extracellular matrix composed of ground substance and fibers. In contrast to other connective tissues, its components are calcified, making it a hard tissue. It has a high tensile strength due to the collagen fibers and also a high compression strength due to the mineral components. Moreover, bone is a light material whose structure is hierarchically organized [1] both macroscopically and microscopically, in such a way that its strength is maximized with the minimum weight.

In spite of these properties, bone injury is quite common, usually caused by a sudden load that exceeds bone strength or the cyclic activity of loads (well below bone strength) that gradually accumulates damage at a rate that cannot be repaired. Depending on the type of injury, a prosthesis or fixation can be used. A prosthesis is normally used to replace damaged or injured joints, while a fixator is used to treat bone fractures, allowing a proper alignment of the fracture being removed after bone healing. The performance of this kind of implant comprises two components, the response of the bone to the implant and the behavior of the material in the host bone. Therefore, it has to meet mechanical and biological requirements to fulfill the objectives specified in its design. From a mechanical point of view, a prosthesis or implant has to be capable of providing mechanical support as the original bone. For example, in the case of a prosthesis, it has to present enough strength to resist overloads, cyclic loads (fatigue), and wear. For fixations, it is necessary to get the appropriate

Biomechanics of Hard Tissues: Modeling, Testing, and Materials.
Edited by Andreas Öchsner and Waqar Ahmed

ISBN: 978-3-527-32431-6

reduction of the fracture fragments and the stabilization of the fracture combined with enough strength to support the loads during the period when they are acting.

In both cases, the surgical implantation of orthopedic implants can result in a strong modification of the load transfer mechanism, over- and underloading the bone in specific sites. This effect causes long-term alterations in bone structure around the implant. These changes can include increased bone formation in some regions and bone loss in others. Increased bone formation can be caused by biological bone induction mechanisms associated with the surgical trauma. Bone resorption can occur by mechanobiological adaptation to the altered loading pattern transmission produced when an implant is incorporated ("stress shielding"). Other causes may also produce bone resorption, such as osteolysis due to wear particles, micromotions in the interface, or poor vascularization. To analyze the functional adaptation that can occur around implants, computer simulations based on finite element analysis (FEA) are normally used. In this chapter, dedicated to the application of mechanobiological models to bone implant design, we shall show some examples related to the analysis of this effect after implantation of different types of prostheses and implants.

4.2 Biological and Mechanobiological Factors in Bone Remodeling and Bone Fracture Healing

4.2.1 Bone Remodeling

Bone is one of the most active tissues in the body, being constantly rebuilt in a process named as *bone adaptation* or *bone remodeling*. Depending on the surface where bone is added or removed, it is usual to distinguish between two different adaptative processes: internal remodeling (or simply remodeling) on trabecular surfaces in the cancellous bone and Haversian systems in the cortical bone, and external remodeling (or modeling) in the periosteum and endosteum. Bone remodeling only occurs at bone surfaces where the effector bone cells are located [2]. The main bone cells responsible for bone remodeling are osteoblasts, osteoclasts, and bone lining cells located on bone surfaces and osteocytes inside the bone matrix [2]. Osteoblasts produce bone. They differentiate from mesenchymal stem cells, which can come from the periosteum layer or from the stromal tissue of the bone marrow. Osteoclasts remove the bone, demineralizing it with acid and dissolving collagen with enzymes. The origin of these cells is the bone marrow. Bone lining cells are inactive osteoblasts that escaped from being buried in the new bone and remained on the bone surface when bone formation stopped. They can be reactivated to osteoblasts in response to various chemical and mechanical stimuli [3]. Osteocytes are, like bone lining cells, former osteoblasts that became buried in the newly formed bone. Many authors [4–9] have proposed that osteocytes are the mechanosensor cells that control bone remodeling. Bone remodeling is the final result of the activity of cellular packets (named as basic multicellular units, BMU) of osteoblasts and osteoclasts that act together in refilling and eliminating bone in a coupled way [10]. In situations of underloading or disuse, bone resorption occurs, whereas under overloading bone formation is the net resulting process. In the latter case, bone is likely to develop microcracks that eventually may collapse and form a macroscopic crack, resulting in fatigue failure or stress bone fracture. In contrast, modeling can be resorptive or formative, always happening through an uncoupled process between osteoblasts and osteoclasts [2]. The role of bone remodeling is multiple and complex [11]. From the biochemical point of view, bone remodeling regulates minerals and hormones, influencing pathologies as important as osteoporosis. From the mechanical point of view, bone remodeling is responsible for the mechanical adaptation of bone tissue, determining its mechanical capacity during the entire life. In fact, there are many experiments that corroborate the effect of mechanical loads on bone adaptation [12–15]. For example, it is widely assumed that remodeling is the mechanism that allows bone to repair damage, reducing the risk of fracture [16–21]. On the other hand, both modeling and remodeling are hypothesized to optimize the stiffness and strength with minimum weight [22–24]. Indeed, it has been determined that peak periosteal strains are similar across species during vigorous activities and rarely exceed 2000–3000 $\mu\varepsilon$ in long bones, which suggests that animal skeleton is continuously "redesigned" to control strain [25, 26].

4.2.2 Bone Fracture Healing

A bone fracture involves the disruption of the resistant properties of bone. However, in contrast to other nonliving materials, bone as a living tissue is able to regenerate without scar formation. This regeneration process consists of the formation and evolution of different tissue types in the fracture site with different mechanical properties and spatial distribution, which are able to restore the original stiffness, strength, and shape of the fractured bone. This process is really complex since the bone callus forms at different stages that overlap over time [27]. Nevertheless, it is important to mention that not all fractures heal correctly: around 5–10% result in a delayed union or a nonunion [28].

Many factors influence the bone healing process, such as genetic, cellular, and biochemical factors, age, type of fracture, interfragmentary motion, oxygen tension, electrical environment, and fracture geometry [29–31]. Biochemical factors, such as, soft tissue coverage and blood supply [31], are important to provide the biological environment for the bone healing process. Several growth factors, such as the transforming growth factor beta (TGF-β), fibroblastic growth factors (FGFs), and platelet-derived growth factors (PDGF) [32], appear and interact in the fracture site. Much interest in this research field has focused on mechanical factors. For example, rigid fixation and moderate gap size can result in primary fracture repair [33], while secondary healing is more likely to occur under flexible fixations [34]. It is generally accepted that some amount of axial movement between fracture fragments stimulates callus formation and favors the quality and quantity of the new tissue, thus increasing the rate at which fracture heals [35, 36]. In contrast, excessive interfragmentary movement delays the healing process and may result in nonunion of bone fragments causing, in some cases, pseudoarthrosis [37]. Another important mechanical factor is the type of load to which the fracture site is subjected (shear [38, 39], torsion [40], compresion, or tension [41]) which results in different fracture healing outcomes.

4.3
Phenomenological Models of Bone Remodeling

For many years, many theoretical models have been proposed to explain how the mechanical environment influences bone remodeling, the "Wolff's law" being the most well-known [42]. In fact, Wolff proposed that there is a dependence between bone structure and the load that it supports. However, he did not provide any idea about the possible processes responsible of this effect. It was Roux in 1881 who proposed that "bone adapted itself in order to support stresses in an optimal way with minimum mass" [43]. Both theories in combination with the development of computational tools and a huge battery of experimental results have motivated the current tremendous development of computational remodeling models. These models may be classified into two main categories depending on the assumptions on which they are based: phenomenological and mechanistic.

Phenomenological models are able to predict bone remodeling through direct relationships between mechanical stimulus and bone response, following known experimental and clinical evidences, but no actual cell processes are considered. At the same time, we can classify this kinds of models into three types:

- Models based on global optimality criteria, in which different mechanical criteria are proposed to be optimal [22, 23, 44].
- Models based on achieving a homeostatic value for a certain mechanical stimulus. These models admit the existence of a certain mechanical stimulus that produces bone apposition or resorption such that, by this process, the stimulus tends to a certain uniform physiological level (homeostasis) in the whole tissue [45–50].
- Models based on damage repair, based on the assumption that bone tries to optimize its strength and stiffness regulating the local damage generated by fatigue or creep [16, 17].

These mathematical models are particularly useful to predict the adaptative bone changes regulated by mechanical factors to improve implant design or treat some patients, as it is shown in Section 4.5.

4.4
Mechanistic Models of Bone Remodeling

Mechanistic models are more complex since they try to characterize, understand, and unravel the role of the mechanical environment in the biological mechanisms involved in bone remodeling. These models are interested not only in the prediction of the long-term behavior of the bone under loading, but also in its rate of adaptation and the specific role of bone cells at each stage of the mechanosensation and adaptation processes. In this direction, some attempts have been made toward its modeling [18–20, 51–55].

In this sense, a first mechanistic model of bone remodeling is the one proposed by Huiskes *et al.* [51]. In this mathematical model, they proposed that bone resorption (by osteoclasts) and bone formation (by osteoblasts) coupled activity is exclusively regulated by mechanical factors. In fact, they considered the strain energy density rate as the regulator mechanical stimulus while the osteocytes' spatial distribution defines the mechanosensor system attenuated by the distance. The osteoclast activation is considered to be regulated by disuse and microdamage, while bone formation is activated when the mechanical stimulus at a surface location exceeds a threshold value.

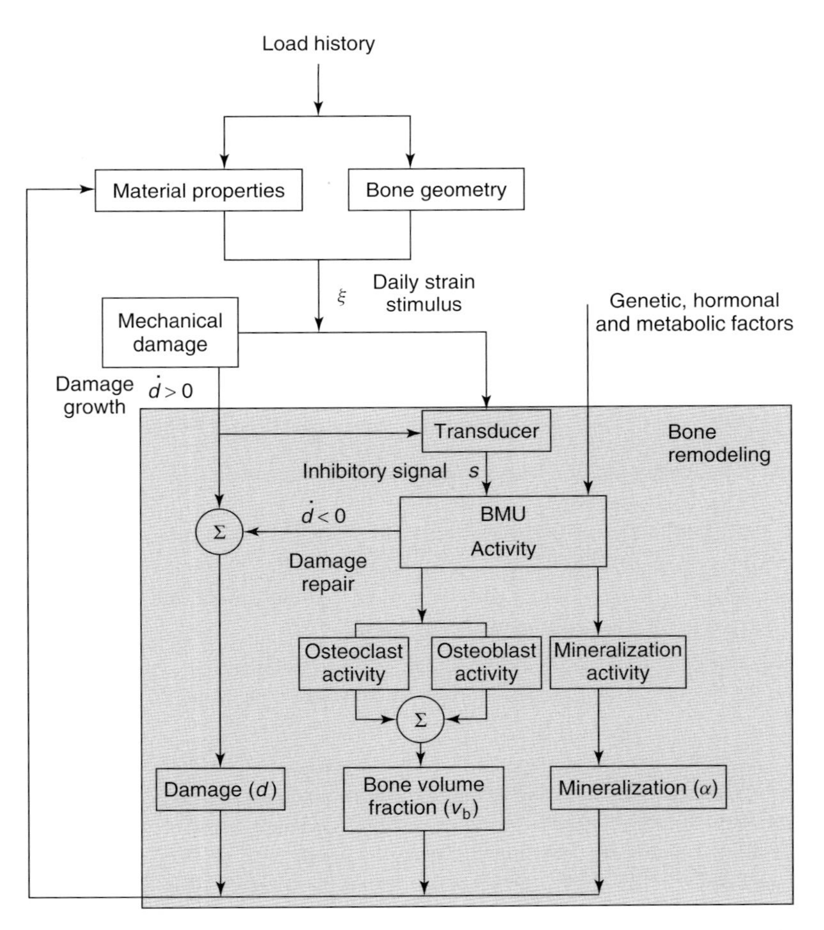

Figure 4.1 Scheme of the bone remodeling theory described in García-Aznar *et al.* [20].

On the other hand, the approach proposed by Hazelwood *et al.* [19] also assumes that bone remodeling is activated by disuse and microdamage, considering osteoclast/osteoblast coupling due to biochemical factors through the BMU activity. Another biological proposal was developed by Hernández *et al.* [18, 56] that includes biological and metabolical influences on bone adaptation. They analyzed the effect of several biological factors directly related to BMU activity, such as mineralization period, focal bone balance, and activation frequency. However, these last works do not take into account the degradation that damage induces into the mechanical properties of the bone and the subsequent modification of the mechanical stimuli, thus making them incomplete for a full understanding of this coupling mechanism between damage accumulation (purely mechanical) and biological bone response. In fact, the determination of the rates of damage accumulation and repair are crucial to prevent stress fractures and to control net bone mass production.

More recently, García-Aznar *et al.* [20] presented a theorical model to analyze the coupling effect between fatigue damage removal/accumulation and biological events that occur in bone remodeling (see Figure 4.1). With this aim, they proposed a continuum damage theory that is capable of predicting damage growth and the subsequent degradation of the elastic modulus of bone under cyclic loading. This approach is coupled with a BMU-based bone remodeling theory similar to the previous existing ones [18, 19, 56]. This combined theory has been used to analyze the influence of different mechanobiological factors. For example, under overloading conditions, bone porosity decreases unless the damage rate is so high that it causes resorption or the occurrence of a "stress fracture." Moreover, this model has also been used to analyze the relevant role that bone fatigue damage accumulation has in the calcium homeostasis [57]. Currently, this model has been extended incorporating the anisotropic behavior of bone during the advance of the BMUs [55].

4.5
Examples of Application of Bone Remodeling Models to Implant Design

According to the 1983 review of Huiskes and Chao [58], the first application of FEA in orthopedics was in 1972 [59]. Since then, this method has become an important tool in orthopedics design, being widely used in biomechanics [60] with many different purposes. The following are among them:

1) Design and preclinical analysis of implants: although FEA does not allow accurate quantitative analysis because of anthrophometric differences, comparative analysis can be performed between different implant designs. Actually, many designs of joint replacement prostheses have been studied using finite element models, either by the manufacturer or by external research groups.
2) Obtainment of fundamental biomechanical knowledge about musculoskeletal structures: analysis of bones, cartilage, ligaments, tendons, and their relationships.
3) Simulation of different adaptive biological processes allowing testing of different constitutive laws for tissue growth, adaptation, and degeneration.

Computational simulations can also be used to predict tissue behavior in response to biomechanical factors.

Nowadays, finite element models are normally used to design implants, analyzing the influence of different factors such as implant variables, like size, shape, position, material stiffness, coating and ingrowth conditions, debonding, or lack of fixation; bone variables, such as geometry, bone density and anisotropy distributions, and response to different loading conditions. In this section, we perform a brief revision of some examples of the application of bone remodeling models to analyze the effect of different design factors in implants.

One of the first works in which a computational theory of bone remodeling was applied to study the influence of a prosthesis implantation was developed by Huiskes *et al.* [48]. Although 2D and axisymmetric models were used, the isotropic theory proposed by Huiskes in this work was used to simulate bone response (elastic modulus adaptation) due to prosthesis implantation. In fact, Huiskes and coworkers carried out the most extensive development of these finite element simulations combined with bone remodeling theories. For example, Huiskes *et al.* [61] studied the dependence of the stress-shielding on the stiffness of the stem analyzing a 3D model of a titanium stem (100 GPa) versus a more flexible stem (20 GPa). They used a 3D FE mesh where the initial apparent density distribution was determined from CT-density values; subsequently, they used an isotropic bone remodeling theory to predict the apparent density distribution after total hip replacement (THR). They concluded that stiffer stems produce more bone resorption than flexible ones. This conclusion has also been shown in animal experiments [62, 63] and in clinical radiographic studies [64].

Weinans *et al.* [65] used 3D FEA for studying the effect of bonded non-cemented total hip arthroplasties in dogs, comparing computational and experimental results through cross-sectional measurements of the canine femurs after two years of follow-up. This comparison showed that long-term changes in the bone around femoral components of THRs can be fully explained with bone remodeling theories. Weinans *et al.* [66] also analyzed non-cemented prosthesis and various situations of prosthesis–bone bonding with finite elements, using apparent density as the variable that quantifies bone remodeling. They checked the influence of the coating conditions (fully, partial, or noncoated) and the fitting characteristics (press fitted or overreamed), concluding that partial coating can reduce bone atrophy relative to fully coated stems. For smooth press-fit stems, computer predictions showed that the amount of bone loss was lower than that for one-third proximally coated or fully coated stems.

van Rietbergen *et al.* [67] also performed 3D finite element simulations for press-fit prostheses implanted into the femurs of dogs, for which CT scan geometric data were available. This model included not only internal remodeling by modification of apparent density but also external remodeling by modifying the boundary geometry. The results of this work confirmed the value of such approaches for preclinical testing of implants.

Kuiper and Huiskes [68] predicted the eventual loss of bone mass from the initial patterns of elastic energy deviation around hip replacements, combining

FEA and the isotropic remodeling theory developed by Huiskes and coworkers [48]. They checked the influence of the strain energy threshold on the simulation of the remodeling process. Finally, they estimated that the difference between several thresholds implied a 4% of bone mass loss.

Kerner *et al.* [69] carried out bone remodeling simulations of patient-specific 3D FE models of bone specimens obtaining their bone densitometry (DEXA). This work again showed the potential of these techniques to study the interaction between the implant and the bone, getting prediction of bone loss in accordance to the DEXA measurements on the specimens.

Prendergast and Taylor [70] used anatomical finite element models to preclinically test the effect of hip prosthesis design. They performed these simulations using a bone remodeling theory based on damage accumulation. They studied the importance of two features of intramedullary prosthesis design on bone adaptation: elastic modulus and the presence of a prosthesis collar. They concluded, as Huiskes *et al.* [61] did, that a low elasticity modulus reduced the bone loss, whereas the presence of a collar had no significant effect.

The isotropic bone remodeling theory developed by Beaupré *et al.* [45] was also applied with 2D FEA to simulate bone adaptation around porous-coated implants [71] in the proximal femur and tibia. The predicted bone density distribution around implanted prostheses was consistent with clinical and experimental findings of other researchers. Similar works were performed in this sense, with this same remodeling theory, to predict bone apparent density alteration around non-cemented acetabular components in 2D models [72].

Weinans *et al.* [73] studied the influence of different parameters on stress shielding that are normally used in these computer simulations, such as loading conditions and bone material properties. They analyzed four different loading conditions and two different bone density–elastic modulus relationships, obtaining differences up to 20% in specific regions due to changes in loading conditions and of 10% after changing the relationship between apparent density and elastic modulus. From these results, they concluded that although bone remodeling is different for each individual, there is no important difference between individuals.

Doblaré and García [23] developed an anisotropic bone remodeling theory, which was used to analyze the influence of anisotropy on bone remodeling after a THR with an Exeter hip prosthesis [24]. They predicted that bone anisotropy changes after replacement, tending to a more isotropic distribution (see Figure 4.2).

Cegoñino *et al.* [74] also analyzed the mechanical stability and bone remodeling adaptation of several distal femoral fractures treated with an intramedullary nail (DFN: distal femoral nail) and with an extramedullary plate (LISS: less invasive stabilization system). Both types of implants achieve a correct stabilization of the fracture, while the LISS plate causes more resorption in the distal region, specifically in the fracture site, due to the bridge effect of the plate in this region.

More recently, Pérez *et al.* [75] applied a bone remodeling model [24] in order to predict the bone remodeling of a resurfacing prosthesis with different cement

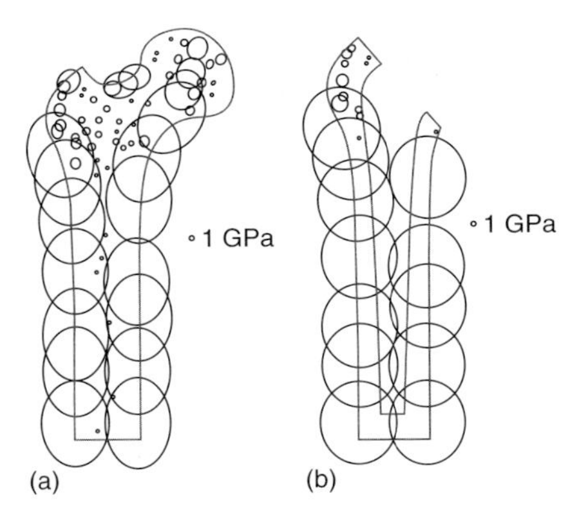

Figure 4.2 Anisotropy distribution: intact femur (a), after total hip replacement (b) (300 analysis steps), [24].

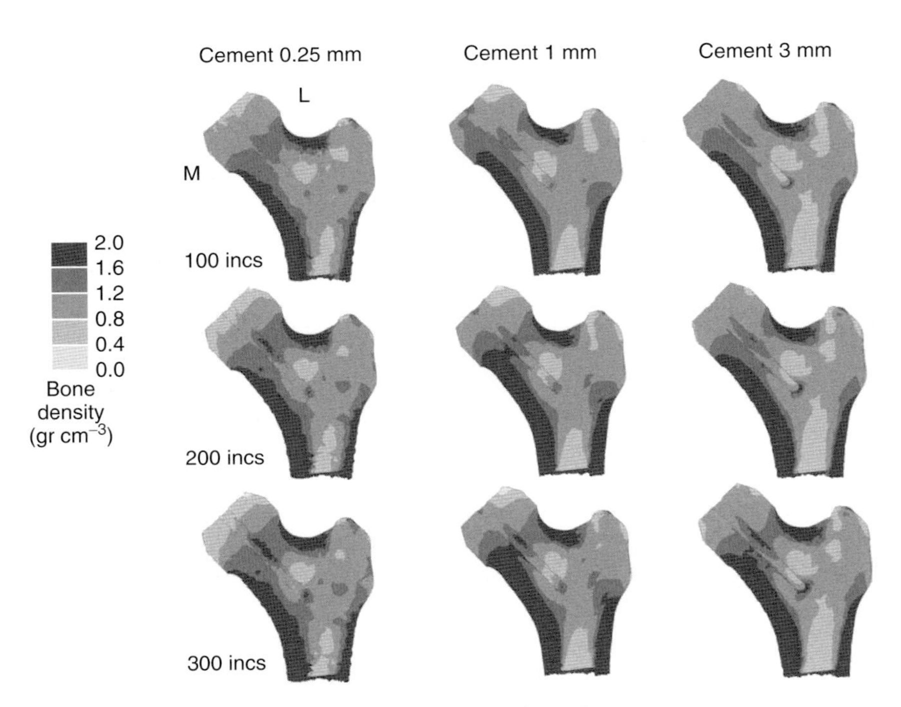

Figure 4.3 Bone density distribution in a resurfaced femoral head at different time intervals for three cement mantle thicknesses (0.25, 1, and 3 mm) (mid-frontal view, M-medial, L-lateral) [75].

mantle thickness (Figure 4.3). Short-term performance of this kind of implants has obtained good results, although the long-term performance remains uncertain. Clinically and numerically, hip resurfacing prostheses resulted in bone resorption at the superior femoral head and in bone apposition around the stem and at the stem tip. From the results predicted, the 1 mm cement mantle thickness may be an appropriate cement configuration.

Most of these examples correspond to the application of phenomenological bone remodeling models to the study of bone adaptation after prosthesis or fixation implantation. As commented above, these models are able to predict the long-term bone response, being an effective tool to evaluate the effect of different design parameters. However, these models present some limitations. For example, most of them are not interested in the evaluation of the bone remodeling rate, an effect that is directly predicted with a mechanistic model. As an example, the mechanistic bone remodeling model developed by García-Aznar *et al.* [20] was used to study the bone adaptation after THR with an Exeter prosthesis. As shown in Figure 4.4, the amount and location of bone mass loss are in good agreement with clinical results [76]. Moreover, these bone changes mainly take place in the first year of THR in the numerical simulation [20], a result that is very similar to clinical observations [77].

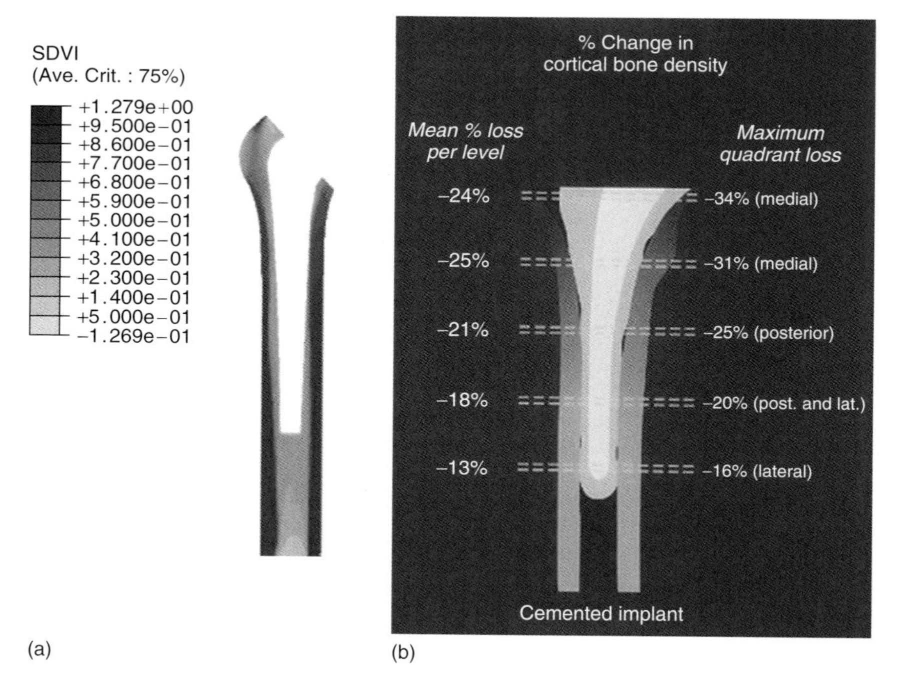

Figure 4.4 (a) Bone volume fraction distribution in the femur after 990 days of prosthesis implantation [20]; (b) change in cortical bone density after THR [76].

4.6
Models of Tissue Differentiation. Application to Bone Fracture Healing

Mechanistic models aim to understand how the mechanical signals are transferred to tissues, how cells sense these signals, and how they are translated into the cascade of biochemical reactions that stimulate cell expression and cell differentiation [43, 78].

Pauwels [79] probably formulated the first mechanistic tissue differentiation theory. This theory regarded the strain and stress invariants as the mechanical stimuli that guided the differentiation process. Deviatoric strain was assumed the stimulus for the formation of collagen fibers, and octahedral strain for the formation of cartilage. Pauwels formulated a set of differentiation rules, which are summarized in Figure 4.5(a), which allowed prediction of the appearance of different tissues in the fracture site. This theory was based on the clinical observation and experience.

Later, Perren and Cordey [80, 81], on the basis of their clinical studies, proposed the interfragmentary movement as the stimulus that guided differentiation in the fracture site. The interfragmentary strain (IS) was defined as the ratio between the relative displacement of the fracture fragments and the gap size. A tissue that ruptures at a certain strain level cannot be formed in a region experiencing strains greater than this (Figure 4.5b). This theory only considers longitudinal or axial strains. Contributions of the other strains, which could also be important, are neglected.

Carter *et al.* [47, 82, 83] proposed a differentiation theory, based on the ideas of Pauwels, aimed at analyzing how mechanical history affects tissue differentiation. The main hypothesis is that growth, maintenance, and remodeling of the tissue are achieved by the transfer of mechanical energy to the energy needed for chemical reactions. They considered that tissue differentiation was dependent on the stress level, and mainly on the invariants of the stress tensor (octahedral and deviatoric components). The studies of Carter *et al.* [84] were the first to use finite element models to explore the relationship between local stress/strain distribution and differentiated tissue types, and thus the first to quantify the level of mechanical stimulus. This theory has been used to study fracture healing [84, 85] and distraction osteogenesis [86].

Claes *et al.* [35, 87, 88] in a multidisciplinary work proposed a new tissue differentiation theory. They simulated the fracture callus in a finite element model, at different stages of bone healing. They established as differentiation hypothesis that the hydrostatic pressure and the principal strains determine the differentiation pathway. These works were based not only on the FEA but also on data from animal experiments and cell cultures. This theory has been used to study fracture healing at fixed times [89] and, also combined with fuzzy logic rules, to predict the evolution of tissues in the fracture callus [90, 91].

Prendergast *et al.* [92, 93] also formulated a model of tissue differentiation. They developed a biphasic poroelastic finite element model of tissues and the validation was based on experiments at a loaded implant interface. They proposed two biophysical stimuli to control differentiation: solid shear deviatoric strain and

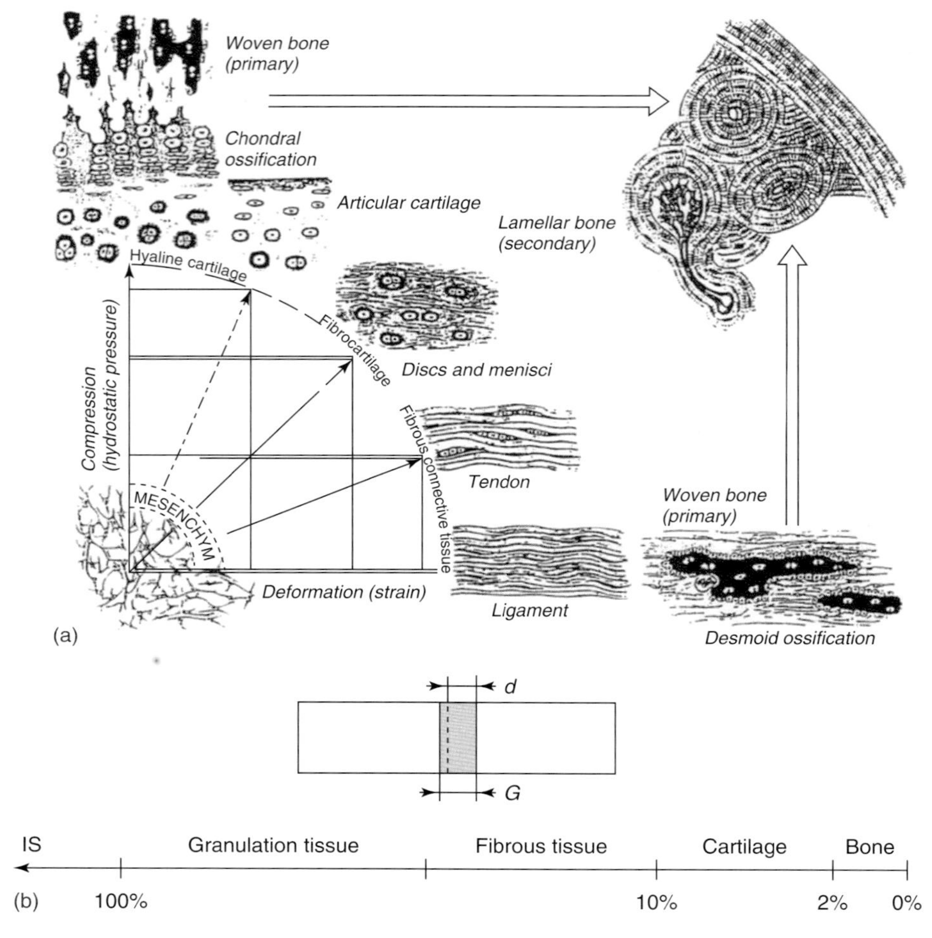

Figure 4.5 Sheme of the differentiation theories proposed by (a) Pauwels [79]; (b) Perren [80].

fluid flow in the interstitial fluid phase. High magnitudes of both favor fibrous tissue formation and low-magnitude ossification. This differentiation theory has been widely applied to simulate fracture healing [94–96], distraction osteogenesis [97], a rabbit bone chamber [98], osteochondral defect repair [99], and in random walk models of cell dispersal [100].

4.7 Mechanistic Models of Bone Fracture Healing

In general, bone healing models are presented in two different ways. On the one hand, there are static biomechanical models in which bone healing is analyzed at specific times of the healing process. On the other hand, there are evolutive models that predict the evolution of tissues in the fracture site through time. In this section, we focus on the second type of models.

One of the first analytical models of fracture healing was developed by Davy and Connolly [101] on the basis of their experimental works of healing canine ribs and radii. In that model they simulated the callus in simple traversal fractures, using concepts of continuum mechanics to determine the stiffness and strength of the fracture site through the healing process. They assumed the complex bone-callus as a variable section beam composed by different materials with elasto-plastic behavior and performed a parametric study.

Logvenkov [102] proposed a mathematical model of the fracture callus. The main elements of that model were osteoprogenitor and cartilage cells, osteoblasts, extracellular matrix, and blood vessels. The derived equations took into account matrix production, dependent on the stress level in the tissues, and cell differentiation dependent on the oxygen concentration. The callus material was considered as a growing elastic body with growth coefficients dependent on a structural tensor and calcium concentration. The structural tensor represented the anisotropy of fibrils' orientation that took place during deformation. The concentrations of oxygen and calcium were connected with the density of blood vessels whose propagation was related to the diffusion of a hypothetical substance. The formulated boundary problem allowed defining changes in the growing zone of the callus and to define its size. Nevertheless, this model did not account for the load history in the healing callus.

Ament and Hofer [103, 104] proposed a theoretical model of tissue healing and differentiation in the fracture callus. They considered both the osteogenic factors determined by the mechanical stimulus represented by the strain energy function. This function was obtained by means of finite element simulation and the effect of vascularity represented by a vascularization factor determined by means of fuzzy logic, from rules defined by medical experts. Similarly, Simon *et al.* [90, 91] proposed a new model based on eight rules of fuzzy logic. They also incorporated the influences of blood supply and mechanical stimulus. The model was able to predict the different tissue evolution in the fracture callus under compression and shear loads.

Lacroix and Prendergast [95] formulated a finite element model considering poroelastic material properties, which included mesenchymal stem cell origin and external bone remodeling based on the differentiation theory proposed by Prendergast *et al.* [93]. The same differentiation theory was used by Isaksson *et al.* [105] to develop a model of cell and tissue differentiation, using a mechanistic approach. The model directly couples cellular mechanisms to mechanical stimulation during bone healing including cell-phenotype-specific activities when modeling tissue differentiation. It was applied to simulate fracture healing under normal and excessive mechanical stimulation and the effect of periosteal stripping.

Bailón Plaza and van der Meulen [106] developed a mathematical framework in which they considered for the first time a differentiation theory based on growth factors. They incorporated chondrogenic and osteogenic growth factors.

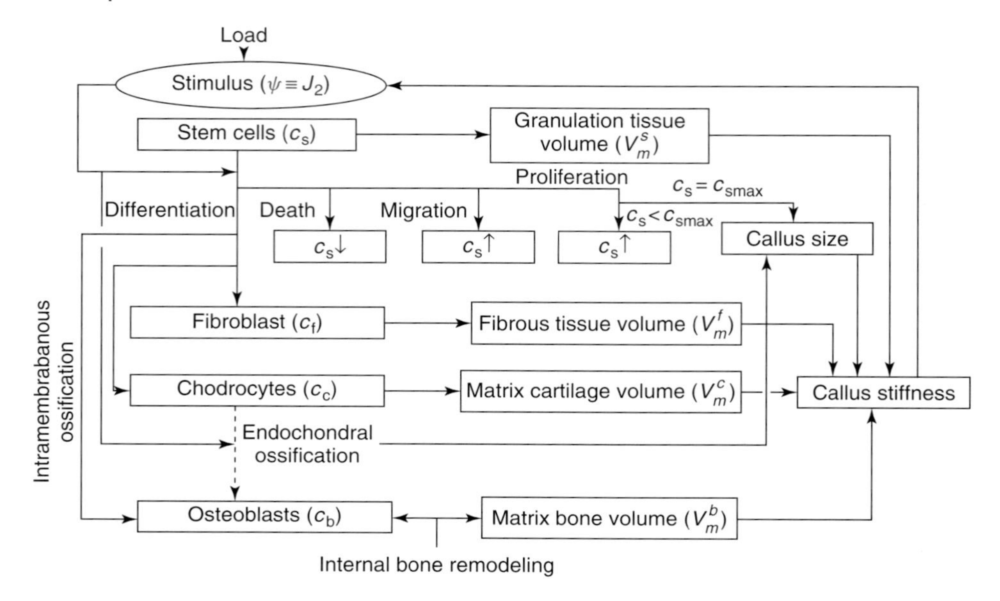

Figure 4.6 Scheme of the mechanobiological model of fracture healing proposed by Gómez-Benito *et al.* [109–111].

They used finite differences to study the healing process on a rat tibia [106]; later, they incorporated the influences of the mechanical factors to the previous growth factor-based model [107]. Recently, Geris *et al.* [108] adopted the model of Bailón-Plaza and van der Meulen incorporating angiogenesis to simulate normal and pathological bone healing with the finite volume method.

Gómez-Benito *et al.* [109–111] developed a continuum mathematical model of the bone healing process (Figure 4.6). It was successfully applied to simulate the influence of gap size [109], amount of interfragmentary movement [111], and external fixator stiffness [110] on the fracture healing process, and distraction osteogenesis [112]. The model simulates tissue regulation and callus growth, taking into account different cellular events (i.e., mesenchymal cell migration and proliferation; mesenchymal cells' differentiation into chondrocytes, fibroblasts, and osteoblasts, death of mesenchymal stem cells, and also the endochondral ossification process), and matrix synthesis, degradation, damage, calcification and remodeling over time. The evolution of the main components of the extracellular matrix of the different tissues (i.e., different collagen types, proteoglycans, minerals, and water) was also analyzed to determine the mechanical properties and permeability according to the particular composition.

In order to define all these processes, the fundamental variables were the concentration (c_i) of each cell type, with subscripts s, b, f, and c indicating stem cells, osteoblasts, fibroblasts, and chondrocytes, respectively, and the mechanical stimulus. Cell concentration can change because of proliferation ($\beta_i^{\mathrm{pr}} \cdot c_i$), migration ($D \cdot \nabla^2 c_i$), and differentiation ($f_{\mathrm{differentiation}}(\psi, t)$) for each specialized cell type

i (osteoblast, chondrocyte, fibroblast, and mesenchymal stem cell) following a similar equation:

$$\frac{Dc_i(\mathbf{x}, t)}{Dt} = \beta_i^{\mathrm{pr}} \cdot c_i - D \cdot \nabla^2 c_i - f_{\mathrm{differentiation}}(\psi, t) \tag{4.1}$$

As a first approach, migration and proliferation of chondrocytes, osteoblasts, and fibroblasts are assumed to be negligible.

Proliferation of stem cells is considered to be proportional to the mechanical stimulus (ψ), which is identified here with the second invariant of the deviatoric strain tensor:

$$\beta_s^{\mathrm{pr}} = \frac{\alpha_{\mathrm{proliferation}} \cdot \psi(\mathbf{x}, t)}{\psi(\mathbf{x}, t) + \psi_{\mathrm{proliferation}}} \tag{4.2}$$

where $\alpha_{\mathrm{proliferation}}$ and $\psi_{\mathrm{proliferation}}$ are constants that define the stem cell proliferation.

They assumed that the differentiation process ($f_{\mathrm{differentiation}}(\psi, t)$) is dependent on the mechanical stimulus and the time that cells need to mature.

Finally, they considered that callus growth is mainly due to mesenchymal cell proliferation ($f^{v}_{\mathrm{proliferation}}(c_s, \psi)$) and chondrocyte hypertrophy during endochondral ossification ($g^{v}_{\mathrm{endochondral}}(\psi, t)$):

$$\mathrm{div}(\mathbf{v}) = f^{v}_{\mathrm{proliferation}}(c_s, \psi) + g^{v}_{\mathrm{endochondral}}(\psi, t) \tag{4.3}$$

It was assumed that the concentration of mesenchymal cells can vary between zero and a maximum or saturation cell density. When the saturation concentration of mesenchymal stem cells is reached, the only way cells can proliferate further is by increasing the callus size at a constant level of cell concentration.

4.8 Examples of Application of Bone Fracture Healing Models to Implant Design

Different factors affecting bone healing have been studied in computational simulations using fracture healing models: the size [95, 109] and type of the fracture [85], the amount of interfragmentary motion [95, 111], the type of load [96], and the stiffness of the external fixator [90, 110, 113]. All these factors could help in the design of new fracture treatments and implants.

Loboa *et al.* [85] analyzed hydrostatic stress and maximum principal tensile strain patterns in regenerating tissue around the site of an oblique fracture and compared the results with the histomorphology of a typical oblique pseudarthrosis. Tissue differentiation predictions were consistent with the characteristic histomorphology of oblique pseudarthrosis. For example, they found that regions of high hydrostatic pressure correlated with locations of periosteal bone resorption. Gómez-Benito *et al.* [109] and Lacroix and Prendergast [95] studied the influence of the gap size on the fracture healing process. In both works, the delay in the healing process with the increase of the gap size was predicted. In the work of Gómez-Benito *et al.* [109], large gap size and intrerfragmentary motions were also analyzed resulting in a nonunion or pseudarthrosis as in experimental works [37].

The influence of the external fixator has also been studied. Lacroix and Prendergast [113] performed a three-dimensional simulation of fracture healing in which a human tibia was attached to an external fixator. They analyzed two different load patterns: healing was successful under the lower load and unsuccessful under the higher load, similar to clinical observations. Gómez-Benito *et al.* [110] studied the influence of the stiffness of the external fixator in the fracture healing pattern. They simulated a simple transverse mid-diaphyseal fracture of an ovine metatarsus. Three different stiffnesses of the external fixator were simulated (2300, 1725 and 1150 N/mm). The model predicted that a lower stiffness of the fixator delays fracture healing and causes a larger callus, similar to what has been observed experimentally. This effect could be observed in the load transfer mechanism between fractured bone and the different fixators through the healing process (Figure 4.7). Simon *et al.* [90] also studied the effect of different stiffness fixators in the fracture healing outcome.

Isaksson *et al.* [96] developed a poroelastic finite element model of an ovine tibia. The influence of torsional and axial loads was studied in their simulations using different mechanoregulatory algorithms. For both torsional and axial loads, they predicted fracture union as was observed in vivo when using the mechanoregulatory algorithm regulated by deviatoric strain and fluid velocity [60].

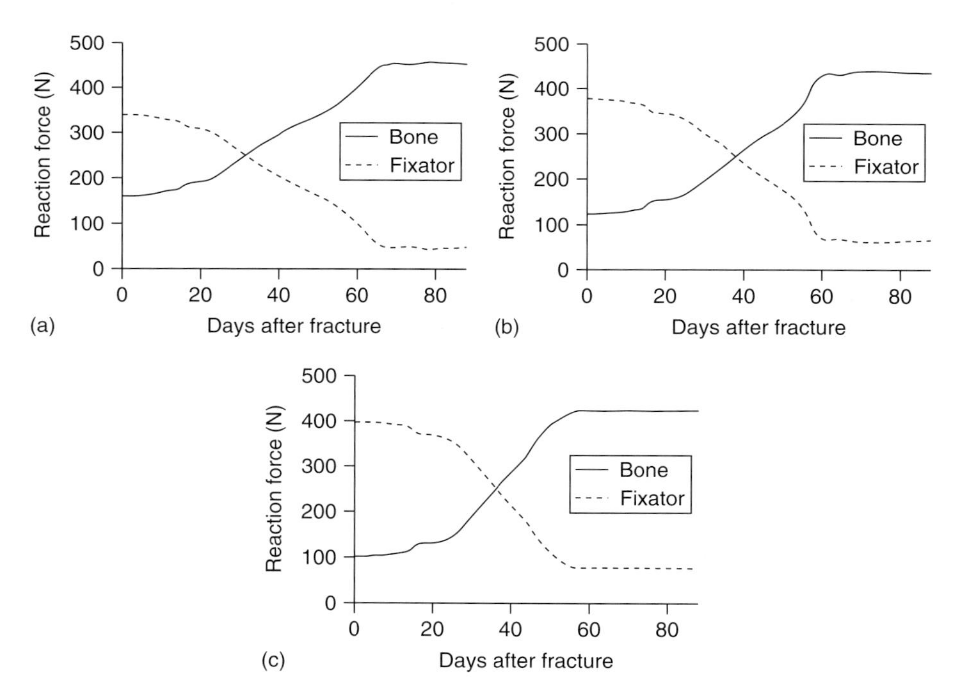

Figure 4.7 Reaction force evolution for different fixator stiffnesses (a) 1150 N/mm^{-1}, (b) 1725 N/mm^{-1}, (c) 2300 N/mm^{-1} [110].

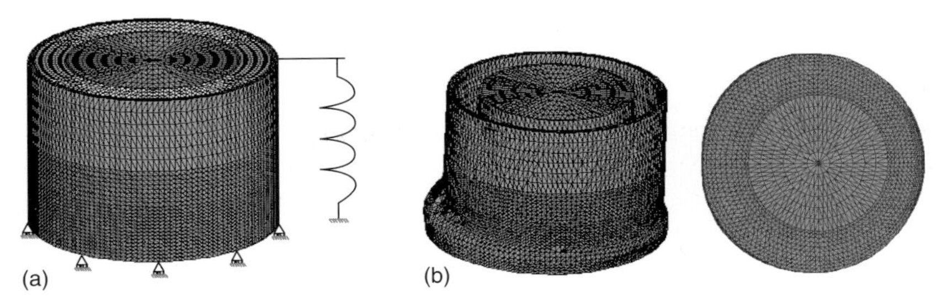

Figure 4.8 (a) Initial finite element mesh and boundary conditions in the fracture stabilized by a unilateral fixator; (b) final finite element mesh: frontal view and transverse plane.

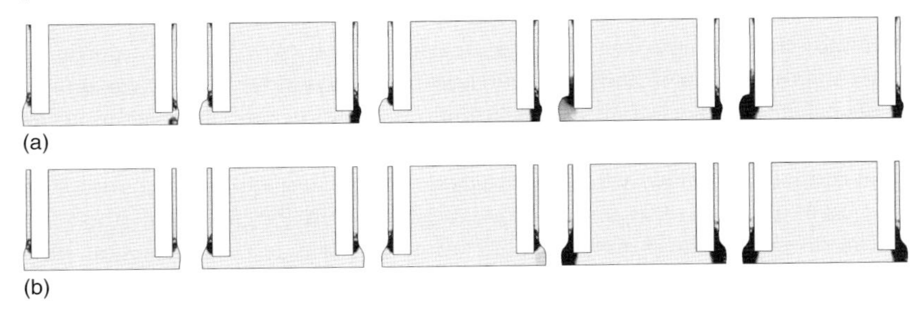

Figure 4.9 Osteoblast evolution (number of cells per cubic millimeter) in (a) the frontal plane; (b) lateral plane for the fracture stabilized by a unilateral fixator.

The influence of bending loads was also analyzed. With this purpose, a mid-diaphyseal fracture of a sheep tibia stabilized by a unilateral fixator was simulated. The unilateral fixator was modeled with a 4 mm radius circular section (Figure 4.8a). The fixator was attached only to one side of the bone by a screw of finite stiffness. The bone was loaded by an axial load of 500 N [114]. In this case, a clear bending effect could be observed. The evolution of bone cells and of the callus geometry was not symmetrical in the screw plane as shown in Figure 4.9 [114, 115].

Bone callus bridged sooner, about six weeks after fracture, on the lateral–frontal side (plane of the fixator, on the side close to the fixator) (Figure 4.9a). On the medial side, the bridge occurred 10 weeks after fracture and a bigger callus resulted. On the contrary, in the lateral plane (Figure 4.9b) the callus was symmetrical not only in geometry, but also in the evolution of the different cells.

This unsymmetrical fixation resulted in high bending movements in addition to the axial interfragmentary movement, causing a nonsymmetrical callus. A similar effect was also observed in similar experimental tests [41].

Recently, bone healing models have been applied to simulate distraction osteogenesis [97, 112, 116]. Distraction osteogeneis is a surgical procedure aimed at producing a large amount of bone; it is widely used in orthopedic and craniofacial surgery. These models have successfully simulated the effect of the distraction rate in the outcome of the process. They have predicted premature union for low distraction rates [97, 112], nonunion at high distraction rates [112], and successful union at moderate rates [97, 112]. The latency period has also been studied in these types of models [116].

4.9 Concluding Remarks

The intrinsic complexity of the events involved in bone tissue after implantation has motivated the development of numerical models with a predictive purpose. Most of this work has focused on the performance of phenomenological models that are capable of estimating the long-term bone behavior and performing comparative analysis to determine the influence of different conditions. However, these models have to be improved in order to better understand the mechanobiological mechanisms that regulate bone adaptation and healing. Nowadays, the power of computational methods is sufficient to check fundamental hypotheses, to understand the implications of biophysical assumptions, and to explore new assumptions that may be later used to design new experiments. In this sense, the development of novel mechanistic models to simulate bone remodeling and healing processes opens a new perspective in research and innovation in the design of implants. This kind of models allows not only performing of comparative analyses, as phenomenological ones, but also better understanding the complex interaction among tissues, cells, and molecular substances through different biophysical events. This modeling requires the use of complex and new numerical methodologies, such as, multiscale, multiphysic, parametric, and stochastic analyses that are in continuous development.

References

1. Rho, J., Kuhn-Spearing, L., and Zioupos, P. (1998) *Med. Eng. Phys.*, **20**(2), 92–102.
2. Carter, D.R. and Beaupré, G.S. (2001) *Skeletal Function and Form: Mechanobiology of Skeletal Development, Aging and Regeneration*, Cambridge University Press.
3. Miller, S.C. and Jee, W.S.S. (1992) *Bone*, vol. 4, pp. 1–19. CRC Press, Boca Raton.
4. Weinbaum, S., Cowin, S.C., and Zeng, Y. (1994) *J. Biomech.*, **27**, 339–360.
5. Rubin, J., Rubin, C., and Jacobs, C.R. (2006) *Gene*, **367**, 1–16.
6. Orr, A.W., Helmke, B.P., Blackman, B.R., and Schwartz, M.A. (2006) *Dev. Cell*, **10**, 11–20.
7. Semeins, C.M., Bronckers, A.L.J.J., Maltha, J.C., Von Den Hoff, J.W., Everts, V., Klein-Nulend, J., Tan, S.D., and Kuijpers-Jagtman, A.M. (2006) *J. Dent. Res.*, **85**, 905–909.
8. Vezeridis, P.S., Semeins, C.M., Chen, Q., and Klein-Nulend, J. (2006) *Biochem. Biophys. Res. Commun.*, **348**, 1082–1088.
9. Chen, A.A., Curtiss, S.B., Martin, R.B., Hazelwood, S.J., Hedgecock, N.L., and Hadi, T. (2007) *Bone*, **40**, 627–637.
10. Frost, H.M. (1963) *Bone Remodeling Dynamics*, Charles C. Thomas, Springfield.
11. Harada, S. and Rodan, G.A. (2003) *Nature*, **423**, 349–355.
12. Vico, L., Moro, L., Hartmann, D., Roth, M., Alexandre, C., Collet, P., and Uebelhart, D. (1997) *Bone*, **20**, 547–551.

13. Oganov, V., Darling, J., Miles, A.W., Owen, G.W., Goodship, A.E., and Cunningham, J.L. (1998) *Acta Austronaut.*, **43**, 65–75.
14. Qin, Y.X., Rubin, C.T., and McLeod, K.J. (1998) *J. Orthop. Res.*, **16**, 482–489.
15. Li, X.J., Jee, W.S., Chow, S.Y., and Woodbury, D.M. (1990) *Anat. Rec.*, **227**, 12–24.
16. Prendergast, P.J. and Taylor, D. (1994) *J. Biomech.*, **27**, 1067–1076.
17. Ramtani, S. and Zidi, M. (2001) *J. Biomech.*, **34**(4), 471–479.
18. Hernandez, C.J., Beaupré, G.S., and Carter, D.R. (2000) *J. Rehabil. Res. Dev.*, **37**(2), 235–244.
19. Hazelwood, S.J., Martin, R.B., Rashid, M.M., and Rodrigo, J.J. (2001) *J. Biomech.*, **34**(3), 299–308.
20. García-Aznar, J.M., Rueberg, T., and Doblaré, M. (2005) *Biomech. Model. Mechanobiol.*, **4**, 147–167.
21. Verborgt, O., Gibson, G.J., and Schaffler, M.B. (2000) *J. Bone Miner. Res.*, **15**, 60–67.
22. Fernandes, P., Rodrigues, H., and Jacobs, C.R. (1999) *Comput. Methods Biomech. Biomed. Eng.*, **2**(2), 125–138.
23. Doblaré, M. and García, J.M. (2002) *J. Biomech.*, **35**(1), 1–17.
24. Doblaré, M. and García, J.M. (2001) *J. Biomech.*, **34**(9), 1157–1170.
25. Lanyon, L.E. and Rubin, C.T. (1982) *J. Exp. Biol.*, **101**, 187–211.
26. Martin, R.B. and Burr, D.B. (1989) *Structure Function and Adaptation of Compact Bone*, Raver Press, New York.
27. Sarmiento, A., Latta, L.L., and Tarr, R.R. (1984) in *Instructional Course Lectures*, A.A.O.S., pp. 83–106.
28. Praemer, A., Furner, S., and Rice, D.P. (1999) *Musculoskeletal Conditions in the United States*, Amer. Acad. of Orthopaedic Surgeons, Data Harbor, Chicago
29. Hadjiargyrou, M., McLeod, K., Ryaby, J.P., and Rubin, C. (1998) *Clin. Orthop.*, **355S**, 216–229.
30. Goodship, A.E., Watkins, P.E., Rigby, H.S., and Kenwright, J. (1993) *J. Biomech.*, **26**(9), 1027–1035.
31. Ruedi, T.P., Buckley, R.E., and Moran, C.G. (2007) *AO Principles of Fracture Management*, AO Publishing - Thieme Verlag.
32. Tsiridis, E., Upadhyay., N., and Giannoudis, P. (2007) *Injury*, **38S**, 11–25.
33. Augat, P., Margevicious, K., Simon, J., Wolf, S., Suger, G., and Claes, L. (1998) *J. Orthop. Res.*, **16**, 475–481.
34. McKibbin, B. (1978) *J. Bone Joint Surg.*, **60B**(2), 150–162.
35. Claes, L., Heigele, C.A., Neidlinger-Wilke, C., Kaspar, D., Seidl, W., Margevicius, K., and Augat, P. (1998) *Clin. Orthop.*, **355S**, 132–147.
36. Goodship, A.E. (1992) *Ann. Rheum. Dis.*, **51**, 4–6.
37. Claes, L., Augat, P., Suger, G., and Wilke, H.J. (1997) *J. Orthop. Res.*, **15**(4), 577–584.
38. Park, S.H., O'Connor, K., McKellop, H., and Sarmiento, A. (1998) *J. Bone Joint Surg.*, **80A**(6), 868–878.
39. Augat, P., Burger, J., Schorlemmer, S., Henke, T., Peraus, M., and Claes, L. (2003) *J. Orthop. Res.*, **21**, 1011–1017.
40. Bishop, N.E., Tami, I., Schneider, E., and Ito, K. (2002) *Acta Bioeng. Biomech.*, **4S1**, 754–755.
41. Hente, R., Füchtmeier, B., Schlegel, U., Ernstbergert, A., and Perren, S.M. (2004) *J. Orthop. Res.*, **22**, 709–715.
42. Wolff, J. (1892) *The Law of Bone Remodelling*. Das Gesetz derTransformation der Knochen, Kirschwald. Translated by Maquet, P. and Furlong, R.
43. Roux, W. (1881) *Der Kamp der Teile im Organismus*, Engelmann, Leipzig.
44. Adachi, T., Tsubota, K.I., Tomita, Y., and Hollister, S.J. (2001) *J. Biomech. Eng.*, **123**, 403–409.
45. Beaupré, G.S., Orr, T.E., and Carter, D.R. (1990) *J. Orthop. Res.*, **8**, 551–651.
46. Cowin, S.C. and Hegedus, D.H. (1976) *J. Elast.*, **6**, 313–326.
47. Carter, D.R., Fyhrie, D.P., and Whalen, R.T. (1987) *J. Biomech.*, **20**(8), 1095–1109.
48. Huiskes, R., Weinans, H., Grootenboer, H.J., Dalstra, M., Fudala, B., and Sloof, T.J. (1987) *J. Biomech.*, **20**(11/12), 1135–1150.
49. Cowin, S.C., Sadegh, A.M., and Luo, G.M. (1992) *J. Biomech. Eng.*, **114**, 129–136.

50. Fyhrie, D.P. and Schaffler, M.B. (1995) *J. Biomech.*, **28**, 135–146.
51. Huiskes, R., Ruimerman, R., van Lenthe, G.H., and Janssen, J.D. (2000) *Nature*, **405**(6787), 704–706.
52. Hernandez, C.J. (2001) *Simulation of bone remodeling during the development and treatment of osteoporosis*, PhD Thesis, Stanford University, Palo Alto California
53. Taylor, D. and Lee, T.C. (2003) *J. Anat.*, **203**, 203–211.
54. Taylor, D., Casolari, E., and Bignardi, C. (2004) *J. Orthop. Res.*, **22**(3), 487–494.
55. Martínez-Reina, J., García-Aznar, J.M., Domínguez, J., and Doblaré, M. (2009) *Biomech. Model. Mechanobiol*, **8**(2), 111–127.
56. Hernandez, C.J., Beaupré, G.S., Marcus, R., and Carter, D.R. (2001) *Bone*, **29**(1), 511–516.
57. Martínez-Reina, J., García-Aznar, J.M., Domínguez, J., and Doblaré, M. (2008) *J. Theor. Biol*, **254**(3), 704–712.
58. Huiskes, R. and Chao, E.Y.S. (1983) *J. Biomech.*, **16**, 385–409.
59. Brekelmans, W.A.M., Poort, H.W., and Sloof, T.J.H. (1972) *Acta Orthop. Scand.*, **43**, 301–317.
60. Prendergast, P.J. (1997) *Clin. Biomech.*, **12**(6), 343–366.
61. Huiskes, R., Weinans, H., and van Rietbergen, B. (1992) *Clin. Orthop.*, **274**, 124–134.
62. Goto, H., Krygier, J.J., Miller, J.E., Brooks, C.E., Bobyn, J.D., and Glassman, A.H. (1990) *Clin. Orthop.*, **261**, 196.
63. Turner, T.M., Sumner, D.R., Urban, R.M., Igloria, R., and Galante, J.O. (1997) *J. Bone Joint Surg. Am.*, **79**, 1381–1390.
64. Engh, C.A. and Bobyn, J.D. (1988) *Clin. Orthop. Relat. Res.*, **231**, 7–28.
65. Weinans, H., Huiskes, R., van Rietbergen, B., Sumner, D.R., Turner, T.M., and Galante, J.O. (1993) *J. Orthop. Res.*, **11**, 500–513.
66. Weinans, H., Huiskes, R., and Grootenboer, H.J. (1994) *J. Biomech. Eng.*, **116**, 393–400.
67. van Rietbergen, B., Huiskes, R., Weinans, H., Sumner, D.R., Turner, T.M., and Galante, J.O. (1993) *J. Biomech.*, **26**, 369–382.
68. Kuiper, J.H. and Huiskes, R. (1997) *J. Biomech. Eng.*, **119**, 228–231.
69. Kerner, J., Huiskes, R., van Lenthe, G.H., Weinans, H., van Rietbergen, B., Engh, C.A., and Amis, A.A. (1999) *J. Biomech.*, **32**(1), 695–703.
70. Prendergast, P.J. and Taylor, D. (1992) *J. Biomed. Eng.*, **14**, 499–503.
71. Orr, T.E., Beaupré, G.S., Carter, D.R., and Schurman, D.J. (1990) *J. Arthroplasty*, **5**, 191–200.
72. Levenston, M.E., Beaupré, G.S., Schurman, D.J., and Carter, D.R. (1993) *J. Arthroplasty*, **8**, 595–605.
73. Weinans, H., Sumner, D.R., Igloria, R., and Natarajan, R.N. (2000) *J. Biomech.*, **33**, 809–817.
74. Cegoñino, J., García, J.M., Seral, B., Doblaré, M., Palanca, D., and Seral, F. (2004) *Comput. Methods Biomech. Biomed. Eng.*, **7**(5), 245–256.
75. Pérez, M.A., Desmarais-Trepanier, C., Nuño, N., Vendittoli, P.-A., Lavigne, M., García-Aznar, J.M., and Doblaré, M. (2008) *J. Biomech.*, under review.
76. Harris, W.H., Maloney, W.J., and Schmalzried, T. (2002) *Clin. Orthop.*, **405**, 70–78.
77. Jurvelin, J.S., Miettinen, H.J., Suomalainen, O.T., Alhava, E.M., Venesmaa, P.K., and Kroger, H.P. (2003) *Acta Orthop. Scand.*, **74**(1), 31–36.
78. van der Meulen, M.C.H. and Huiskes, R. (2002) *J. Biomech.*, **35**, 401–414.
79. Pauwels, F. (1941) Grundriβ einer Biomechanik der Frakturheilung, in *34th Kongress der Deutschen Orthopädischen Gessellschaft*, Ferdinand Enke Verlag, Stuttgart, (Biomechanics of the Locomotor Apparatus) translated by P. Maquet and R. Furlong (eds.), 1980, Springer, Berlin, 375–407.
80. Perren, S.M. and Cordey, J. (1980) *Currect Concepts of Internal Fixation of Fractures*, pp. 63–77. Springer-Verlarg, Berlin.
81. Perren, S.M. (1979) *Clin. Orthop. Relat. Res.*, **138**, 175–196.
82. Beaupré, G.S., Giori, N.J., Blenman-Fyhrie, P.R., and Carter,

D.R. (1992) 4th Conference of the ISFR. Book of Abstracts, pp. 1–11.

83. Carter, D.R., Beaupré, G.S., Giori, N.J., and Helms, J.A. (1998) *Clin. Orthop.*, **355S**, 41–55.

84. Blenman, P.R., Carter, D.R., and Beaupré, G.S. (1989) *J. Orthop. Res.*, **7**, 398–407.

85. Loboa, E.G., Beaupré, G.S., and Carter, D.R. (2001) *J. Orthop. Res.*, **19**(6), 1067–1072.

86. Morgan, E.F., Longaker, M.T., and Carter, D.R. (2006) *Matrix Biol.*, **25**(2), 94–103.

87. Claes, L. and Heigele, C.A. (1996) Proceedings of the 5th Meeting of the International Society for Fracture Repair, volume 16, Ottawa, Canada,

88. Claes, L.E. and Heigele, C.A. (1999) *J. Biomech.*, **32**, 255–266.

89. Gardner, T.N. and Mishra, S. (2003) *Med. Eng. Phys.*, **25**, 455–464.

90. Simon, U., Augat, P., Forster, E., Breuer, H., and Claes, L. (2004) Sixth International Symposium on Computer Methods in Biomechanics & Biomedical Engineering.

91. Shefelbine, S.J., Augat, P., Claes, L., and Simon, U. (2005) *J. Biomech.*, **38**(12), 2440–2450.

92. Weinans, H. and Prendergast, P.J. (1996) *Bone*, **19**, 143–149.

93. Prendergast, P.J., Huiskes, R., and Søballe, K. (1997) *J. Biomech.*, **6**, 539–548.

94. Lacroix, D. and Prendergast, P.J. (1999) 23rd Annual Meeting of ASB.

95. Lacroix, D. and Prendergast, P.J. (2002) *J. Biomech.*, **35**, 1163–1171.

96. Isaksson, H., van Donkelaar, C.C., Huiskes, R., and Ito, K. (2006) *J. Orthop. Res.*, **24**(5), 898–907.

97. Isaksson, H., Comas, O., Mediavilla, J., Wilson, W., van Donkelaar, C.C., Huiskes, R., and Ito, K. (2007) *J. Biomech.*, **40**(9), 2002–2011.

98. Geris, L., Andreykiv, A., Osterwyck, H.V., Stolen, J.V., Keulen, F.F., Duyck, J., and Naert, I. (2004) *J. Biomech.*, **37**, 763–769.

99. Kelly, D.J. and Prendergast, P.J. (2005) *J. Biomech.*, **38**(7), 1413–1422.

100. Pérez, M.A. and Prendergast, P.J. (2007) *J. Biomech.*, **40**(1), 2244–2253.

101. Davy, D.T. and Connolly, J.F. (1982) *J. Biomech.*, **15**, 235–247.

102. Logvenkov, S.A. (1996) 10th Conference ESB, Leuven, p. 153.

103. Ament, Ch. and Hofer, E.P. (2000) *J. Biomech.*, **33**, 961–968.

104. Ament, Ch. and Hofer, E.P. (1996) Proceedings of the 5th Meeting of the International Society for Fracture Repair.

105. Isaksson, H., van Donkelaar, C.C., Huiskes, R., and Ito, K. (2008) *J. Theor. Biol.*, **252**(2), 230–246.

106. Bailón-Plaza, A. and van der Meulen, M.C.H. (2001) *J. Theor. Biol.*, **212**, 191–209.

107. Bailón-Plaza, A. and van der Meulen, M.C.H. (2003) *J. Biomech.*, **36**(8), 1069–1077.

108. Geris, L., Gerisch, A., Vander Sloten, J., Weiner, R., and Van Oosterwyck, H. (2008) *J. Theor. Biol.*, **251**(1), 137–158.

109. Gómez-Benito, M.J., García-Aznar, J.M., Kuiper, J.H., and Doblaré, M. (2005) *J. Theor. Biol.*, **235**(1), 105–119.

110. Gómez-Benito, M.J., García-Aznar, J.M., Kuiper, J.H., and Doblaré, M. (2006) *J. Biomech. Eng.*, **128**(3), 290–299.

111. García-Aznar, J.M., Kuiper, J.H., Gómez-Benito, M.J., Doblaré, M., and Richardson, J.B. (2007) *J. Biomech.*, **40**(7), 1467–1476.

112. Reina-Romo, E., Gómez-Benito, M.J., García-Aznar, J.M., Domínguez, J., and Doblaré, M. (2010) *Biomech. Model. Mechanobiol.*, **9**(1), 103–115.

113. Lacroix, D. and Prendergast, P.J. (2002) *Comput. Methods Biomech. Biomed. Eng.*, **5**(5), 369–376.

114. Duda, G.N., Eckert-Hübner, K., Sokiranski, R., Kreutner, A., Miller, R., and Claes, L. (1998) *J. Biomech.*, **31**(3), 201–210.

115. Yang, L., Nayagam, S., and Saleh, M. (2003) *Clin. Biomech.*, **18**(2), 166–172.

116. Boccaccio, A., Prendergast, P., Pappalettere, C., and Kelly, D. (2008) *Med. Biol. Eng. Comput.*, **46**(3), 283–298. doi: 10.1007/s11517-007-0247-1.

5
Biomechanical Testing of Orthopedic Implants; Aspects of Tribology and Simulation

Yoshitaka Nakanishi

5.1
Introduction

Orthopedic implants have been the subject of considerable research in recent years. They are generally made of biocompatible materials. They are considered to be artificial and external components that compensate for a movement disturbance under load. The most representative implants are total knee- or hip-joint prosthesis, where two weightbearing surfaces suffer from a variety of tribological problems, namely, friction, wear, and lubrication. In addition to wear damage on the bearing material, the wear debris are also of concern because they can induce tissue reaction. The prosthesis loosening produced from such a reaction is of major consideration in the long-term survival of contemporary joint prosthesis, *in vivo*.

Although joint simulators investigate the performance and wear rate of joint prostheses under *in vivo* kinematic conditions prior to clinical trials, such simulators are complex and expensive to construct [1–6]. Furthermore, the multifactorial nature of each joint involves many complex factors that influence the development of wear, *in vivo*. Simple configuration wear testing, such as the pin-on-plate wear test, has been used extensively for screening novel materials for use in joint prostheses prior to more complex joint simulation tests. Such a simple configuration of the test allows for better control of individual tribological variables, leading to a better fundamental understanding of how these factors independently influence wear. However, these simple configuration wear devices underestimate the kinematic and physiological conditions found in the joint.

In this chapter, the physiological and kinematic conditions of joint implants *in vitro* are evaluated in an attempt to simulate *in vivo* wear conditions.

5.2
Tribological Testing of Orthopedic Implants

Relative movement between two bearing surfaces may be classified into two main groups, sliding and rolling. However, when there is a difference in severity of

Biomechanics of Hard Tissues: Modeling, Testing, and Materials.
Edited by Andreas Öchsner and Waqar Ahmed

ISBN: 978-3-527-32431-6

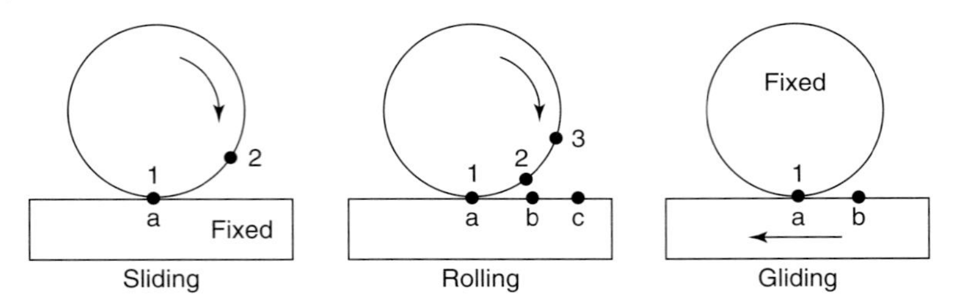

Figure 5.1 Kinematic conditions of contact for the various test configurations: sliding, rolling, and gliding. Adapted from McGloughlin and Kavanagh [7].

friction between two bearing surfaces, the sliding is sometimes classified into further two groups, sliding and gliding [7, 8]. The kinematic conditions are represented graphically as shown in Figure 5.1. Sliding is a condition where the contact point on the lower bearing surface remains stationary relative to the contact point on the upper bearing surface which moves. Rolling is a condition where the relative contact point velocities of both the lower and the upper bearing surfaces are equal. Finally gliding is a condition where the contact point on the upper surface is stationary and the contact point on the lower surface moves.

The primary function of a lubricant is to reduce friction or wear (or both) between moving bearing surfaces in contact with each other. There is a variety of lubricants including the synovial fluid in human joints. The following types of lubrication are considered in order of decreasing film thickness [9]:

- Hydrodynamic lubrication
- Elastohydrodynamic lubrication
- Mixed lubrication
- Boundary lubrication.

In hydrodynamic lubrication, the load is supported by the pressure developed due to the relative motion and geometry of the bearing surfaces. In hydrodynamic or fluid-film lubrication, there is no contact between the bearing surfaces. Friction arises purely from shearing of the viscous lubricant.

In contact situations involving high loads such as gears, ball bearings, and other high-contact-stress geometries, there are two additional requirements. The first is that the surface deforms elastically leading to a localized change in the geometry which favors lubrication. The second is that the lubricant becomes more viscous under the high pressure that exists in the contact zone. Here, the lubricant pressures existing in the contact zone approximate those of dry contact Herzian stress. This is the definition of elastohydrodynamic lubrication, sometimes abbreviated as EHL or EHD.

Although the most important function of a lubricant is to prevent contact, a better fundamental understanding about the transition from hydrodynamic and elastohydrodynamic lubrication to boundary lubrication is required. This is the region where lubrication goes from the desirable hydrodynamic condition of no contact to the less acceptable "boundary" condition, where increased contact usually leads to higher friction and wear. This condition is sometimes referred to as *mixed lubrication.*

Boundary lubrication is often described as a condition of lubrication in which the friction and wear between two surfaces in relative motion are determined by the surface properties of the solids and the chemical nature of the lubricant rather than its viscosity. According to another common definition, boundary lubrication occurs when the surface of the bearing solid is separated by a film whose thickness is on the molecular level.

This concept is illustrated by the Stribeck curve derived from bearing friction experiments. When the coefficients of friction for bearing, reported by Stribeck, were plotted against a dimensionless grouping, now known as the *Gumbel number*, of the form:

$$\text{viscosity} \times \text{speed/load} \tag{5.1}$$

it took the form shown in Figure 5.2. This relationship is widely used to indicate the mode of lubrication in various bearing forms, since it exhibits little variation for different configurations.

Boundary lubrication prevails in the left-hand side of the graph, where the coefficient of friction is essentially independent of both speed and load. Friction thus obeys the laws of dry friction, with coefficients of friction being ≈ 0.1.

In the mixed lubrication region, where the total friction arises from a combination of viscous shearing of the lubricant and direct asperity contact, the coefficient of friction falls as the speed of sliding is increased or the load is decreased. In this region, the coefficient of friction varies rapidly as the operating parameters

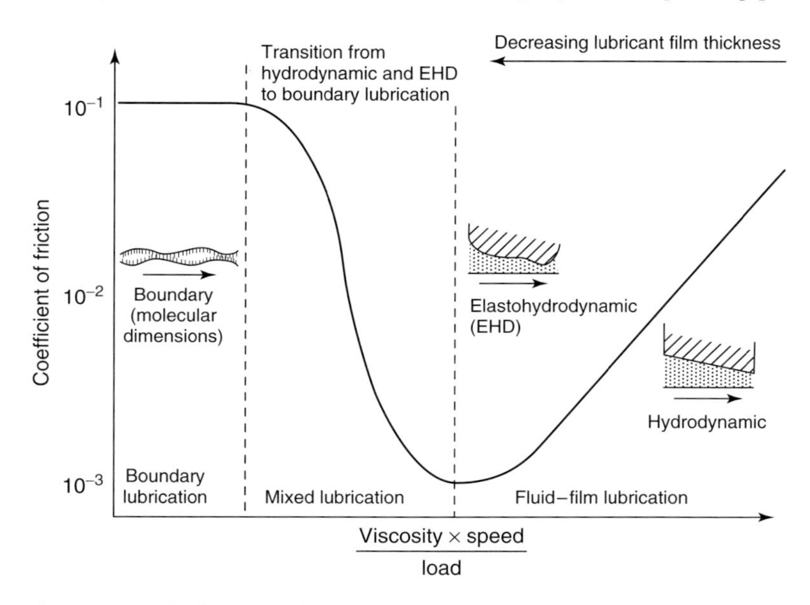

Figure 5.2 Stribeck curve and lubrication models.

change, from the boundary lubrication value of 0.1 to the best value achieved in hydrodynamic lubrication which approximates 0.001.

The region of full fluid-film lubrication is shown on the right-hand side of Figure 5.2. It is characterized by a rising coefficient of friction as the speed increases or the load decreases. If the coefficient of friction is recorded in a bearing as the speed and load are changed, the results are similar to a Stribeck curve (Figure 5.2).

Generation of bearing wear debris or surface damage is studied using the boundary and mixed lubrication regimes, where direct contact occurs between two bearing surfaces at asperity levels [10]. Wear mechanisms of an ultrahigh molecular weight polyethylene (UHMWPE), commonly used in bearing material of joint prostheses, have been well established qualitatively, by observation of the polyethylene surface and debris formation under high magnification. Three classical mechanisms of wear have been established for the sliding of a hard counterface on polyethylene. These are (i) adhesive wear; (ii) counterface or third-body abrasive wear (Figure 5.3); and (iii) fatigue wear derived from plowing friction (Figure 5.4). Particles such as bone and bone cement have been found embedded beneath and on the surface of the polyethylene components after clinical retrieval and these particles can cause third-body debris (abrasive grains). Some researchers have concluded that third-body debris is the main cause of increased polyethylene wear and premature failure *in vivo* and suggest that the use of bone cement in joint replacement can increase the probability of increased counterface roughness.

As can be seen from the examples of testing protocol of the American Society for Testing and Materials (ASTM) F732-82 [11], the pin-on-plate wear test is one of the most widespread and useful configuration tests. Despite the development and extensive use of simple pin-on-plate tests, there have been relatively few studies concerning the effect of the pin geometry on wear rate (Figure 5.5) [12].

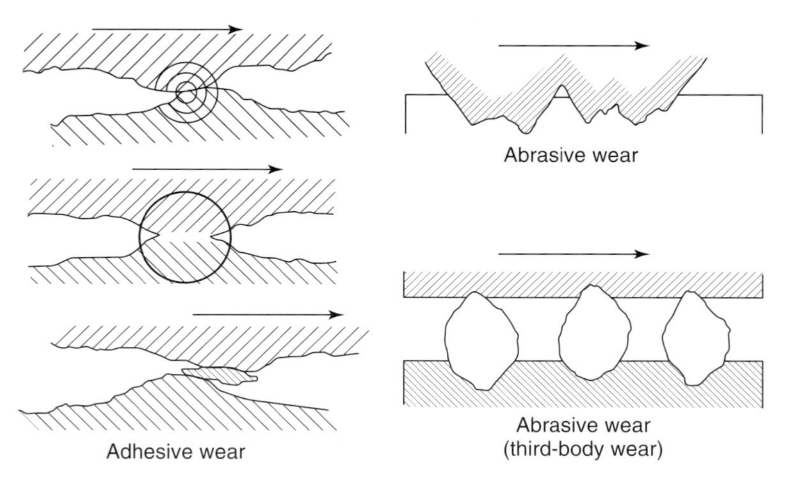

Figure 5.3 Wear mechanisms.

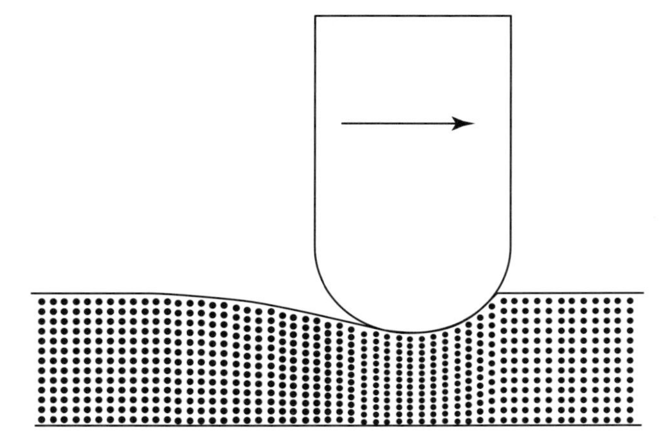

Figure 5.4 Plowing friction resulting in fatigue wear of bearing material.

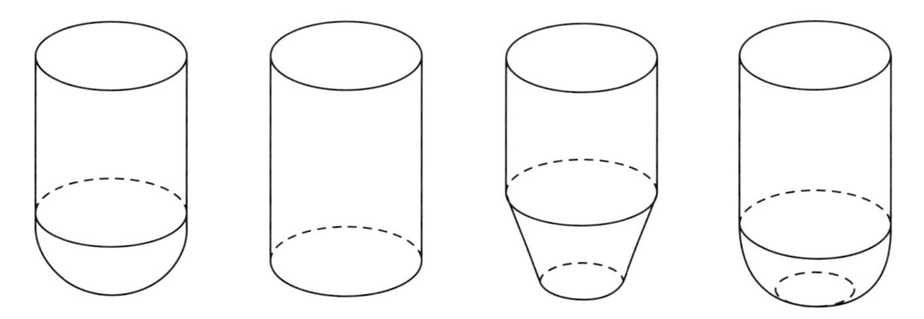

Figure 5.5 Various surface profiles of the pin used for pin-on-disc configuration tests. Adapted from Besong *et al.* [12].

There are clear differences in the contact stress between the pin and plate configuration, and misalignment between the pin and the plate may raise the stress within the tribological system. The plowing friction shown in Figure 5.4 is also an important consideration, because the increase in friction and wear on the lower bearing material is enhanced by plowing.

For the past 20 years, laboratory studies on polyethylene wear have been conducted using unidirectional or reciprocating linear wear-type testing machines consisting of a polymer pin on a metallic or ceramic flat. In recent years, multidirectional motions have been introduced to more realistically address *in vivo* kinematics (Figure 5.6) [13–15]. Unidirectional reciprocating motion causes the polyethylene surface to appear to become oriented and strain hardened. It appears that the multidirectional motion due to physiological gait patterns causes the surface layer to be constantly redrawn and reoriented at acute angles. This leads to shearing of polyethylene particles from the surface, producing wear.

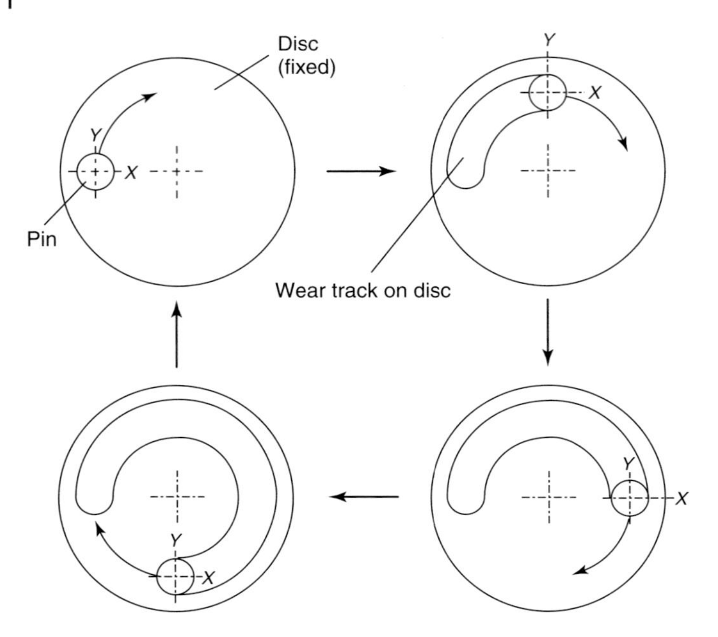

Figure 5.6 Example of multidirectional motion of pin on disc. Adapted from Nakanishi *et al.* [15].

It has been recognized for a number of years that water does not produce adequate boundary lubrication in joint simulations of *in vivo* wear. It has also been confirmed that natural synovial fluid used as a lubricant in joint simulations affects the tribological characteristic of bearing materials in joint prostheses [14].

The synovial fluid found in the joint cavity can be regarded as a dialysate of blood. The large molecules in blood cannot penetrate the synovial membrane and are excluded from the joint cavity. The polysaccharide hyaluronic acid (HA) is the main component of the joint fluid. It is produced by the synovial membrane and forms a high molecular mass complex [16, 17]. Both aging and arthritis can reduce both the concentration and the relative molecular weight of HA, and thus reduce the viscosity of the synovial fluid.

Bovine serum is the most commonly used lubricant in *in vitro* wear tests. Boundary lubricants that occur in bovine serum include proteins of various molecular weights and phospholipids [18, 19]. In some cases, a water-based liquid that contains the principal constituents of synovial fluid is used for wear tests in order to reduce considerable individual variability of bovine serum, which may affect the wear results (Table 5.1). The wear of the polyethylene in the pin-on-plate tests with water as a lubricant are highly variable, however, and depend on the formation of a polymer transfer film on the counterface. The wear factors in the bovine serum lubricated tests are more consistent and transfer onto the counterface does not occur. Bovine serum is, therefore, considered to be more representative of *in vivo* conditions [20, 21].

Table 5.1 Composition of simulated synovial fluid.

Solvent	Additive	Concentration ($g\ dl^{-1}$)
0.01 $mol\ l^{-1}$ phosphate-buffered saline	Albumin (human serum protein)	2.0
	γ-Globulin (human serum protein)	1.0
	Phospholipid L α-DPPC (dipalmitoylphosphatidylcholine)	0.2
	Cholesterol	0.1
	Sodium azide	0.3

Adapted from Nakanishi and Higaki [1].

The boundary lubricating properties of proteins and phospholipids are changed by the length of exposure time and contact pressure. Thus, experimental results should be analyzed with these factors in mind. Research on how the exposure time affects boundary film formation on bearing surfaces shows that the specific wear rate of a polyethylene pin sliding on a metallic plate is increased by a decrease in the exposure time, even if the contact pressure and the sliding speed are held constant (Figure 5.7) [22, 23]. Wear joint simulation test conducted using bovine serum lubrication showed that the dependence of friction on contact stress for the polyethylene socket was similar to that of semicrystalline polymers under dry

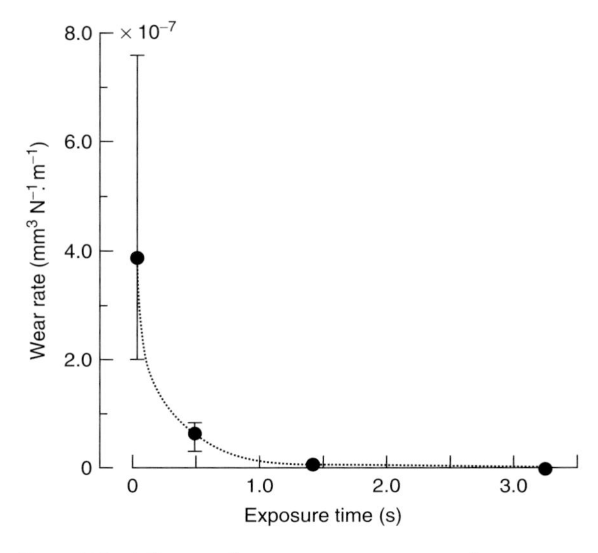

Figure 5.7 Influence of exposure time on wear of UHMWPE. Adapted from Nakanishi *et al.* [23].

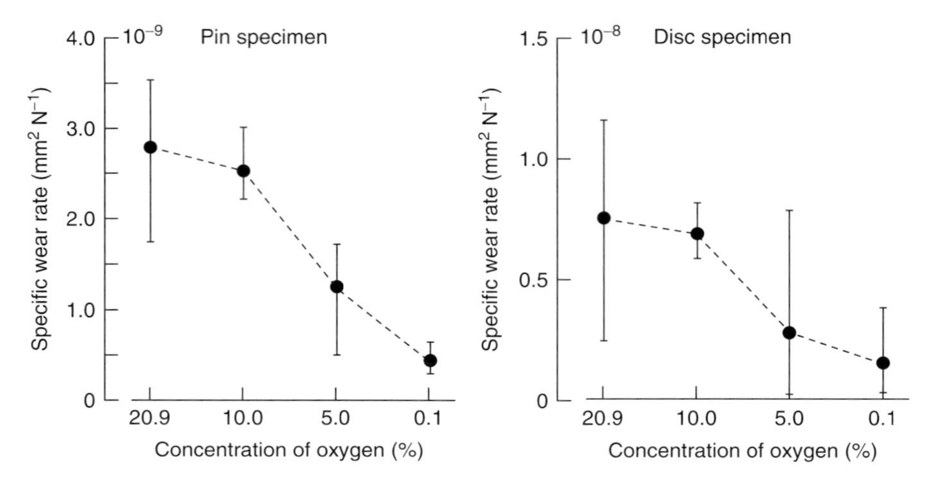

Figure 5.8 Influence of dissolved oxygen on wear of Co–Cr–Mo alloy. Typically, pin-on-disc tests were performed. Pin and disc, Co–Cr–Mo alloys; lubrication, simulated synovial fluid. Adapted from Nakanishi *et al.* [26].

sliding. This suggests that partial dry contact at asperity levels occurred in the metal–polyethylene ball-in-socket joint that used serum bovine lubrication [24].

As with other experiments, oxygen concentration in our experimental system is an important factor to be considered. A metallic surface is composed of three layers including a superficial layer of oxide film, the work-affected layer, and the base metal. The oxide film generally improves corrosion resistance and decreases wear, but sufficient oxygen is needed to reform the oxide film after damage from friction. However, the damaged oxide film in the human body may be difficult to restore, because the oxygen concentration *in vivo* is low (Figure 5.8) [25, 26].

The oxygen concentration in human joints will also affect the material and wear properties of polyethylene. Oxidative degradation of polyethylene is known to cause increased wear rate of the polymer in total joint replacement, leading to failure of these devices [27, 28].

5.3
Tribological Testing of Tissue from a Living Body

When performing tribological testing on living tissue, it is important to safeguard against tissue desiccation prior to testing. Living tissues generally contain a significant amount of fluid, and as a result, material and tribological properties will vary in accordance with the degree of desiccation. Furthermore, the dampness or softness of the surface and the individual variations in the geometry of the tissues also complicate testing as well as introduce challenges in the design of appropriate methods of fastening the tissue to the testing device. The geometric configuration

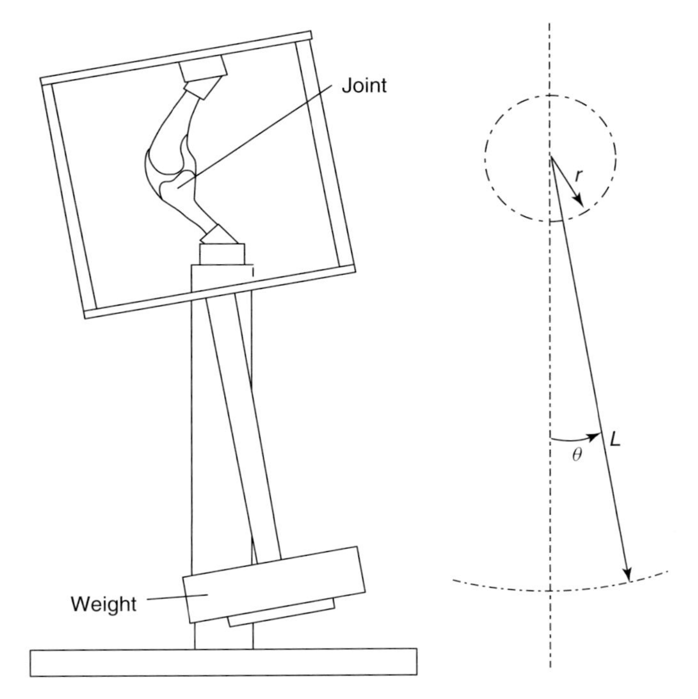

Figure 5.9 Schematic showing the pendulum friction tester. The coefficient of friction is calculated by the formula $f = L\Delta\theta/4r$, where r is the radius of the bearing in joint, L is the distance between the center of gravity and the center of rotation of the friction tester, and $\Delta\theta$ is the change in damping amplitude. Adapted from Kawano *et al.* [16].

of the articular surfaces of hip, knee, and shoulder joints is quite complicated. The bearing surface is not perfectly spheroidal and it is difficult to slide the two bearing surfaces past each other without causing large deformations, which require high energy for relative movement. To avoid this artifact, a sphere-on-flat sliding test is recommended. For this test, a spherical section composed of the convex portion of the joint will be used, if a glass plate with a hydrophilic nature similar to articular cartilage can be adapted to provide the counterface material [29]. If the tribological characteristics of the joint's original shape is required, pendulum testing is thought to be a more useful simulation (Figure 5.9).

Using the assumption that all of the potential energy loss is consumed by friction and that the friction due to air resistance is negligible, the coefficient of friction is calculated by a simple formula [16, 30]

$$f = L\Delta\theta/4r \tag{5.2}$$

where r is the radius of the bearing in joint, L is the distance between the center of gravity and the center of rotation of the friction tester, and $\Delta\theta$ is the change in damping amplitude.

The work (force × distance) becomes an important parameter for estimating the adhesion formation on the injured flexor tendon that cannot withstand the reciprocating motions of a pendulum test [17].

5.4 Theoretical Analysis for Tribological Issues

Theoretical analyses of the tribology of natural and artificial joints can be classified as either an estimation of theoretical film thickness between two bearing surfaces [31] or a calculation of contact pressure distributions in the bearing material [32].

The elastohydrodynamic lubrication analysis, where the governing equations consist of Reynolds and elasticity equations, is used for an estimate of the film thickness, and the lubricant representing the synovial fluid is assumed to be Newtonian, isoviscous, and incompressible. However, the main component of synovial fluid is the polysaccharide HA, which is non-Newtonian (Figure 5.10). Issues concerning the viscosity of the lubricant will inevitably arise in more precise analyses [33].

The contact pressure distribution in the bearing material is generally computed by means of a three-dimensional finite element method (FEM). The FEM is a revolutionary technology that is widely used in industry; however, it cannot be overemphasized that the precision of the analysis is dependent upon its boundary conditions. Neither the influence of friction at the contact point between two objects nor the influence of wear or wear debris on FEM analysis has yet to be established.

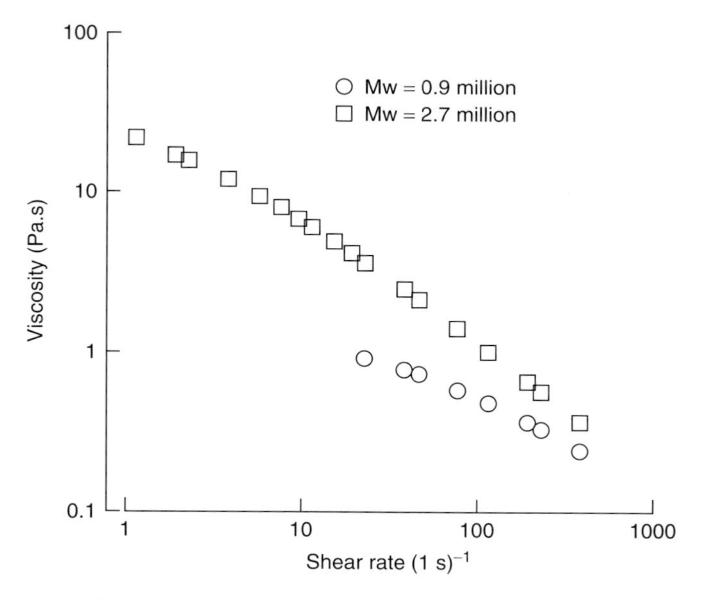

Figure 5.10 Non-Newtonian liquid of water solution of hyaluronic acid (HA).

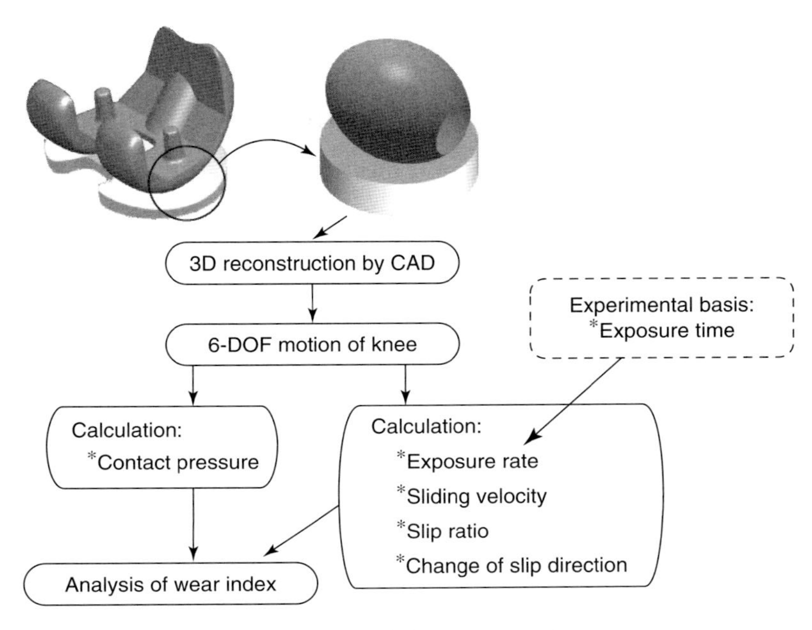

Figure 5.11 Flow chart of experimentally based numerical analysis for total knee-joint prostheses.

The tribological phenomena discussed in this chapter have not been entirely elucidated, since this would require a mathematical representation of the entire tribological system, which is not feasible. A more reasonable solution to these problems is through experiment-based numerical analysis (Figure 5.11) [34]. Numerical analyses using simplified configuration wear tests have identified individual factors that affect tribological phenomena and have established individual numeric models and correlated functions. The results from the integration of these individual numeric models or correlated functions, however, should be compared to the results from simulated body environments such as *in vitro* joint simulators. The new numerical models derived from these simulations could then be used to develop a more precise representation of the actual tribological phenomena. Considering the range of factors that influence the wear rate of a bearing material, further theoretical and experimental modeling is required in order to predict the *in vivo* behavior of bearing materials used in human joint replacements.

References

1. Nakanishi, Y. and Higaki, H. (2007) *Jpn. J. Tribol.*, **52**(4), 401–409.
2. ISO 14242-1 (2002) *Implants for Surgery – Wear of Total Hip-Joint Prostheses - Part1: Loading and Displacement Parameters for Wear-Testing Machines and Corresponding Environmental Conditions for Test.*
3. ISO 14242-2 (2000) *Implants for Surgery – Wear of Total Hip-Joint Prostheses - Part 2: Methods of Measurement.*

4. ISO 14243-3 (2004) *Implants for Surgery – Wear of Total Hip-Joint Prostheses - Part 3: Loading and Displacement Parameters for Wear-Testing Machines with Displacement Control and Corresponding Environmental Conditions for Test.*
5. ISO 14243-1 (2002) *Implants for Surgery – Wear of Total Knee-Joint Prostheses - Part 1: Loading and Displacement Parameters for Wear-Testing Machines with Load Control and Corresponding Environmental Conditions for Test.*
6. ISO 14243-2 (2000) *Implants for Surgery – Wear of Total Knee-Joint Prostheses - Part 2: Methods of Measurement.*
7. McGloughlin, T.M. and Kavanagh, A.G. (2000) *Proc. Inst. Mech. Eng. H*, **214**, 349–359.
8. Cornwall, G.B., Bryant, J.T., and Hansson, C.M. (2001) *Proc. Inst. Mech. Eng. H*, **215**, 95–106.
9. Furey, M.J. (2000) in *Biomechanics Principles and Applications*, Chapter 1 (eds D.J. Schneck and J.D. Bronzino), CRC Press LLC, Boca Raton, FL, pp. 73–97.
10. Bowden, F.P. (1950) *Nature*, **166**, 330–334.
11. ASTM F732-82 (1991) *Standard Practice for Reciprocating Pin-on-Flat Evaluation of Friction and Wear Properties of Polymeric Materials for Use in Total Joint Prosthese*, vol. 12, American Society for Testing Materials, Philadelphia, PA, pp. 262–269.
12. Besong, A.A., Jin, Z.M., and Fisher, J. (2001) *Proc. Inst. Mech. Eng. H*, **215**, 605–610.
13. Bragdon, C.R., O'Connor, D.O., Lowenstein, J.D., Jasty, M., and Syniuta, W.D. (1996) *Proc. Inst. Mech. Eng. H*, **210**, 157–165.
14. Joyce, T.J., Vandelli, C., Cartwright, T., and Unsworth, A. (2001) *Wear*, **250**, 206–211.
15. Nakanishi, Y., Higaki, H., Umeno, T., Miura, H., and Iwamoto, Y. (2007) *Jpn. J. Clin. Biomech.*, **28**, 205–212.
16. Kawano, T., Miura, H., Mawatari, T., Moro-Oka, T., Nakanishi, Y., Higaki, H., and Iwamoto, Y. (2003) *Arthritis Rheum.*, **48**(7), 1923–1929.
17. Moro-Oka, T., Miura, H., Mawatari, T., Kawano, T., Nakanishi, Y., Higaki, H., and Iwamoto, Y. (2005) *J. Orthop. Res.*, **18**(5), 835–840.
18. Wimmer, M.A., Sprecher, C., Hauert, R., Tager, G., and Fischer, A. (2003) *Wear*, **255**, 1007–1014.
19. Nakanishi, Y., Murakami, T., Higaki, H., and Miyagawa, H. (1999) *JSME Int. J. C*, **42**(3), 481–486.
20. Derbyshire, B., Fisher, J., Dowson, D., Hardaker, C., and Brummitt, K. (1994) *Med. Eng. Phys.*, **16**, 229–236.
21. Bell, J., Tipper, J.L., Ingham, E., Stone, M.H., and Fisher, J. (2001) *Proc. Inst. Mech. Eng. H*, **215**, 259–263.
22. Nakanishi, Y., Miyagawa, H., Higaki, H., Miura, H., and Iwamoto, Y. (2003) *J. Syn. Lubr.*, **19**(4), 273–282.
23. Nakanishi, Y., Higaki, H., Umeno, T., Miura, H., and Iwamoto, Y. (2007) *Jpn. J. Clin. Biomech.*, **28**, 213–218.
24. Wang, A., Essner, A., and Klein, R. (2001) *Proc. Inst. Mech. Eng. H*, **215**, 133–139.
25. Nakanishi, Y., Hoshino, R., Takashima, T., Higaki, H., Umeno, T., Miura, H., and Iwamoto, Y. (2006) Influence of dissolved oxygen in lubricating liquid on tribological characteristics of metallic bearing for artificial joints, 5th World Congress of Biomechanics (Ed. D. Liepsch), Medimond International Proceedings, 2006, pp. 151–156.
26. Nakanishi, Y., Higaki, H., Takashima, T., Umeno, T., Shimoto, K., Miura, H., and Iwamoto, Y. (2008) *Jpn. J. Clin. Biomech.* **1**, 29.
27. Sakoda, H., Fisher, J., Lu, S., and Buchanan, F. (2001) *J. Mater. Sci. – Mater. M.*, **12**, 1043–1047.
28. Costa, L., Luda, M.P., Trossarelli, L., Brach del Prever, E.M., Crova, M., and Gallinaro, P. (1998) *Biomaterials*, **19**, 659–668.
29. Higaki, H., Murakami, T., and Nakanishi, Y. (1997) *JSME Int. J. C*, **40**(4), 776–781.
30. Murakami, T., Higaki, H., Sawae, Y., Ohtsuki, N., Moriyama, S., and Nakanishi, Y. (1998) *Proc. Inst. Mech. Eng. H*, **212**, 23–35.

31. Liu, F., Jin, Z.M., Hirt, F., Rieker, C., Roberts, P., and Grigoris, P. (2005) *Proc. Inst. Mech. Eng. H*, **219**, 319–328.
32. Cosmi, F., Hoglievina, M., Fancellu, G., and Martinelli, B. (2006) *Proc. Inst. Mech. Eng. H*, **220**, 871–879.
33. Nakanishi, Y., Murakami, T., and Higaki, H. (2000) *Proc. Inst. Mech. Eng. H*, **214**, 181–192.
34. Miura, H., Higaki, H., Nakanishi, Y., Mawatari, T., Moro-Oka, T., Murakami, T., and Iwamoto, Y. (2002) *J. Arthoplasty*, **17**(6), 760–766.

6
Constitutive Modeling of the Mechanical Behavior of Trabecular Bone – Continuum Mechanical Approaches

Andreas Öchsner and Seyed Mohammad Hossein Hosseini

6.1
Introduction

Typical uniaxial stress–strain curves for different types of bones are shown in Figure 6.1 for the compressive regime. For all the curves, an initial linear elastic behavior can be observed. The common approach is to describe this elastic part on the basis of Hooke's law, cf. Sections 6.3.1–6.3.4. This elastic range is followed by a strong nonlinear behavior of almost constant stress (so-called stress plateau). At higher strains, some curves show a strong increase in the stress where densification begins.

These macroscopic stress–strain curves are similar to the behavior known from completely different types of materials such as cellular polymers and metals or even concrete. Although the deformation mechanism on the microlevel can be completely different, a common approach is to use the constitutive equations of metal plasticity to describe the nonelastic behavior, cf. [1–3]. As we see in this chapter, bones have some kind of cellular or porous structure, and the classical equations of full dense metals (e.g., von Mises or Tresca) must be extended by at least the hydrostatic pressure to account for the fact that such a material is even in the plastic range compressible. This general theory of a yield or failure surface based on stress invariants is introduced in Section 6.3.5. Many extensions of this theory are known for bones. However, the main focus is to thoroughly introduce the concept of a yield and limit surface so that possible extensions (e.g., by damage variables or the consideration of anisotropy) are easier to incorporate.

Many different approaches to derive new constitutive equations are known. Nowadays, the finite element method is the standard tool in computational engineering and advanced analysis tools (e.g., μCT) allow an extremely detailed imaging of bone structure. More and more powerful computer hardware (RAM and CPU) enables and supports this trend. However, there are approaches based on simplified model structures that reveal some advantages compared to these highly computerized approaches. Thus, some classical model structures are presented in the second part of the chapter. These simpler models are, in many cases, able to consider the major physical effect and may finally yield a mathematical equation to

Biomechanics of Hard Tissues: Modeling, Testing, and Materials.
Edited by Andreas Öchsner and Waqar Ahmed

ISBN: 978-3-527-32431-6

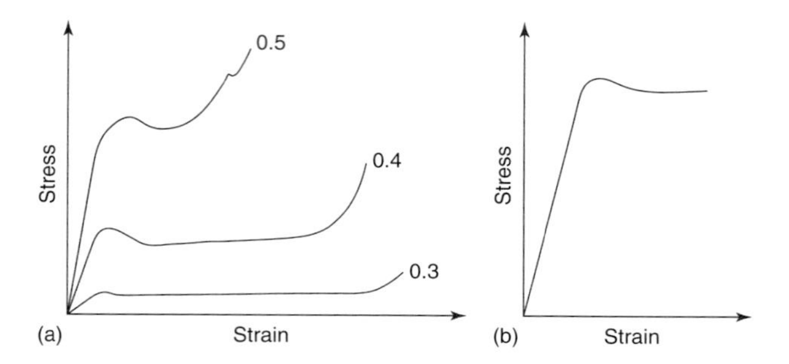

Figure 6.1 Uniaxial stress–strain curves in compression: (a) trabecular bone with different relative densities (after [4]); (b) compact bone (after [5]).

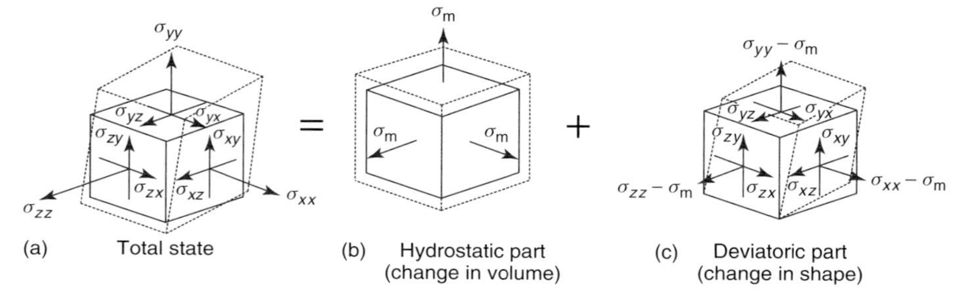

Figure 6.2 Decomposition of the stress tensor into its spherical and deviatoric parts. (a) totale state, (b) hydrostatic part (change in volume), and (c) deviatoric part (change in shape).

describe the material behavior. However, incorporated material parameters should be obtained from well-defined experimental investigations.

6.2 Summary of Elasticity Theory and Continuum Mechanics

6.2.1 Stress Tensor and Decomposition

It is of great importance in the framework of limit or failure surfaces of isotropic materials to decompose the stress tensor σ_{ij} into a pure volume changing (spherical or hydrostatic) tensor σ_{ij}^{o} and a pure shape changing (deviatoric) stress tensor s_{ij} (cf. Figure 6.2)[1]:

$$\sigma_{ij} = \sigma_{ij}^{\text{o}} + s_{ij} = \sigma_{\text{m}}\delta_{ij} + s_{ij} \tag{6.1}$$

1) It should be noted that in the case of anisotropic materials, a hydrostatic stress state may result in a shape change, [6].

In Eq. (6.1), $\sigma_{\mathrm{m}} = \frac{1}{3}(\sigma_{xx} + \sigma_{yy} + \sigma_{zz})$ denotes the mean normal stress[2] and δ_{ij} the Kronecker tensor (δ_{ij} is equal to 1 if $i = j$, and 0 if $i \neq j$). Furthermore, Einstein's summation convention was used [7].

Equation (6.1) can be written with components in the following way as

$$\underbrace{\begin{bmatrix} \sigma_{xx} & \sigma_{xy} & \sigma_{xz} \\ \sigma_{xy} & \sigma_{yy} & \sigma_{yz} \\ \sigma_{xz} & \sigma_{yz} & \sigma_{zz} \end{bmatrix}}_{\text{Stress tensor } \sigma_{ij}} = \underbrace{\begin{bmatrix} \sigma_{\mathrm{m}} & 0 & 0 \\ 0 & \sigma_{\mathrm{m}} & 0 \\ 0 & 0 & \sigma_{\mathrm{m}} \end{bmatrix}}_{\text{Hydrostatic tensor } \sigma_{ij}^{\mathrm{o}}} + \underbrace{\begin{bmatrix} s_{xx} & s_{xy} & s_{xz} \\ s_{xy} & s_{yy} & s_{yz} \\ s_{xz} & s_{yz} & s_{zz} \end{bmatrix}}_{\text{Deviatoric tensor } s_{ij}} \tag{6.2}$$

It can be seen that the elements outside the main diagonal, that is, the shear stresses, are the same for the stress and the deviatoric stress tensors:

$$s_{ij} = \sigma_{ij} \quad \text{for} \quad i \neq j \tag{6.3}$$

$$s_{ij} = \sigma_{ij} - \sigma_{\mathrm{m}} \quad \text{for} \quad i = j \tag{6.4}$$

The hydrostatic part of σ_{ij} has in the case of metallic materials (full dense materials), for temperatures approximately under $0.3 \cdot T_{\mathrm{kf}}$ (T_{kf}: melting temperature), nearly no influence on the occurrence of inelastic strains since dislocations slip only under the influence of shear stresses. On the other hand, the hydrostatic stress has a considerable influence on the failure or yielding behavior in the case of cellular materials, in soil or damage mechanics.

6.2.2 Invariants

To ensure the independence of a stress-based description of physical phenomena from the chosen coordinate system (objectiveness), it is meaningful to use a set of independent tensor invariants instead of the stress tensor components σ_{ij}. These invariants are independent of the orientation of the coordinate system and represent the physical content of the stress tensor. The so-called characteristic equation[3]

$$\det\left(\sigma_{ij} - \lambda\delta_{ij}\right) = 0 \tag{6.5}$$

or in components

$$\begin{vmatrix} \sigma_{xx} - \lambda & \sigma_{xy} & \sigma_{xz} \\ \sigma_{xy} & \sigma_{yy} - \lambda & \sigma_{yz} \\ \sigma_{xz} & \sigma_{yz} & \sigma_{zz} - \lambda \end{vmatrix} = 0 \tag{6.6}$$

leads to the cubic equation

$$\lambda^3 - I_1(\sigma_{ij})\lambda^2 + I_2(\sigma_{ij})\lambda - I_3(\sigma_{ij}) = 0 \tag{6.7}$$

2) Also called the hydrostatic stress. In the context of soil mechanics, also the pressure $p = -\sigma_{\mathrm{m}}$ is used. Be aware that some finite element codes use this definition.

3) det(...) denotes the determinant of a matrix or tensor.

for the definition of the three scalar principal stress invariants I_1, I_2 and I_3. The three roots λ_i of Eq. (6.7) are the principal stresses: $\sigma_{\mathrm{I}} = \max(\lambda_1, \lambda_2, \lambda_3)$, $\sigma_{\mathrm{III}} = \min(\lambda_1, \lambda_2, \lambda_3)$, and $\sigma_{\mathrm{II}} = (I_1 - \sigma_{\mathrm{I}} - \sigma_{\mathrm{III}})$. It may be noted here that the principal stress state is the state that has no shear components, that is, only the stress components are the principal stresses σ_i ($i = \mathrm{I, II, III}$) on the main diagonal of the stress tensor. Another interpretation of the principal invariants is given by

- I_1 = trace[4] of σ_{ij}:

$$I_1 = \sigma_{xx} + \sigma_{yy} + \sigma_{zz} \tag{6.8}$$

- I_2 = sum of the two-row main subdeterminants of σ_{ij}:

$$I_2 = \begin{vmatrix} \sigma_{xx} & \sigma_{xy} \\ \sigma_{xy} & \sigma_{yy} \end{vmatrix} + \begin{vmatrix} \sigma_{yy} & \sigma_{yz} \\ \sigma_{yz} & \sigma_{zz} \end{vmatrix} + \begin{vmatrix} \sigma_{xx} & \sigma_{xz} \\ \sigma_{xz} & \sigma_{zz} \end{vmatrix} \tag{6.9}$$

- I_3 = determinant of σ_{ij}:

$$I_3 = \begin{vmatrix} \sigma_{xx} & \sigma_{xy} & \sigma_{xz} \\ \sigma_{xy} & \sigma_{yy} & \sigma_{yz} \\ \sigma_{xz} & \sigma_{yz} & \sigma_{zz} \end{vmatrix} \tag{6.10}$$

In addition to these principal invariants, there is also often another set of invariants used. This set is included in the principal invariants I_i and called *basic invariants* J_i:

$$J_1 = I_1 \tag{6.11}$$

$$J_2 = \tfrac{1}{2} I_1^2 - I_2 \tag{6.12}$$

$$J_3 = \tfrac{1}{3} I_1^3 - I_1 I_2 + I_3 \tag{6.13}$$

The definition of both principal and basic invariants is summarized and compared in Table 6.1.

It can be seen from Table 6.1 that the spherical tensor σ_{ij}^{o} is completely characterized by its first invariant, because the second and third invariants are its powers. The stress deviator tensor s_{ij} is completely characterized by its second and third invariants. Therefore, the physical contents of the stress state σ_{ij} can be completely described either by the three invariants or, if we use the decomposition in its spherical and deviatoric parts, by the first invariant of the spherical tensor and the second and third invariants of the stress deviator tensor. It should be noted here that this statement is valid for both, that is, the principal and basic invariants. In the following, we will only use the basic invariants, and thus, the physical content of stress state will be described by the following set of invariants:

$$\sigma_{ij} \rightarrow J_1^{\mathrm{o}}, J_2', J_3' \tag{6.14}$$

Table 6.2 summarizes formulae based on the general stress components σ_{ij} for the practical calculation of the basic stress invariants.

Finally, it should be mentioned here that it is also possible to derive the same sets of invariants from the strain tensor ε_{ij}.

4) The trace of a tensor is the sum of the diagonal terms.

Table 6.1 Definition of principal and basic invariants.

	Stress tensor	
σ_{ij}	σ^{o}_{ij}	s_{ij}
	Principal invariants	
I_1, I_2, I_3	I^o_1, I^o_2, I^o_3	I'_1, I'_2, I'_3
$I_1 = \sigma_{ii}$	$I^o_1 = \sigma_{ii}$	$I'_1 = 0$
$I_2 = \frac{1}{2}(\sigma_{ii}\sigma_{jj} - \sigma_{ij}\sigma_{ji})$	$I^o_2 = \frac{1}{3}(\sigma_{ii})^2$	$I'_2 = -\frac{1}{2} s_{ij}s_{ji}$
$I_3 = \frac{1}{3}\left(\frac{1}{2}\sigma_{ii}\sigma_{jj}\sigma_{kk} + \sigma_{ij}\sigma_{jk}\sigma_{ki} - \frac{3}{2}\sigma_{ij}\sigma_{ji}\sigma_{kk}\right)$	$I^o_3 = \frac{1}{27}(\sigma_{ii})^3$	$I'_3 = \frac{1}{3} s_{ij}s_{jk}s_{ki}$
$\Rightarrow I_1, I_2, I_3$	$\Rightarrow I^o_1$	$\Rightarrow I'_2, I'_3$
	Basic invariants	
J_1, J_2, J_3	J^o_1, J^o_2, J^o_3	J'_1, J'_2, J'_3
$J_1 = \sigma_{ii}$	$J^o_1 = \sigma_{ii}$	$J'_1 = 0$
$J_2 = \frac{1}{2}\sigma_{ij}\sigma_{ji}$	$J^o_2 = \frac{1}{6}(\sigma_{ii})^2$	$J'_2 = \frac{1}{2} s_{ij}s_{ji}$
$J_3 = \frac{1}{3}\sigma_{ij}\sigma_{jk}\sigma_{ki}$	$J^o_3 = \frac{1}{9}(\sigma_{ii})^3$	$J'_3 = \frac{1}{3} s_{ij}s_{jk}s_{ki}$
$\Rightarrow J_1, J_2, J_3$	$\Rightarrow J^o_1$	$\Rightarrow J'_2, J'_3$

Table 6.2 Basic invariants in terms of σ_{ij}.

Invariants	**General stress values**
	Stress tensor
J_1	$\sigma_{xx} + \sigma_{yy} + \sigma_{zz}$
J_2	$\frac{1}{2}\left(\sigma^2_{xx} + \sigma^2_{yy} + \sigma^2_{zz}\right) + \sigma^2_{xy} + \sigma^2_{xz} + \sigma^2_{yz}$
J_3	$\frac{1}{3}(\sigma^3_{xx} + \sigma^3_{yy} + \sigma^3_{zz} + 3\sigma^2_{xy}\sigma_{xx} + 3\sigma^2_{xy}\sigma_{yy} + 3\sigma^2_{xz}\sigma_{xx} + 3\sigma^2_{xz}\sigma_{zz} + 3\sigma^2_{yz}\sigma_{yy} + 3\sigma^2_{yz}\sigma_{zz} + 6\sigma_{xy}\sigma_{xz}\sigma_{yz})$
	Spherical tensor
J^o_1	$\sigma_{xx} + \sigma_{yy} + \sigma_{zz}$
J^o_2	$\frac{1}{6}(\sigma_{xx} + \sigma_{yy} + \sigma_{zz})^2$
J^o_3	$\frac{1}{9}(\sigma_{xx} + \sigma_{yy} + \sigma_{zz})^3$
	Stress deviator tensor
J'_1	0
J'_2	$\frac{1}{6}\left[(\sigma_{xx} - \sigma_{yy})^2 + (\sigma_{yy} - \sigma_{zz})^2 + (\sigma_{zz} - \sigma_{xx})^2\right] + \sigma^2_{xy} + \sigma^2_{yz} + \sigma^2_{zx}$
J'_3	$s_{xx}s_{yy}s_{zz} + 2\sigma_{xy}\sigma_{yz}\sigma_{zx} - s_{xx}\sigma^2_{yz} - s_{yy}\sigma^2_{zx} - s_{zz}\sigma^2_{xy}$
with	$s_{xx} = \frac{1}{3}(2\sigma_{xx} - \sigma_{yy} - \sigma_{zz})$
	$s_{yy} = \frac{1}{3}(-\sigma_{xx} + 2\sigma_{yy} - \sigma_{zz})$
	$s_{zz} = \frac{1}{3}(-\sigma_{xx} - \sigma_{yy} + 2\sigma_{zz})$

6.3
Constitutive Equations

Figure 6.3 shows the emplacement of the constitutive equation in the framework of solid mechanical modeling of a material particle. The constitutive equation combines the kinematic and equilibrium equations by indicating the dependence of strains on stresses.

According to the level of description, the constitutive equations are distinguished as microscopic, stochastic, and macroscopic [8]. Microscopic constitutive equations describe the material behavior based on variables (e.g., dislocation density) taken from materials physics. A transfer of the microscopic material behavior on the macroscopic level has not been successful till now. Stochastic constitutive equations also describe the material behavior on the micro level. For that purpose, probabilistic processes are applied. The transfer on the macroscopic level is due to its possible mathematical structure. Macroscopic constitutive equations are also called *phenomenological models*. Their field of application is the continuum mechanical modeling of components and structures. The material is idealized as a homogeneous continuum, which constitution is described by a few variables such as strain, temperature, and occasionally by additional so-called internal variables. In the case of multiphase or inhomogeneous materials, such as cellular or fiber-reinforced materials, this assumption does not hold in the first place. By averaging the different material properties, an approximate homogeneity can be assumed, cf. Figure 6.4.

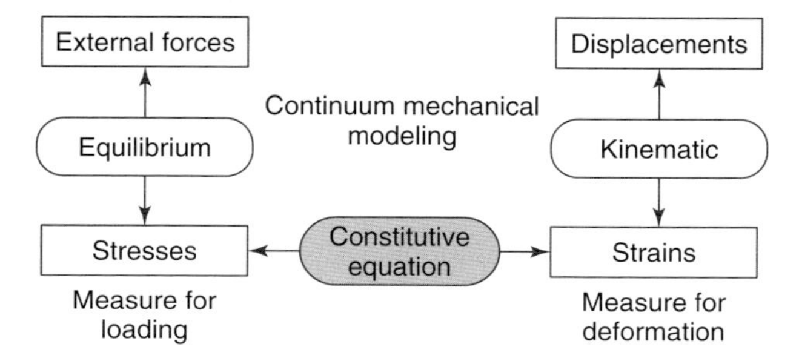

Figure 6.3 Solid mechanical modeling of engineering materials.

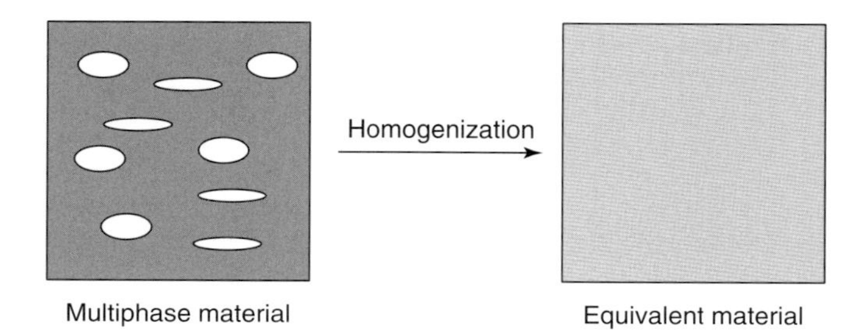

Figure 6.4 Averaging the materials properties to obtain an equivalent material.

6.3.1
Linear Elastic Behavior: Generalized Hooke's Law for Isotropic Materials

In general Cauchy elasticity, the stress tensor σ_{ij} is a nonlinear tensor function of the strain tensor ε_{kl}:

$$\sigma_{ij} = f_{ij}(\varepsilon_{kl}) \tag{6.15}$$

whereas in linear elasticity of homogeneous materials the Cauchy stress tensor is, in extension to Hooke's law ($\sigma = E\varepsilon$) of the year 1678, proportional to the Cauchy strain tensor through the linear transformation:

$$\sigma_{ij} = C_{ijkl}\varepsilon_{kl} \tag{6.16}$$

It may be noted here that Eq. (6.16) is the simplest generalization of the linear dependence of stress and strain observed by Hooke in a simple tension test, and consequently Eq. (6.16) is referred to as the *generalized Hooke's law*. The fourth-order tensor C_{ijkl} has 81 independent components in total and is called the *elasticity tensor*. In the case of linear elasticity, the components of the elasticity tensor are constant values. Owing to symmetry of σ_{ij} and ε_{kl}, the pairs of indices ij and kl in C_{ijkl} can be permutated. Hence, the number of independent constants of the elasticity tensor is reduced to 36. A further reduction of constants is obtained if we assume the existence of a strain energy potential $w(\varepsilon_{ij})$ from which the stresses are derived by differentiation as

$$\sigma_{ij} = \frac{\partial w}{\partial \varepsilon_{kl}} \tag{6.17}$$

After partial differentiation of Eq. (6.16) with respect to ε_{ij} and consideration of Eq. (6.17), one obtains the components of the elasticity tensor as

$$C_{ijkl} = \frac{\partial \sigma_{ij}}{\partial \varepsilon_{kl}} = \frac{\partial^2 w}{\partial \varepsilon_{ij}\partial \varepsilon_{kl}} \tag{6.18}$$

There is an additional symmetry due to the commutation of the order of the partial derivative with respect to ε_{ij} and ε_{kl}, and the number of independent constants is reduced to 21. A material for which such a constitutive relation is assumed is also called *Green elastic material* [9]. By considering the material symmetries, (e.g., crystal systems), a further reduction in the number of constants can be achieved. For an isotropic material, the mechanical properties are the same for all directions and it can be shown [10] that the components of the elasticity tensor are determined by two independent constants, the so-called Lamé's constants λ and μ and the stress–strain relationship is given for this case by

$$\sigma_{ij} = 2\mu\varepsilon_{ij} + \lambda\varepsilon_{kk}\delta_{ij} \tag{6.19}$$

The characterization of metallic materials is in general based on the engineering constants, Young's modulus E and Poisson's ratio ν. On the other hand, the description of isotropic materials in soil or rock mechanics is based on the shear modulus G and the bulk modulus K rather than E and ν. The conversion of the elastic constants can be done with the aid of the relationships given in Table 6.3.

Table 6.3 Conversion of elastic constants.

	λ, μ	E, ν	μ, ν	E, μ	K, ν	G, ν	K, G
λ	λ	$\frac{\nu E}{(1+\nu)(1-2\nu)}$	$\frac{2\mu\nu}{1-2\nu}$	$\frac{\mu(E-2\mu)}{3\mu-E}$	$\frac{3K\nu}{1+\nu}$	$\frac{2G\nu}{1-2\nu}$	$K-\frac{2G}{3}$
μ	μ	$\frac{E}{2(1+\nu)}$	μ	μ	$\frac{3K(1-2\nu)}{2(1+\nu)}$	μ	μ
K	$\lambda+\frac{2}{3}\mu$	$\frac{E}{3(1-2\nu)}$	$\frac{2\mu(1+\nu)}{3(1-2\nu)}$	$\frac{\mu E}{3(3\mu-E)}$	K	$\frac{2G(1+\nu)}{3(1-2\nu)}$	K
E	$\frac{\mu(3\lambda+2\mu)}{\lambda+\mu}$	E	$2\mu(1+\nu)$	E	$3K(1-2\nu)$	$2G(1+\nu)$	$\frac{9KG}{3K+G}$
ν	$\frac{\lambda}{2(\lambda+\mu)}$	ν	ν	$\frac{E}{2\mu}-1$	ν	ν	$\frac{3K-2G}{2(3K+G)}$
G	μ	$\frac{E}{2(1+\nu)}$	μ	G	$\frac{3K(1-2\nu)}{2(1+\nu)}$	G	G

Table 6.4 Reference values of typical engineering materials (all values are given near room temperature).

Material	E (GPa)	ν in –	G (GPa)	Density (g cm^{-3})
Aluminum	70	0.35	26	2.70
Iron	211	0.29	82	7.874
Magnesium	45	0.290	17	1.738
Titanium	116	0.32	44	4.506
Human bone (trabecular)	0.186 ⋮ 0.724			0.25 ⋮ 0.50

Mechanical and physical reference values[5] of some typical engineering materials are summarized in Table 6.4. The measured and predicted values for trabecular bone comprise a wide range and are influenced by many factors (e.g., age, wet or dry etc.). In order to provide some typical values for trabecular bone and relate them to these classical engineering materials, the power law by Hodgskinson *et al.* [11] was applied, which describes Young's modulus over a large density range.

In the following description, the tensor notation is abandoned and the engineering notation (so-called Voigt notation) is consistently introduced and observed, that is, the components of the second-order strain tensor ε_{ij} and the stress tensor σ_{ij} are arranged into column vectors

$$\varepsilon_{ij} = \begin{bmatrix} \varepsilon_x & \varepsilon_{xy} & \varepsilon_{xz} \\ \varepsilon_{yx} & \varepsilon_y & \varepsilon_{yz} \\ \varepsilon_{zx} & \varepsilon_{zy} & \varepsilon_z \end{bmatrix} \rightarrow \boldsymbol{\varepsilon} = \{\varepsilon_x \, \varepsilon_y \, \varepsilon_z \, 2\varepsilon_{xy} \, 2\varepsilon_{yz} \, 2\varepsilon_{xz}\}^{\mathrm{T}} \tag{6.20}$$

5) Conversion from GPa to MPa: value times 1000; conversion from g cm^{-3} to kg m^{-3}: value times 1000; 1 Pa = 1 N m^{-2}.

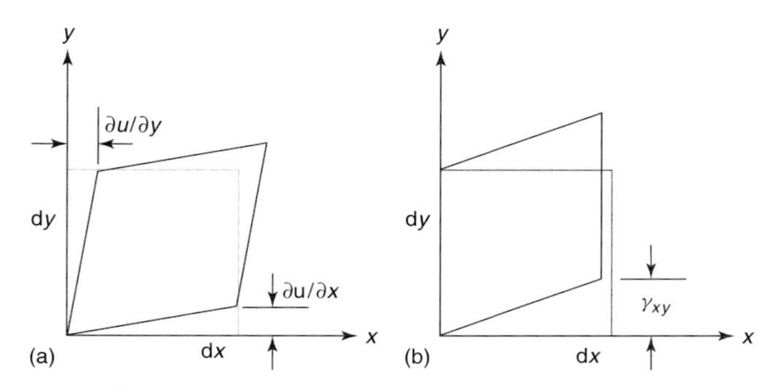

Figure 6.5 Definition of shear strain: (a) tensor definition as the average $\varepsilon_{xy} = \varepsilon_{yx} = (\partial v/\partial x + \partial u/\partial y)/2$; (b) engineering definition as the total $\gamma_{xy} = \partial v/\partial x + \partial u/\partial y$.

and

$$\sigma_{ij} = \begin{bmatrix} \sigma_x & \sigma_{xy} & \sigma_{xz} \\ \sigma_{yx} & \sigma_y & \sigma_{yz} \\ \sigma_{zx} & \sigma_{zy} & \sigma_z \end{bmatrix} \rightarrow \boldsymbol{\sigma} = \{\sigma_x \, \sigma_y \, \sigma_z \, \sigma_{xy} \, \sigma_{yz} \, \sigma_{xz}\}^{\mathrm{T}} \tag{6.21}$$

and the fourth-order elasticity tensor C_{ijkl} is represented by a square (6×6)-matrix **C**. This formalism is closer to actual computer implementations than the tensorial notation [12]. Furthermore, we use the unit vector $\mathbf{1} = \{1\,1\,1\,0\,0\,0\}^{\mathrm{T}}$ and the diagonal matrix $\mathbf{L} = \lceil 1\,1\,1\,2\,2\,2 \rfloor$, which is not a unit matrix because the strain vector is expressed using the engineering definition of shear strain (e. g., $\gamma_{xy} = 2\varepsilon_{xy}$) rather than the tensor definition, cf. Figure 6.5.

Now, the generalized Hooke's law for isotropic materials based on Lamé's constants can be written in matrix form as

$$\boldsymbol{\sigma} = (2\mu \mathbf{L}^{-1} + \lambda \mathbf{1} \otimes \mathbf{1})\boldsymbol{\varepsilon} \tag{6.22}$$

or in explicit form with all components as

$$\begin{Bmatrix} \sigma_x \\ \sigma_y \\ \sigma_z \\ \sigma_{xy} \\ \sigma_{yz} \\ \sigma_{xz} \end{Bmatrix} = \begin{bmatrix} \lambda + 2\mu & \lambda & \lambda & 0 & 0 & 0 \\ \lambda & \lambda + 2\mu & \lambda & 0 & 0 & 0 \\ \lambda & \lambda & \lambda + 2\mu & 0 & 0 & 0 \\ 0 & 0 & 0 & \mu & 0 & 0 \\ 0 & 0 & 0 & 0 & \mu & 0 \\ 0 & 0 & 0 & 0 & 0 & \mu \end{bmatrix} \cdot \begin{Bmatrix} \varepsilon_x \\ \varepsilon_y \\ \varepsilon_z \\ 2\,\varepsilon_{xy} \\ 2\,\varepsilon_{yz} \\ 2\,\varepsilon_{xz} \end{Bmatrix} \tag{6.23}$$

The dyadic product $\otimes$ used in Eq. (6.22) is in general defined for two vectors **a** and **b** by

$$\mathbf{a} \otimes \mathbf{b} = \begin{Bmatrix} a_1 \\ a_2 \\ \vdots \\ a_n \end{Bmatrix} \otimes \begin{Bmatrix} b_1 \\ b_2 \\ \vdots \\ b_n \end{Bmatrix} = \begin{bmatrix} a_1 b_1 & a_1 b_2 & \dots & a_1 b_n \\ a_2 b_1 & a_2 b_2 & \dots & a_2 b_n \\ \vdots & \vdots & \ddots & \vdots \\ a_n b_1 & \vdots & \dots & a_n b_n \end{bmatrix} \tag{6.24}$$

where the same result can be obtained by the matrix multiplication $\mathbf{a} \cdot \mathbf{b}^{\mathrm{T}}$.

Alternatively, Eq. (6.22) can be written in the form of Eq. (6.19) as

$$\begin{aligned}\boldsymbol{\sigma} &= 2\mu\mathbf{L}^{-1}\boldsymbol{\varepsilon} + \lambda\mathbf{1}\otimes\mathbf{1}\,(\varepsilon_{\mathrm{m}}\mathbf{1}+\mathbf{e}) = 2\mu\mathbf{L}^{-1}\boldsymbol{\varepsilon} + \lambda\varepsilon_{\mathrm{m}}\underbrace{\mathbf{1}\mathbf{1}^{\mathrm{T}}\mathbf{1}}_{3\mathbf{1}} + \lambda\underbrace{\mathbf{1}\mathbf{1}^{\mathrm{T}}\mathbf{e}}_{\mathbf{0}}\\ &= 2\mu\mathbf{L}^{-1}\boldsymbol{\varepsilon} + \lambda(3\varepsilon_{\mathrm{m}})\mathbf{1}\end{aligned} \tag{6.25}$$

Substituting for λ and μ in terms of E and ν (cf. Table 6.3) gives the alternative formulation of Eq. (6.23) based on the engineering constants as

$$\begin{Bmatrix}\sigma_x\\ \sigma_y\\ \sigma_z\\ \sigma_{xy}\\ \sigma_{yz}\\ \sigma_{xz}\end{Bmatrix} = \frac{E}{(1+\nu)(1-2\nu)}\begin{bmatrix}1-\nu & \nu & \nu & 0 & 0 & 0\\ \nu & 1-\nu & \nu & 0 & 0 & 0\\ \nu & \nu & 1-\nu & 0 & 0 & 0\\ 0 & 0 & 0 & \frac{1-2\nu}{2} & 0 & 0\\ 0 & 0 & 0 & 0 & \frac{1-2\nu}{2} & 0\\ 0 & 0 & 0 & 0 & 0 & \frac{1-2\nu}{2}\end{bmatrix}\cdot\begin{Bmatrix}\varepsilon_x\\ \varepsilon_y\\ \varepsilon_z\\ 2\,\varepsilon_{xy}\\ 2\,\varepsilon_{yz}\\ 2\,\varepsilon_{xz}\end{Bmatrix} \tag{6.26}$$

In a simple tension test, the only nonzero stress component σ_x causes axial strain ε_x and transverse strains $\varepsilon_y = \varepsilon_z$. Thus, Eq. (6.26) yields

$$\varepsilon_x = \frac{\sigma_x}{E} \quad \text{and} \quad \varepsilon_y = -\nu\varepsilon_x = -\frac{\nu\sigma_x}{E} \tag{6.27}$$

By using Eq. (6.27), one can calculate the elastic constants, Young's modulus E and Poisson's ratio ν, from a uniaxial tension or compression test. Introducing the shear modulus G and the bulk modulus K according to Table 6.3 yields a further formulation of Hooke's generalized law as

$$\begin{Bmatrix}\sigma_x\\ \sigma_y\\ \sigma_z\\ \sigma_{xy}\\ \sigma_{yz}\\ \sigma_{xz}\end{Bmatrix} = \begin{bmatrix}K+\frac{4}{3}G & K-\frac{2}{3}G & K-\frac{2}{3}G & 0 & 0 & 0\\ K-\frac{2}{3}G & K+\frac{4}{3}G & K-\frac{2}{3}G & 0 & 0 & 0\\ K-\frac{2}{3}G & K-\frac{2}{3}G & K+\frac{4}{3}G & 0 & 0 & 0\\ 0 & 0 & 0 & G & 0 & 0\\ 0 & 0 & 0 & 0 & G & 0\\ 0 & 0 & 0 & 0 & 0 & G\end{bmatrix}\cdot\begin{Bmatrix}\varepsilon_x\\ \varepsilon_y\\ \varepsilon_z\\ 2\,\varepsilon_{xy}\\ 2\,\varepsilon_{yz}\\ 2\,\varepsilon_{xz}\end{Bmatrix} \tag{6.28}$$

The quantity $K + \frac{4}{3}G$ in Eq. (6.28) is also known as the *constraint modulus* [13]. Decomposing the stress and strain vectors into spherical and deviatoric components decouples the volumetric response from the distortional response, and Hooke's law can be expressed in terms of the volumetric and deviatoric strains in the following form:[6)]

$$\boldsymbol{\sigma} = \boldsymbol{\sigma}^{\mathrm{o}} + \mathbf{s} = \sigma_{\mathrm{m}}\mathbf{1} + \mathbf{s} = 3K\varepsilon_{\mathrm{m}}\cdot\mathbf{1} + 2G\mathbf{L}^{-1}\mathbf{e} \tag{6.29}$$

From the point of view of continuum mechanics, a state under pure shear stress (result: shear modulus G) and a state under pure hydrostatic stress (result: bulk modulus K) should aim to determine the elastic constants, since the constants are independent of each other in this case. However, the experimental determination of the elastic constants is mostly based on a simple realizable tension or

6) This form is also called the canonical form [14].

Table 6.5 Formulations of generalized isotropic Hooke's law based on different elastic constants (elastic stiffness form).

Hooke's law
$\boldsymbol{\sigma} = \mathbf{C}\boldsymbol{\varepsilon} \quad \vert \quad \sigma_{ij} = C_{ijkl}\,\varepsilon_{kl}$
Formulation based on Lamé's constants
$\boldsymbol{\sigma} = (\lambda\mathbf{1}\otimes\mathbf{1} + 2\mu\mathbf{L}^{-1})\boldsymbol{\varepsilon} \quad \vert \quad \sigma_{ij} = \lambda\varepsilon_{kk}\delta_{ij} + 2\mu\varepsilon_{ij}$
Formulation based on Young's modulus and Poisson's ratio
$\boldsymbol{\sigma} = \frac{E}{1+\nu}\cdot\left(\frac{\nu}{1-2\nu}\mathbf{1}\otimes\mathbf{1} + \mathbf{L}^{-1}\right)\boldsymbol{\varepsilon} \quad \vert \quad \sigma_{ij} = \frac{E}{1+\nu}\cdot\left(\varepsilon_{ij} + \frac{\nu}{1-2\nu}\,\delta_{ij}\varepsilon_{kk}\right)$
Formulation based on bulk modulus and shear modulus
$\boldsymbol{\sigma} = \left(K\mathbf{1}\otimes\mathbf{1} + 2G(\mathbf{L}^{-1} - \frac{1}{3}\mathbf{1}\otimes\mathbf{1})\right)\boldsymbol{\varepsilon} \quad \vert \quad \sigma_{ij} = 2G\cdot\left(\varepsilon_{ij} + \frac{3K-2G}{6G}\,\delta_{ij}\varepsilon_{kk}\right)$
Decomposition in volumetric and deviatoric parts
$\boldsymbol{\sigma} = \underbrace{3K\varepsilon_{\mathrm{m}}\mathbf{1}}_{\text{Volumetric response}} + \underbrace{2G\mathbf{L}^{-1}\mathbf{e}}_{\text{Deviatoric response}} \quad \vert \quad \sigma_{ij} = \underbrace{K\varepsilon_{kk}\delta_{ij}}_{\text{Volumetric response}} + \underbrace{2Ge_{ij}}_{\text{Deviatoric response}}$

compression test, from which Young's modulus E and Poisson's ratio ν can be obtained.

A summary of the different formulations of Hooke's law in matrix form is given in Table 6.5 and compared with the corresponding tensor form given in common literature, for example, [2, 10, 15, 16].

The decomposition of the stress vector into spherical and deviatoric components can also be performed based on projection tensors. The deviatoric stress vector $\mathbf{s}$ is obtained by subtracting the spherical state of stress from the actual state of stress. Thus, we can write

$$\mathbf{s} = \boldsymbol{\sigma} - \sigma_{\mathrm{m}}\mathbf{1} = \boldsymbol{\sigma} - \frac{1}{3}\,\mathbf{1}\mathbf{1}^{\mathrm{T}}\boldsymbol{\sigma} = \left(\mathbf{I} - \frac{1}{3}\,\mathbf{1}\mathbf{1}^{\mathrm{T}}\right)\boldsymbol{\sigma} = \left(\mathbf{I} - \frac{1}{3}\,\mathbf{1}\otimes\mathbf{1}\right)\boldsymbol{\sigma} = \mathbf{I}'\boldsymbol{\sigma} \quad (6.30)$$

and define the deviatoric projection tensor

$$\mathbf{I}' = \mathbf{I} - \frac{1}{3}\,\mathbf{1}\otimes\mathbf{1} \quad (6.31)$$

to transform the actual stress state in its deviatoric part. Respecting $\mathbf{1}^{\mathrm{T}}\mathbf{1} = 3$, we can indicate the following properties of the deviatoric projection tensor:

- $$\mathbf{I}'\mathbf{1} = \mathbf{0} \tag{6.32}$$

Proof:

$$\mathbf{I}'\mathbf{1} = (\mathbf{I} - \tfrac{1}{3}\,\mathbf{1}\mathbf{1}^{\mathrm{T}})\mathbf{1} = \mathbf{I}\mathbf{1} - \tfrac{1}{3}\mathbf{1}\,\underbrace{\mathbf{1}^{\mathrm{T}}\mathbf{1}}_{3} = \mathbf{1} - \mathbf{1} = \mathbf{0}$$

- $$\mathbf{I}'\mathbf{I}' = \mathbf{I}' \ \text{ or in general } \ \left(\mathbf{I}'\right)^{n} = \mathbf{I}' \ \text{ for any } \ n \in \mathbb{N} \tag{6.33}$$

Proof:

$$\begin{aligned}
\mathbf{I}'\mathbf{I}' &= (\mathbf{I} - \frac{1}{3}\mathbf{1}\otimes\mathbf{1})(\mathbf{I} - \frac{1}{3}\mathbf{1}\otimes\mathbf{1}) \\
&= \underbrace{\mathbf{I}\mathbf{I}}_{\mathbf{I}} - \tfrac{1}{3}\underbrace{\mathbf{I}\mathbf{1}\mathbf{1}^{\mathrm{T}}}_{\mathbf{1}\mathbf{1}^{\mathrm{T}}} - \tfrac{1}{3}\underbrace{\mathbf{1}\mathbf{1}^{\mathrm{T}}\mathbf{I}}_{\mathbf{1}\mathbf{1}^{\mathrm{T}}} + \tfrac{1}{9}\mathbf{1}\,\underbrace{\mathbf{1}^{\mathrm{T}}\mathbf{1}}_{3}\,\mathbf{1}^{\mathrm{T}} \\
&= \mathbf{I} - \tfrac{2}{3}\mathbf{1}\mathbf{1}^{\mathrm{T}} + \tfrac{1}{3}\mathbf{1}\mathbf{1}^{\mathrm{T}} \\
&= \mathbf{I} - \tfrac{1}{3}\mathbf{1}\mathbf{1}^{\mathrm{T}}
\end{aligned}$$

This relationship expresses the fact that the deviatoric part of the stress cannot be changed by any other deviatoric projection and that the spherical part is equal to zero. A corresponding projection tensor can be derived for the spherical part of the actual state of stress:

$$\mathbf{I}^{\mathrm{o}} = \mathbf{I} - \mathbf{I}' = \frac{1}{3}\mathbf{1}\otimes\mathbf{1} = \frac{1}{3}\mathbf{1}\mathbf{1}^{\mathrm{T}} \tag{6.34}$$

Thus, we can rewrite Eq. (6.29) based on the transformation tensors as

$$\boldsymbol{\sigma} = \boldsymbol{\sigma}^{\mathrm{o}} + \mathbf{s} = \mathbf{I}^{\mathrm{o}}\boldsymbol{\sigma} + \mathbf{I}'\boldsymbol{\sigma} \tag{6.35}$$

In the following, the strain energy per unit volume w will be derived in its canonical form of two decoupled contributions, that is, its spherical and deviatoric parts:

$$w = \frac{1}{2}\boldsymbol{\sigma}\boldsymbol{\varepsilon} = \frac{1}{2}\left(\sigma_{\mathrm{m}}\mathbf{1}^{\mathrm{T}}\varepsilon_{\mathrm{m}}\mathbf{1} + \mathbf{s}^{\mathrm{T}}\boldsymbol{\varepsilon}\right) \tag{6.36}$$

By using Eq. (6.29), we can substitute σ_{m} and $\mathbf{s}$ and express the strain energy as

$$w = \frac{1}{2}\left(3K\varepsilon_{\mathrm{m}}\mathbf{1}^{\mathrm{T}}\varepsilon_{\mathrm{m}}\mathbf{1} + 2G\mathbf{L}^{-1}\mathbf{e}^{\mathrm{T}}\mathbf{e}\right) = \frac{1}{2}\left(9K\varepsilon_{\mathrm{m}}^{2} + 2G\mathbf{e}^{\mathrm{T}}\mathbf{L}^{-1}\mathbf{e}\right) \tag{6.37}$$

In Eq. (6.37), the positive strain energy argument delimits the range of possible values of Poisson's ratio to $-1 \leq \nu \leq 0.5$ (cf. Table 6.3).

In the last part of this section, we will provide relationships between the strain vector and the stress vector. Equation (6.16) can be written in the matrix form as

$$\boldsymbol{\varepsilon} = \mathbf{D}\boldsymbol{\sigma} \tag{6.38}$$

where the elastic compliance matrix $\mathbf{D}$ is given by the inverse of the elasticity matrix $\mathbf{C}$:

$$\mathbf{D} = \mathbf{C}^{-1} \tag{6.39}$$

Table 6.6 Formulations of generalized isotropic Hooke's law based on different elastic constants (elastic compliance form).

Hooke's law	
$\boldsymbol{\varepsilon} = \mathbf{C}^{-1}\,\boldsymbol{\sigma}$	$\varepsilon_{ij} = C^{-1}_{ijkl}\,\sigma_{kl}$
Formulation based on Lamé's constants	
$\boldsymbol{\varepsilon} = \frac{1}{2\mu}\cdot\left(\mathbf{L} - \frac{\lambda}{2\mu+3\lambda}\,\mathbf{1}\otimes\mathbf{1}\right)\boldsymbol{\sigma}$	$\varepsilon_{ij} = \frac{1}{2\mu}\cdot\left(\sigma_{ij} - \frac{\lambda}{2\mu+3\lambda}\,\sigma_{kk}\delta_{ij}\right)$
Formulation based on Young's modulus and Poisson's ratio	
$\boldsymbol{\varepsilon} = \frac{1+\nu}{E}\cdot\left(\mathbf{L} - \frac{\nu}{1+\nu}\,\mathbf{1}\otimes\mathbf{1}\right)\boldsymbol{\sigma}$	$\varepsilon_{ij} = \frac{1+\nu}{E}\cdot\left(\sigma_{ij} - \frac{\nu}{1+\nu}\,\sigma_{kk}\delta_{ij}\right)$
Formulation based on bulk modulus and shear modulus	
$\boldsymbol{\varepsilon} = \frac{1}{2G}\cdot\left(\mathbf{L} - \frac{3K-2G}{9K}\,\mathbf{1}\otimes\mathbf{1}\right)\boldsymbol{\sigma}$	$\varepsilon_{ij} = \frac{1}{2G}\cdot\left(\sigma_{ij} - \frac{3K-2G}{9K}\,\sigma_{kk}\delta_{ij}\right)$
Decomposition in volumetric and deviatoric parts	
$\boldsymbol{\varepsilon} = \frac{1}{3K}\,\sigma_{\mathrm{m}}\mathbf{1} + \frac{1}{2G}\,\mathbf{Ls}$	$\varepsilon_{ij} = \frac{1}{3K}\,\sigma_{\mathrm{m}}\delta_{ij} + \frac{1}{2G}\,s_{ij}$

This inversion can be done using Sherman–Morrison formula [17, 18] given in general form for a matrix **a** and two vectors **u** and **v** (α, β scalars) as

$$\left(\alpha\mathbf{A} + \beta(\mathbf{u}\otimes\mathbf{v})\right)^{-1} = \frac{1}{\alpha}\left(\mathbf{A}^{-1} - \frac{\beta\left(\mathbf{A}^{-1}\mathbf{u}\right)\otimes\left(\mathbf{v}^{\mathrm{T}}\cdot\mathbf{A}^{-1}\right)^{\mathrm{T}}}{\alpha + \beta\mathbf{v}^{\mathrm{T}}\cdot\mathbf{A}^{-1}\mathbf{u}}\right) \tag{6.40}$$

as long as $\frac{\beta}{\alpha}\,\mathbf{v}^{\mathrm{T}}\cdot\mathbf{A}^{-1}\mathbf{u} \neq -1$

A summary of the different formulations of Hooke's law in compliance form is given in Table 6.6 and compared with the corresponding tensor form given in common literature, for example, [2, 10, 15, 16].

6.3.2 Linear Elastic Behavior: Generalized Hooke's Law for Orthotropic Materials

Although most materials can be treated as approximately isotropic, strictly speaking, all materials are anisotropic to some extent and the material properties are not the same in every direction. In the following, we consider the important case of an orthotropic material, that is, a material with three principal, mutually orthogonal axes. These axes are also called the *material principal axes*. Let us assume in the following that the principal axes are coincident with the global coordinate system

(x, y, z). We will start with the compliance form of Hooke's law:

$$\begin{Bmatrix} \varepsilon_x \\ \varepsilon_y \\ \varepsilon_z \\ 2\varepsilon_{xy} \\ 2\varepsilon_{yz} \\ 2\varepsilon_{xz} \end{Bmatrix} = \begin{bmatrix} \frac{1}{E_x} & -\frac{\nu_{yx}}{E_y} & -\frac{\nu_{zx}}{E_z} & 0 & 0 & 0 \\ -\frac{\nu_{xy}}{E_x} & \frac{1}{E_y} & -\frac{\nu_{zy}}{E_z} & 0 & 0 & 0 \\ -\frac{\nu_{xz}}{E_x} & -\frac{\nu_{yz}}{E_y} & \frac{1}{E_z} & 0 & 0 & 0 \\ 0 & 0 & 0 & \frac{1}{G_{xy}} & 0 & 0 \\ 0 & 0 & 0 & 0 & \frac{1}{G_{yz}} & 0 \\ 0 & 0 & 0 & 0 & 0 & \frac{1}{G_{xz}} \end{bmatrix} \cdot \begin{Bmatrix} \sigma_x \\ \sigma_y \\ \sigma_z \\ \sigma_{xy} \\ \sigma_{yz} \\ \sigma_{xz} \end{Bmatrix} \tag{6.41}$$

Here, E_x, E_y, and E_z are Young's moduli for the principal axes, and ν_{ij} are Poisson's ratios for these axes. The Poisson's ratio ν_{xy}, for example, represents the ratio between strains ε_x and ε_y when the material is subjected to uniaxial stress in the y-direction, that is

$$\nu_{xy} = -\frac{\varepsilon_x}{\varepsilon_y}\bigg|_{\sigma_y=\sigma} \tag{6.42}$$

The coefficients G_{xy}, G_{yz}, and G_{xz} represent the shear moduli for the x-y, y-z, and x-z planes, respectively. The compliance matrix is symmetric, and therefore the following relations must be satisfied (observe the fact that Poisson's ratios are not symmetric):

$$\frac{\nu_{xy}}{E_x} = \frac{\nu_{yx}}{E_y}, \quad \frac{\nu_{yz}}{E_y} = \frac{\nu_{zy}}{E_z}, \quad \frac{\nu_{xz}}{E_x} = \frac{\nu_{zx}}{E_z} \tag{6.43}$$

From the previous equation, it follows that the compliance matrix comprises nine independent material constants and that the following equation holds[7]:

$$\nu_{xy}\nu_{yz}\nu_{zx} = \nu_{xz}\nu_{yx}\nu_{zy} \tag{6.44}$$

The stiffness form of Hooke's law is obtained by inverting the compliance matrix as

$$\begin{Bmatrix} \sigma_x \\ \sigma_y \\ \sigma_z \\ \sigma_{xy} \\ \sigma_{yz} \\ \sigma_{xz} \end{Bmatrix} = \begin{bmatrix} \frac{1-\nu_{yz}\nu_{zy}}{E_y E_z D} & \frac{\nu_{yx}+\nu_{zx}\nu_{yz}}{E_y E_z D} & \frac{\nu_{zx}+\nu_{yx}\nu_{zy}}{E_y E_z D} & 0 & 0 & 0 \\ \frac{\nu_{yx}+\nu_{zx}\nu_{yz}}{E_y E_z D} & \frac{1-\nu_{xz}\nu_{zx}}{E_x E_z D} & \frac{\nu_{zy}+\nu_{xy}\nu_{zx}}{E_x E_z D} & 0 & 0 & 0 \\ \frac{\nu_{zx}+\nu_{yx}\nu_{zy}}{E_y E_z D} & \frac{\nu_{zy}+\nu_{xy}\nu_{zx}}{E_x E_z D} & \frac{1-\nu_{xy}\nu_{yx}}{E_x E_y D} & 0 & 0 & 0 \\ 0 & 0 & 0 & G_{xy} & 0 & 0 \\ 0 & 0 & 0 & 0 & G_{yz} & 0 \\ 0 & 0 & 0 & 0 & 0 & G_{xz} \end{bmatrix} \cdot \begin{Bmatrix} \varepsilon_x \\ \varepsilon_y \\ \varepsilon_z \\ 2\varepsilon_{xy} \\ 2\varepsilon_{yz} \\ 2\varepsilon_{xz} \end{Bmatrix} \tag{6.45}$$

where

$$D = \frac{1 - \nu_{xy}\nu_{yx} - \nu_{xz}\nu_{zx} - \nu_{yz}\nu_{zy} - 2\nu_{yx}\nu_{xz}\nu_{zy}}{E_x E_y E_z} \tag{6.46}$$

is the determinant of the compliance matrix in Eq. (6.41) multiplied by $G_{xy}G_{yz}G_{xz}$.

7) Used to simplify the expression for the determinant of Eq. (6.41).

6.3.3 Linear Elastic Behavior: Generalized Hooke's Law for Orthotropic Materials with Cubic Structure

If the properties of an orthotropic material are identical in all three directions (x, y, and z), the material is said to have a cubic structure. A huge number of materials have cubic symmetry, for example, all the face centered cubic (FCC) and body centered cubic (BCC) metals. Thus, if we have

$$E_x = E_y = E_z = E \tag{6.47}$$

$$G_{xy} = G_{yz} = G_{xz} = G \tag{6.48}$$

$$\nu_{xy} = \nu_{yx} = \cdots = \nu \tag{6.49}$$

the compliance form of Hooke's law for an orthotropic material with cubic structure is reduced to

$$\begin{Bmatrix} \varepsilon_x \\ \varepsilon_y \\ \varepsilon_z \\ 2\varepsilon_{xy} \\ 2\varepsilon_{yz} \\ 2\varepsilon_{xz} \end{Bmatrix} = \frac{1}{E} \cdot \begin{bmatrix} 1 & -\nu & -\nu & 0 & 0 & 0 \\ -\nu & 1 & -\nu & 0 & 0 & 0 \\ -\nu & -\nu & 1 & 0 & 0 & 0 \\ 0 & 0 & 0 & \frac{E}{G} & 0 & 0 \\ 0 & 0 & 0 & 0 & \frac{E}{G} & 0 \\ 0 & 0 & 0 & 0 & 0 & \frac{E}{G} \end{bmatrix} \cdot \begin{Bmatrix} \sigma_x \\ \sigma_y \\ \sigma_z \\ \sigma_{xy} \\ \sigma_{yz} \\ \sigma_{xz} \end{Bmatrix} \tag{6.50}$$

where E, ν, and G are the three *independent* material constants. The stiffness form of Hooke's law is obtained by inverting the compliance matrix as

$$\begin{Bmatrix} \sigma_x \\ \sigma_y \\ \sigma_z \\ \sigma_{xy} \\ \sigma_{yz} \\ \sigma_{xz} \end{Bmatrix} = \begin{bmatrix} \frac{E(\nu-1)}{2\nu^2+\nu-1} & -\frac{E\nu}{2\nu^2+\nu-1} & -\frac{E\nu}{2\nu^2+\nu-1} & 0 & 0 & 0 \\ -\frac{E\nu}{2\nu^2+\nu-1} & \frac{E(\nu-1)}{2\nu^2+\nu-1} & -\frac{E\nu}{2\nu^2+\nu-1} & 0 & 0 & 0 \\ -\frac{E\nu}{2\nu^2+\nu-1} & -\frac{E\nu}{2\nu^2+\nu-1} & \frac{E(\nu-1)}{2\nu^2+\nu-1} & 0 & 0 & 0 \\ 0 & 0 & 0 & G & 0 & 0 \\ 0 & 0 & 0 & 0 & G & 0 \\ 0 & 0 & 0 & 0 & 0 & G \end{bmatrix} \cdot \begin{Bmatrix} \varepsilon_x \\ \varepsilon_y \\ \varepsilon_z \\ 2\varepsilon_{xy} \\ 2\varepsilon_{yz} \\ 2\varepsilon_{xz} \end{Bmatrix} \tag{6.51}$$

6.3.4 Linear Elastic Behavior: Generalized Hooke's Law for Transverse Isotropic Materials

Special classes of orthotropic materials are those that have the same properties in one plane (e.g., the x–y plane) and different properties in the direction normal to this plane (e.g., the z-axis). This implies that the solid can be rotated with respect to the loading direction about one axis without measurable effect on the solid's response. Such materials are called *transverse isotropic*, and they are described by five independent elastic constants, instead of nine for fully orthotropic ones.

Examples of transversely isotropic materials include hexagonal close-packed crystals, some piezoelectric materials (e.g. PZT-4, barium titanate), and fiber-reinforced composites, where all fibers are in parallel.

By convention, the five elastic constants in transverse isotropic constitutive equations are Young's modulus and Poisson's ratio in the x–y symmetry plane (index 'p'), E_p and ν_p, Young's modulus and Poisson's ratio in the z-direction, E_{pz} and ν_{pz}, and the shear modulus in the z-direction, G_{zp}.

The compliance form of Hooke's law takes the form

$$\begin{Bmatrix} \varepsilon_x \\ \varepsilon_y \\ \varepsilon_z \\ 2\varepsilon_{xy} \\ 2\varepsilon_{yz} \\ 2\varepsilon_{xz} \end{Bmatrix} = \begin{bmatrix} \frac{1}{E_p} & -\frac{\nu_p}{E_p} & -\frac{\nu_{zp}}{E_z} & 0 & 0 & 0 \\ -\frac{\nu_p}{E_p} & \frac{1}{E_p} & -\frac{\nu_{zp}}{E_z} & 0 & 0 & 0 \\ -\frac{\nu_{pz}}{E_p} & -\frac{\nu_{pz}}{E_p} & \frac{1}{E_z} & 0 & 0 & 0 \\ 0 & 0 & 0 & \frac{1+\nu_p}{E_p} & 0 & 0 \\ 0 & 0 & 0 & 0 & \frac{1}{G_{pz}} & 0 \\ 0 & 0 & 0 & 0 & 0 & \frac{1}{G_{pz}} \end{bmatrix} \cdot \begin{Bmatrix} \sigma_x \\ \sigma_y \\ \sigma_z \\ \sigma_{xy} \\ \sigma_{yz} \\ \sigma_{xz} \end{Bmatrix} \tag{6.52}$$

where Poisson's ratios are not symmetric, but satisfy

$$\frac{\nu_{pz}}{E_p} = \frac{\nu_{zp}}{E_z} \tag{6.53}$$

The stiffness form of Hooke's law is obtained by inverting the compliance matrix as

$$\begin{Bmatrix} \sigma_x \\ \sigma_y \\ \sigma_z \\ \sigma_{xy} \\ \sigma_{yz} \\ \sigma_{xz} \end{Bmatrix} = \begin{bmatrix} \frac{1-\nu_{pz}\nu_{zp}}{E_p E_z D} & \frac{\nu_p+\nu_{zp}\nu_{pz}}{E_p E_z D} & \frac{\nu_{zp}+\nu_p\nu_{zp}}{E_p E_z D} & 0 & 0 & 0 \\ \frac{\nu_p+\nu_{zp}\nu_{pz}}{E_p E_z D} & \frac{1-\nu_{pz}\nu_{zp}}{E_p E_z D} & \frac{\nu_{zp}+\nu_p\nu_{zp}}{E_p E_z D} & 0 & 0 & 0 \\ \frac{\nu_{zp}+\nu_p\nu_{zp}}{E_p E_z D} & \frac{\nu_{zp}+\nu_p\nu_{zp}}{E_p E_z D} & \frac{1-\nu_p^2}{E_p^2 D} & 0 & 0 & 0 \\ 0 & 0 & 0 & \frac{E_p}{1+\nu_p} & 0 & 0 \\ 0 & 0 & 0 & 0 & G_{pz} & 0 \\ 0 & 0 & 0 & 0 & 0 & G_{pz} \end{bmatrix} \cdot \begin{Bmatrix} \varepsilon_x \\ \varepsilon_y \\ \varepsilon_z \\ 2\varepsilon_{xy} \\ 2\varepsilon_{yz} \\ 2\varepsilon_{xz} \end{Bmatrix} \tag{6.54}$$

where

$$D = \frac{(1-\nu_p)(1-\nu_p-2\nu_{pz}\nu_{zp})}{E_p^2 E_z} \tag{6.55}$$

Table 6.7 summarizes the different formulations of Hooke's law and the assigned number of independent variables.

6.3.5
Plastic Behavior, Failure, and Limit Surface

The three essential ingredients of plastic analysis are the yield criterion, the flow rule, and the hardening rule, cf. [19]. The yield criterion relates the state of stress to the onset of yielding. The flow rule relates the state of stress σ_{ij} to the corresponding increments of plastic strain $d\varepsilon_{ij}^{p}$ when an increment of plastic

Table 6.7 Generalized linear elastic Hooke's law and independent material constants.

Type	Number constants
Anisotropic	21
Orthotropic	9
Transverse isotropic	5
Orthotropic, cubic structure	3
Isotropic	2

flow occurs. The hardening rule describes how the yield criterion is modified by straining beyond the initial yield. In the following section, the mathematical and graphical representations of the initial yield criterion are discussed in detail. The initial yield criterion can generally be expressed for isotropic and anisotropic materials as

$$F = F(\sigma_{ij}) \tag{6.56}$$

where the state of stress σ_{ij} can be split into its spherical part σ_{ij}^{o} and deviatoric part σ'_{ij}. For an isotropic material, the stress state can then be expressed in terms of combinations of three independent stress invariants. In the following, the set of the so-called *basic* invariants, $J_1^{\text{o}}, J_2', \text{and } J_3'$, is used, where J_1^{o} is the first invariant of the spherical stress tensor (σ_{ij}^{o}) and J_2' and J_3' are the second and third invariants of the deviatoric tensor (σ'_{ij}), [20, 21]. Further sets of independent invariants can be found, for example, in Altenbach *et al.* [22].

Thus, one can replace Eq. (6.56) for an isotropic material by

$$F = F(J_1^{\text{o}}, J_2', J_3') \tag{6.57}$$

The yield condition $F = 0$ represents a hypersurface in the n-dimensional stress space (in the case of isotropic materials, n is equal to the six independent stress tensor components) and is also called the *yield surface*. A direct *graphical* representation of this yield surface in a Cartesian coordinate system with three coordinates is not possible due to its dimension. However, a reduction of the dimensions is possible if a principal axis transformation is applied to the argument σ_{ij}. The components of the stress tensor are then uniquely reduced for isotropic materials to the principal stresses σ_{I}, σ_{II}, and σ_{III} on the principal diagonal of the stress tensor. In such a principal stress space, it is now possible to graphically represent the yield condition as a three-dimensional surface. A hydrostatic stress state ($\sigma_{\text{I}} = \sigma_{\text{II}} = \sigma_{\text{III}}$) lies in such a principal stress system on the space diagonal (hydrostatic axis). Any plane perpendicular to the hydrostatic axis is called an *octahedral plane*. The particular octahedral plane passing through the origin is the deviatoric plane or π-plane [20]. On the basis of the dependency of the yield condition on the invariants, a descriptive classification can be performed. Yield conditions independent of hydrostatic stress can be represented by the invariants J_2' and J_3'. Stress states with J_2' = const. lie on a circle around the hydrostatic axis in

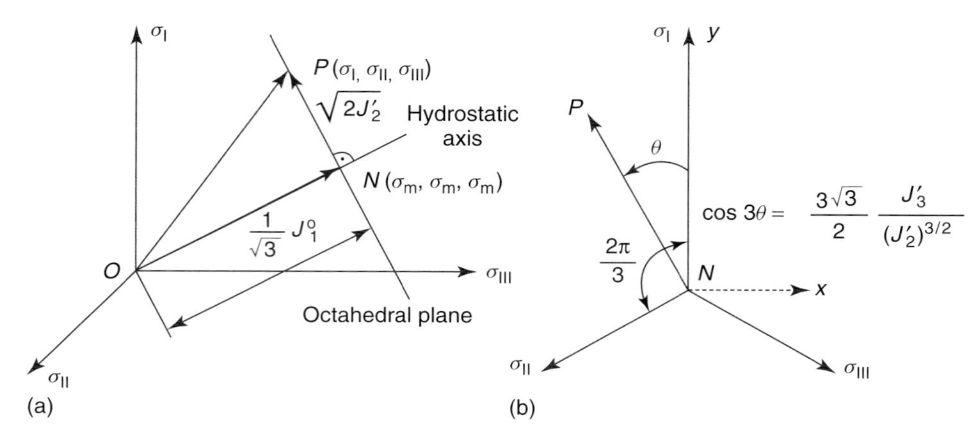

Figure 6.6 Geometrical interpretation of basic stress invariants in relation to an arbitrary stress state P: (a) principal stress space; (b) octahedral plane, J_1^o = const. (view along the hydrostatic axis).

an octahedral plane. A dependency of the yield condition on J_3' results in a deviation from the circle shape. A dependency on J_1^o denotes a size change of the cross section of the yield surface along the hydrostatic axis. However, the shape of the cross section remains similar in the mathematical sense. Therefore, a dependency on J_1^o can be represented by sectional views through planes along the hydrostatic axis. The geometrical interpretation of stress invariants [13] is given in Figure 6.6.

It can be seen that an arbitrary stress state P can be expressed by its position along the hydrostatic axis $(\frac{1}{\sqrt{3}}J_1^o) = \xi$ and its polar coordinates $(\sqrt{2J_2'} = \rho, \theta(J_2', J_3'))$ in the octahedral plane through P. For the set of polar coordinates, the so-called stress Lode angle θ is defined in the range $0 \leq \theta \leq 60^\circ$ as [23]

$$\cos(3\theta) = \frac{3\sqrt{3}}{2} \cdot \frac{J_3'}{(J_2')^{3/2}} \tag{6.58}$$

The set of coordinates $(\xi, \rho, \cos(3\theta))$ is known as the Haigh–Westergaard coordinates.

To investigate the shape of the yield surface, multiaxial stress states must be simulated and the shape within the corresponding octahedral plane has to be drawn according to the following transformation, which projects a principal stress state first in the octahedral plane (angle of transformation: $\cos\vartheta = 1/\sqrt{3}$) and then to the Cartesian coordinate system (x, y) shown in Figure 6.6(b).

$$y = \frac{2}{\sqrt{6}} \cdot (\sigma_{\text{I}} - 0.5 \cdot (\sigma_{\text{II}} + \sigma_{\text{III}})) \tag{6.59}$$

$$x = \frac{1}{\sqrt{2}} \cdot (\sigma_{\text{III}} - \sigma_{\text{II}}) \tag{6.60}$$

In such a case, it is essential whether the plastic behavior is pressure sensitive, that is, depends on J_1^o, or not. If there is no dependency on J_1^o, that is, a constant

Table 6.8 Values of the stress Lode angle for basic tests.

Case	Component	θ according to Eq. (6.58)	θ as given in Figure 6.6
Uniaxial tension	$\sigma_{I} = \sigma$ or	0°	0°
	$\sigma_{II} = \sigma$ or	0°	120° (-60°)
	$\sigma_{III} = \sigma$	0°	240° (60°)
Uniaxial compression	$\sigma_{I} = -\sigma$ or	60°	0° (180°)
	$\sigma_{II} = -\sigma$ or	60°	-60° (120°)
	$\sigma_{III} = -\sigma$	60°	60°
Biaxial tension	$\sigma_{I} = \sigma_{II} = \sigma$ or	60°	60°
	$\sigma_{I} = \sigma_{III} = \sigma$ or	60°	-60°
	$\sigma_{II} = \sigma_{III} = \sigma$	60°	180°
Biaxial compression	$\sigma_{I} = \sigma_{II} = -\sigma$ or	0°	240°
	$\sigma_{I} = \sigma_{III} = -\sigma$ or	0°	120°
	$\sigma_{II} = \sigma_{III} = -\sigma$	0°	0°
Triaxial tension	$\sigma_{I} = \sigma_{II} = \sigma_{III} = \sigma$	–	–
Triaxial compression	$\sigma_{I} = \sigma_{II} = \sigma_{III} = -\sigma$	–	–
Pur shear	$\sigma_{I} = \tau;\, \sigma_{II} = -\tau$ $(\sigma_{xy} = \tau)$	30°	-30°
	$\sigma_{I} = \tau;\, \sigma_{III} = -\tau$ $(\sigma_{xz} = \tau)$	30°	30°
	$\sigma_{II} = \tau;\, \sigma_{III} = -\tau$ $(\sigma_{yz} = \tau)$	30°	90°

shape along the hydrostatic axis, then all evaluated points can be drawn in a single octahedral plane. However, if one expects a dependency, then only stress states with the same hydrostatic stress are allowed to be represented in the same octahedral plane where J_1^{o} = const. holds. As a result, for example, uniaxial tensile ($J_1^{\text{o}} = \sigma_{I}$) and pure shear tests ($J_1^{\text{o}} = 0$) cannot be represented in the same octahedral plane. In order to draw the shape of the yield surface for a pressure-sensitive material, the following differing multiaxial stress states with $J_1^{\text{o}} = 0$, for example, $\sigma_{I} = -\sigma_{II}$ $(\sigma_{III} = 0)$, or $\sigma_{I} = -2\sigma_{II} = -2\sigma_{III}$ $(\sigma_{I} > 0 \vee \sigma_{I} < 0)$, are a possibility to obtain values in the same deviatoric plane. It should be noted here again that the shape changes only its size along the hydrostatic axes but remains similar in the mathematical sense. Typical Lode angles for basic experiments are summarized in Table 6.8, where the evaluation of Eq. (6.58) and the graphical representation in the octahedral plane are given.

For an isotropic material, the labels I, II, III attached to the principal coordinate axes are arbitrary. It follows that the yield condition must have threefold symmetry and it is only required to investigate the sector $\theta = 0^{\circ}$ to 60°, cf. Figure 6.7. The other sectors follow directly from the symmetry. If the uniaxial yield stress is in addition the same in tension and compression, a sixfold symmetry is obtained and only the sector $\theta = 0^{\circ}$ to 30° needs to be investigated, cf. Figure 6.7.

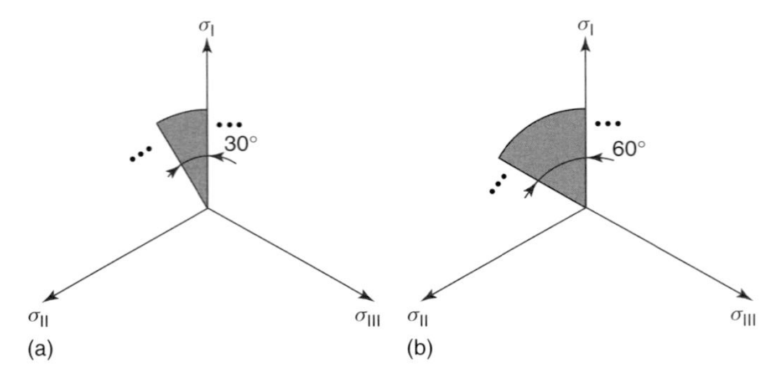

Figure 6.7 Schematic representation of periodic segments in the octahedral plane: (a) 30° symmetry segment; (b) 60° symmetry segment.

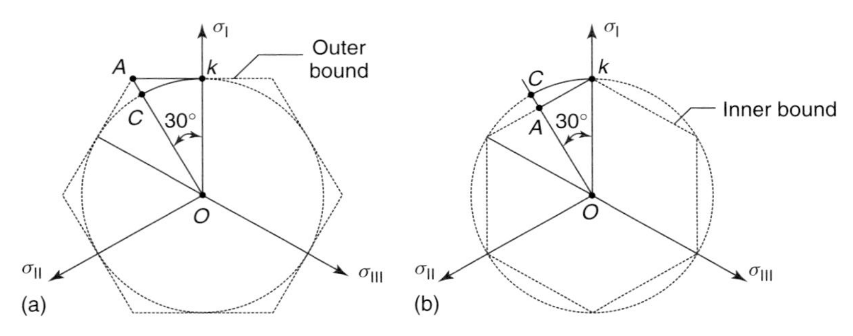

Figure 6.8 Schematic representation of outer (a) and inner (b) bounds of a yield condition in the octahedral space.

Further restrictions for the shape of the yield condition in the octahedral plane are based on the requirement of a convex shape[8], cf. Figure 6.8. The inner bound corresponds to the Tresca yield criterion. An influence of the third invariant would result in a deviation from the circular shape (e.g., von Mises).

The maximum distance between the circle and the outer bound is obtained for $\theta = 30^\circ$ as (cf. Figure 6.8(b))

$$\frac{AC}{k} = \frac{AO - k}{k} = \frac{1}{\cos 30^\circ} - 1 = \frac{2}{\sqrt{3}} - 1 = 15.47\,\% \tag{6.61}$$

of the circle diameter. The maximum distance between the circle and the inner bound is obtained under the same angle as (cf. Fig. 6.8(b))

$$\frac{AC}{k} = \frac{k - AO}{k} = 1 - \cos 30^\circ = 1 - \frac{\sqrt{3}}{2} = 13.40\,\% \tag{6.62}$$

8) The convexity of the yield surface can be derived from Drucker's stability postulate [6, 24].

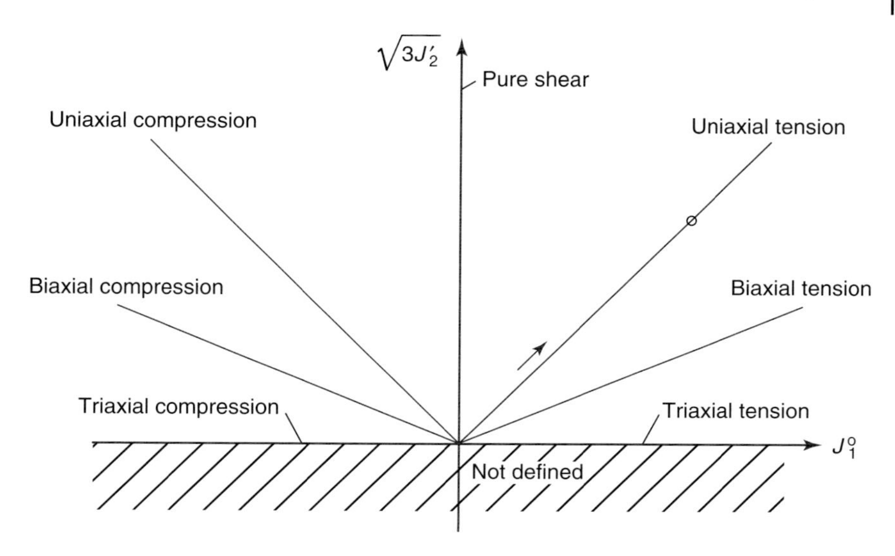

Figure 6.9 Schematic representation of basic tests in the J_1°–$\sqrt{3J_2'}$ invariant space.

In the case that a material has the same uniaxial yield stress in tension and compression, the influence of the third invariant is often disregarded since the maximum error is in the indicated range of the outer and inner bounds, cf. Eqs (6.61) and (6.62). In such a case, the mathematical description can be performed in a J_1°–$\sqrt{3J_2'}$ invariant space as indicated in Figure 6.9.

In the J_1°–$\sqrt{3J_2'}$ invariant space, basic material tests can be identified as lines through the origin as indicated in Figure 6.9. Table 6.9 summarizes the linear equations for these basic experiments. As one can see, a uniaxial tensile test is represented, for example, by the bisecting line in the first quadrant of the Cartesian J_1°–$\sqrt{3J_2'}$ coordinate system. Performing a uniaxial tensile test would mean to "walk" from the origin along this straight line (by monotonically increasing the load) until a material limit, for example, initial yield or failure, is reached. This point (indicated in Figure 6.9 by ∘) makes part of the yield or limit surface and the

Table 6.9 Definition of basic tests in the J_1°– $\sqrt{3J_2'}$ invariant space.

Case	J_1°	$\sqrt{3J_2'}$	**Comment**
Uniaxial tension (σ)	σ	σ	Slope: 1
Uniaxial compression ($-\sigma$)	$-\sigma$	σ	Slope: −1
Biaxial tension (σ)	2σ	σ	Slope: 0.5
Biaxial compression ($-\sigma$)	-2σ	σ	Slope: −0.5
Triaxial tension (σ)	3σ	0	Horizondal axis
Triaxial compression ($-\sigma$)	-3σ	0	Horizondal axis
Pure shear (τ)	0	$\sqrt{3}\,\tau$	Vertical axis

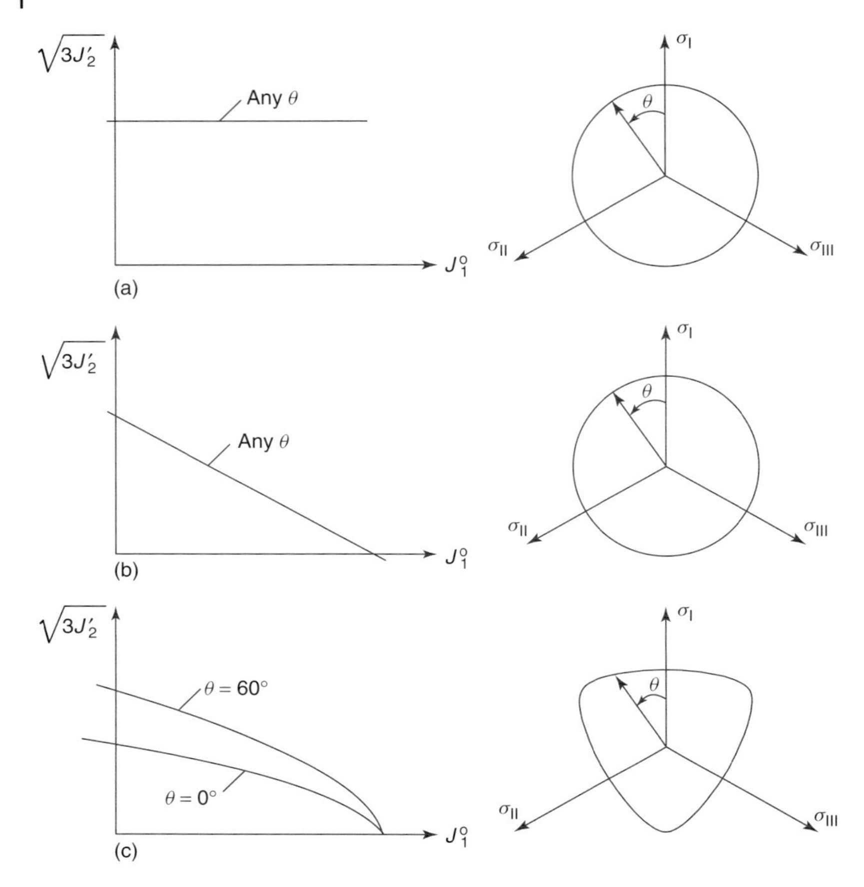

Figure 6.10 Schematic representation of different yield/failure conditions: (a) $F = F(J_2')$; (b) $F = F(J_1^\circ, J_2')$; (c) $F = F(J_1^\circ, J_2', J_3')$.

connection of all such points obtained from different experiments (i.e., from walks along different lines) give the final shape, which must be described in terms of the variables to obtain the expression $F(J_1^\circ, J_2') = 0$.

Some typical representations of yield or limit surfaces are schematically shown in Figure 6.10. A representation that depends only on J_2' is shown in part (a): a constant diameter is obtained along the hydrostatic axis, and a circular shape in the octahedral plane. Incorporating a dependency on the hydrostatic stress (part (b)) changes the diameter along the hydrostatic axis but the octahedral plain still displays a circle. Only the third invariant (part (c)) can deform the circle in the octahedral plane. For such a dependency, sectional views for $\theta = 0°$ ("tensile meridian"), $\theta = 60°$ ("compression meridian"), or $\theta = 30°$ ("shear meridian") are common representations in the J_1°–$\sqrt{3J_2'}$ space. Along these meridians, a hydrostatic stress state is superimposed on the respective stress states.

The flow rule is in general stated in terms of a function Q, which is described in units of stress, and is called a *plastic potential*. With $\mathrm{d}\lambda$, a scalar called *plastic multiplier*, the plastic strain increments are given by

$$\mathrm{d}\varepsilon = \mathrm{d}\lambda \cdot \frac{\partial Q}{\partial \boldsymbol{\sigma}} \tag{6.63}$$

The flow rule given in Eq. (6.63) is said to be associated if $Q = F$, otherwise it is nonassociated. Hardening can be modeled as isotropic (i.e., initial yield surface expands uniformly without distortion and translation as plastic flow occurs) or as kinematic (i.e., initial yield surface translates as a rigid body in stress space, maintaining its size, shape, and orientation), either separately or in combination. With the isotropic hardening parameter κ and the kinematic hardening parameters $\boldsymbol{\alpha} = \{\alpha_x, \ldots, \alpha_{xz}\}^{\mathrm{T}}$, the hardening vector $\mathbf{q} = \{\kappa, \boldsymbol{\alpha}\}^{\mathrm{T}}$ can be composed and the yield condition under consideration of hardening effects results as

$$F = F(\boldsymbol{\sigma}, \mathbf{q}) = 0 \tag{6.64}$$

Equation (6.64) can be further extended, for example, by incorporating damage effects or anisotropy (fabric-dependent criterion), [25, 26]. More details on modeling of plastic material behavior can be found in the standard books [22, 27–29].

6.4 The Structure of Trabecular Bone and Modeling Approaches

At the first glance, the bone may look as a solid dense material. However, most of the bones rather form some kind of composite material where the outer shell is made up of a dense layer, the so-called *compact bone*[9] and a core made up of a cellular structure, the so-called *trabecular bone*[10], cf. Figure 6.11. These two types of bones are distinguished based on their porosity[11] and microstructure. Trabecular bone is much more porous with porosity ranging anywhere from approximately 50 to 90% whereas the cortical bone is much denser with a porosity ranging between 5 and 10%. According to Gibson and Ashby [30], any bone with a porosity larger than 30% is classified as "cancelous." A typical range for the density of trabecular bone is between 94 $\mathrm{kg\,m^{-3}}$ (extremely low density) and 780 $\mathrm{kg\,m^{-3}}$ (moderately high), [11]. Compact bone has a density of about 1800 ... 1900 $\mathrm{kg\,m^{-3}}$, [11, 31].

9) Also known as cortical bone; Latin name: substantia corticalis.

10) Also known as cancellous or spongy bone; Latin name: substantia spongiosa or substantia spongiosa ossium.

11) The porosity is defined as the fraction of the volume of void-space and the total or bulk volume of material, including the solid and void components. The range is between 0 and 1, or as a percentage between 0 and 100%.

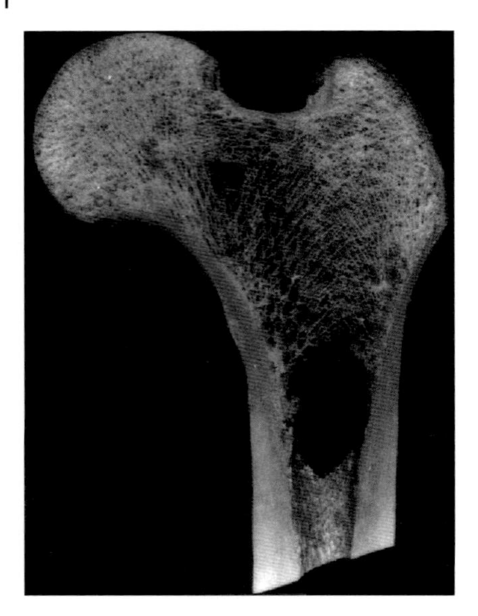

Figure 6.11 Cross-sectional view of the head of a femur.

A typical structure of a vertebral cancellous bone (μCT image) is shown in Figure 6.12 for a healthy state (a) and osteoporotic[12)] state (b). It can be seen that the structure is open and made up of a network of beams or rods (so-called trabeculae). In the case of the diseased structure, a much higher porosity characterized by much bigger holes and spaces compared to the healthy bone can be observed. This loss of density or mass may result in the fracture of the bone even for normal, that is, every day, loads. The density of the trabeculae was investigated by Galante *et al.* [32] based on 63 specimens, and a mean value of 1820 kg m^{-3} was found. This value is exactly in the range of compact bone and this is the reason why some authors [31, 33] assign the physical properties of the compact bone to the beams or rods of trabecular bone in the scope of modeling approaches.

The structure of trabecular bone and its relation to mechanical properties have several important applications in biomedical engineering. The loss of the normal density of bone (osteoporosis) results in a fragile bone. The knowledge about the fracture strength as a function of the bone density can help predict the risk of fracture and recommend appropriate treatments and behavioral rules. The actual density of the bone can be determined by a safe and painless bone mineral density (BMD) test[13)].

12) Osteoporosis, or porous bone, is a disease characterized by low bone mass and structural deterioration of bone tissue, leading to bone fragility and an increased susceptibility to fractures, especially of the hip, spine, and wrist, although any bone can be affected.

13) A routine X-ray can reveal osteoporosis of the bone, which appears much thinner and lighter than normal bones. Unfortunately, by the time X-rays can detect osteoporosis, at least 30% of the bone has already been lost. Major medical organizations are recommending a dual energy X-ray absorptiometry (DXA, formerly known as DEXA) scan for diagnosing osteoporosis.

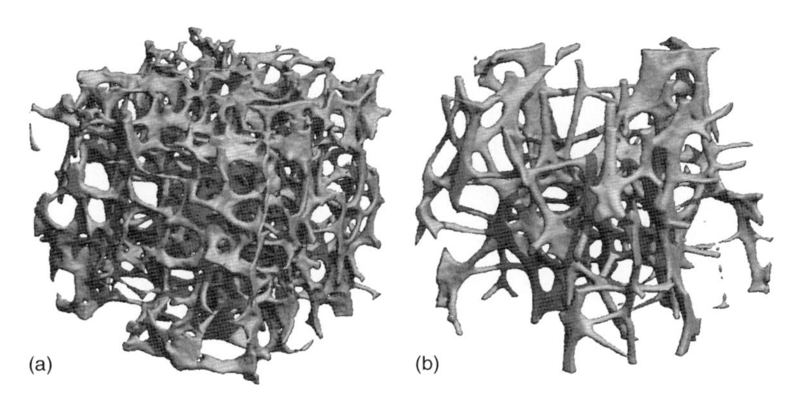

Figure 6.12 Images (μCT) of human bones: (a) healthy vertebra; (b) osteoporotic vertebra, (© by SCANCO Medical AG, Bruttisellen Switzerland).

Total hip joint replacement involves surgical removal of the diseased head ("ball") of a thigh bone, the femur, and a "cup-shaped" bone of the pelvis called the *acetabulum* ("socket") and replacing them with an artificial ball and stem inserted into the femur bone and an artificial plastic cup socket. Knowing the mechanical properties of trabecular bone allows the design and optimization of artificial hips with properties close to the natural complement, which is supposed to prevent the loosening of the implanted prosthesis.

Osteoarthritis[14] is a type of arthritis (damage of the body joints) that is caused by the breakdown and eventual loss of the cartilage of one or more joints. Cartilage is a proteinaceous substance that serves as a "lubricant" or "cushion" between the bones of the joints. The load transmission between the bones and a joint is related to the mechanical properties of all constituents, and thus, changing properties of the trabecular may cause damage to the tribological system.

6.4.1 Structural Analogies: Cellular Plastics and Metals

For a long time, the development of artificial cellular materials has been aimed at utilizing the outstanding properties of biological materials in technical applications. As an example, the geometry of honeycombs was identically converted into aluminum structures, which have been used since the 1960s as cores of lightweight sandwich elements in the aviation and space industries [34, 35]. Nowadays, in particular, foams made of polymeric materials are widely used in all fields of technology. For example, Styrofoam® and hard polyurethane foams are widely used as packaging materials. Other typical application areas are the fields of heat

14) Also known as degenerative arthritis, degenerative joint disease.

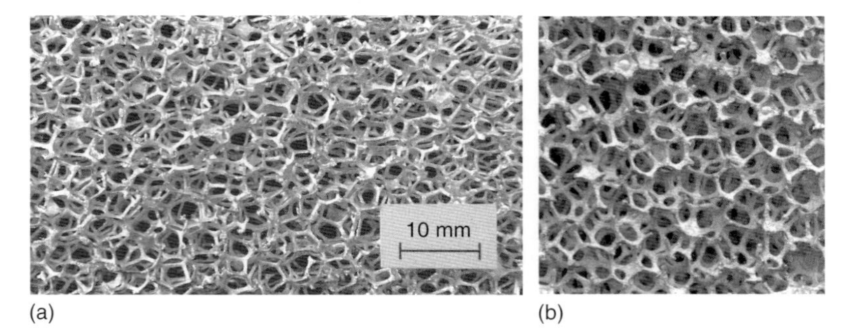

Figure 6.13 Open-cell Al foams: (a) Duocel®; (b) M-Pore®.

and sound absorption. During the last few years, techniques for foaming metals and metal alloys and for manufacturing novel metallic cellular structures have been developed [36]. Figure 6.13 shows two different Al foams that reveal a structure quite similar to that of the bones shown in Figure 6.12. On the basis of these structural analogies, it is possible to model the geometry of natural cellular (trabecular bone) and artificial cellular materials (plastic or metallic foams) with the same mathematical or engineering approaches despite the fact that the base material properties might be quite different.

The approaches found in literature to model the physical behavior of cellular materials and to derive respective constitutive equations to predict their behavior can be broadly divided into four directions:

- the use of phenomenological models,
- the application of equations developed for composite materials [37, 38],
- the analysis of model structures representative of the physical structure,
- the analysis of the real structure, for example, based on μCT images [39–41].

The last approach, that is, consideration of the real geometry, would require the lowest degree of simplifications since the geometry would be exactly represented for a certain volume. However, there are still problems connected with this most advanced approach:

- It is required to consider a representative volume element (RVE) that comprises a certain number of cells in order to be representative for the heterogeneous or random cell structure. In addition, a minimum number of cells is required to avoid any influence of the edge effects occurring at the "free" surface of the volume (which can be the boundary of the scanned volume) in order to represent macroscopic values for a larger arrangement of cells [42]. The precise number depends, however, on the corresponding cell structure and should be examined separately.
- The standard engineering tool to simulate the geometry is the finite element method. The conversion of the geometrical information based on μCT scans [43] is normally automatically done and so-called tetrahedral elements are generated. However, these elements reveal inferior performance in the case of nonlinear

material behavior. Strategies to apply brick (hexaeder) elements by converting voxels into finite elements are given, for example, in Keyak *et al.* [44] and van Rietbergen *et al.* [45]. However, this approach results in unsmoothed surfaces with sharp geometric discontinuities, which may give erroneous results [46], especially in the case of nonlinear material behavior
- The numerical approach based on the finite element method with real geometries implies high demands with respect to the computational hardware and, for example, the detailed representation of the size of a mechanical test specimen still reaches the limits of actual computer memory (RAM). In addition, access to expensive scanning hardware (CT) is required.

In the following, we will concentrate on the third approach, that is, analysis of model structures. In this case, the real geometry is strongly simplified. On the other hand, analytical (e.g., Bernoulli–Euler beam theory) or numerical methods (e.g., finite element method) at a much lower demand may be applied to investigate the physical properties. The major idea is to create a model structure that represents some characteristics of the real geometry (e.g., porosity) and behaves (e.g., deforms) in a similar way as the real structure (e.g., major deformation behavior represented by beams under bending). However, such a modeling approach cannot aim to determine the precise values of the macroscopic material parameters since a dependency on the pore geometry and arrangement exits. Nevertheless, this approach may aim to determine the principal characteristics of the constitutive equation, while the exact values of the respective free material parameters need to be determined experimentally for a given real material. In other words, the approach may be able to derive an equation to describe the material behavior, for example, as a function of porosity, but there are parameters in the equation that must be evaluated through experiments. This approach is, for example, common for classical engineering materials, where ductile metals can be described by the von Mises yield condition (known mathematical equation) whereas the yield stress (free material parameter) must be experimentally determined for each specific ductile metal.

Typical expressions for the average mechanical properties such as stiffness of cellular material are obtained as power laws of density or relative density with the following structure:

$$\text{mechanicalproperty} = C_1 \cdot (\text{density})^{C_2} \tag{6.65}$$

where C_1 and C_2 are real numbers to be fitted by experimental values. In the following, some classical modeling approaches for the geometry will be described.

The so-called ball-and-strut model by Gent and Thomas [47] was applied to represent the cellular morphology of open-cell rubber foams, cf. Figure 6.14(a). It consists of circular struts joined at their intersections by polymer spheres. The connecting elements (spheres) were assumed to be undeformable and the orientation distribution of the struts was taken to be random. In addition, it was assumed that the entire surface of the spherical dead volume is covered by strut elements. This model was later applied by Lederman [49], who assumed that in

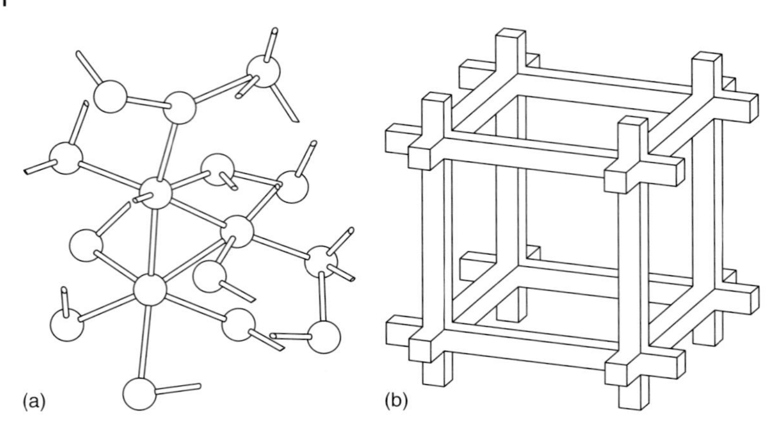

Figure 6.14 Simple idealizations of open-cell structures: (a) the ball-and-strut model [47]; (b) the cubic strut model [48].

general the dead volume does not need to be entirely covered by strut elements. The important geometrical parameter in this model is the ratio of the sphere diameter to the free strut length.

A cubical arrangement of square-section struts (cf. Figure 6.14(b)) was proposed by Gent and Thomas [48], who investigated the elastic behavior and the tensile rupture of open-cell foamed elastic materials. In their analysis, they assumed that the junctions of the struts are essentially undeformable compared with the struts themselves, and they referred to this as *dead volume*. An equivalent model for closed-cell systems was proposed by Matonis [50]. Cubic strut models have been extensively applied by Gibson and Ashby to different types of materials (bone, polymer, and metal foams [30, 31, 51]) to investigate different macroscopic properties (e.g., elastic and plastic behavior) by consideration of beam bending of the struts. Anisotropy can be simply incorporated in this cubical model by uniformly stretching it in one of the principal directions, [52]. It should be noted here that the cubic strut model shown in Figure 6.14(b) behaves in the linear elastic range according to Hooke's law for orthotropic materials with cubic structure, cf. Section 6.3.3. Nevertheless, the derived material properties are in most of the cases assigned to an equivalent isotropic material.

For the cubic plate models shown in Figure 6.15, the primary deformation mode is bending as in the case of the cubic strut model. However, these models were motivated by the cancellous bone of lower porosity, where the structure transforms into a more closed network of plates. Some of these plate elements have small perforations in them resulting in cells that are not entirely closed [53].

Columnar models with a hexagonal cross section are shown in Figure 6.16 for a rodlike structure (a) and a platelike structure (b) of cancellous bone. These models are motivated by bones where the loading is mainly uniaxial (e.g., as the vertebrae) and the trabeculae often develop a columnar structure with cylindrical symmetry [54, 55]. The elastic behavior of such a model is described by generalized Hooke's law for transverse isotropic materials, cf. Section 6.3.4.

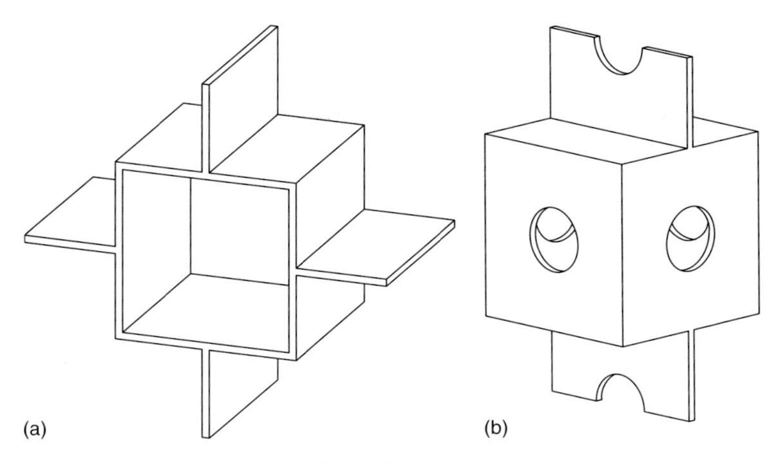

Figure 6.15 (a) Cubic model of plate-like asymmetric structure [31]; (b) perforated plate model [30].

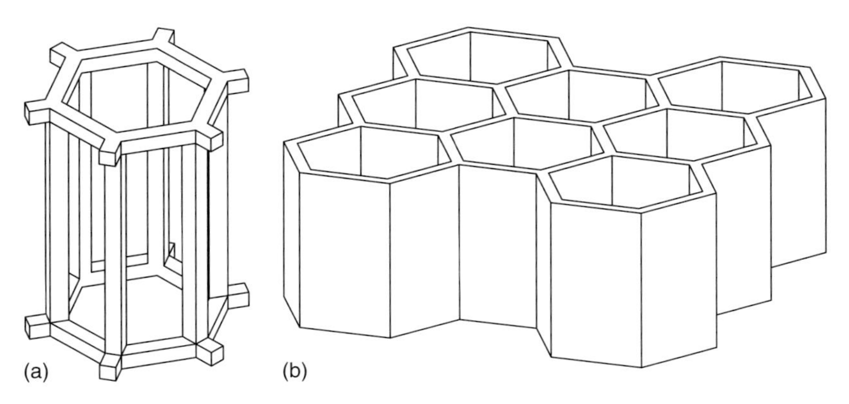

Figure 6.16 (a) Hexagonal model of rodlike columnar structure [31]; (b) prismatic plate model [30].

The cubic block model in Figure 6.17 represents an orthotropic material with cubic structure (cf. Section 6.3.3) and the elastic properties have been investigated by Beaupre and Hayes [56] with the finite element method based on different loading conditions, that is, uniaxial, compressive, and shear loads. A rather transverse isotropic material is represented by the plate–rod structure shown in Figure 6.17(b). Different deformation mechanisms are activated if a load is applied perpendicular to the plates (bending state) or if loads are applied parallel to the plates [30]. A similar model with a regular arrangement of the connecting rods (now an orthotropic structure with two identical principal axes) was applied by Raux *et al.* [57] to model the trabecular architecture of the human patella.[15)]

15) The human patella, also known as the *knee cap or kneepan*, is a thick, triangular bone which articulates with the femur and covers and protects the knee joint.

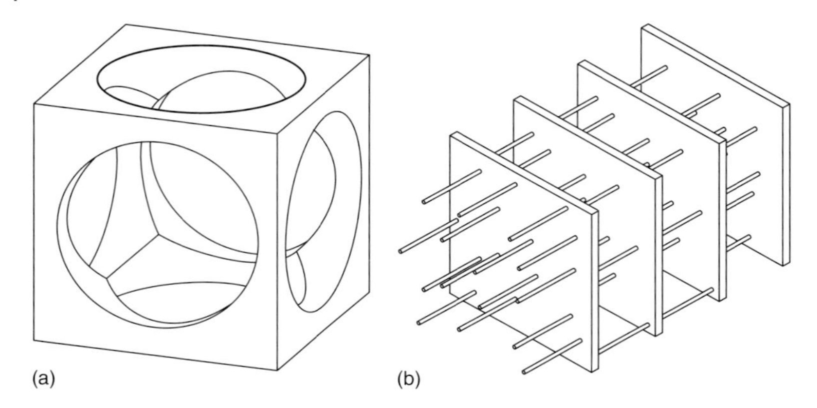

Figure 6.17 (a) Open-celled cubic block [56]; (b) parallel plate model [30].

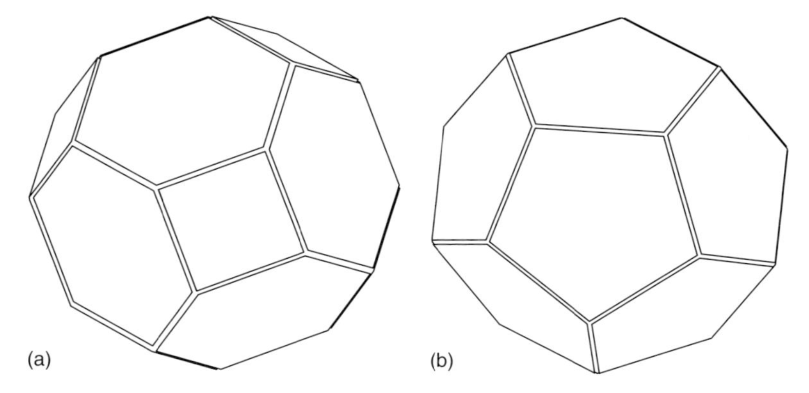

Figure 6.18 (a) Tetrakaidecahedron [30]; (b) pentagonal dodecahedron [30].

A different modeling approach is based on so-called polyhedral cells [30]. Typical representatives are the tetrakaidecahedron, which is composed of six squares and eight hexagons (cf. Figure 6.18(a)) and the pentagonal dodecahedron, which is built up of 12 regular pentagons (cf. Figure 6.18(b)). It should be noted here that only the tetrakaidecahedron is a true space-filling body and the pentagonal dodecahedron does not pack properly unless distorted or combined with other structures.

At the end of this section, let us in addition mention the geometrical model by Ko [58] who considered geometrical shapes of interstices of hexagonal closest packing and FCC closest packing of uniform spheres. His approach can be imagined as a uniform expansion of each sphere so that a contacting surface becomes flat and tangent to a contact point. Thus, each sphere becomes a polyhedron. As an example, the hexagonal closest packing will deform each sphere into trapezo-rhombic dodecahedron with six equilateral trapezoids and six congruent rhombics. Most of the interstices are then squeezed to form 24 edges of such a dodecahedron.

6.5 Conclusions

Continuum mechanical approaches are an important tool to model the mechanical behavior of trabecular bone. The first part of this chapter summarized the common approaches based on generalized Hooke's law in its different forms under symmetry considerations. The second part of the continuum mechanical approaches introduced basic ideas for yield and failure surfaces. The major intention of this part was to collect some kind of basic understanding, which can be extended to more complex approaches, for example, in the form of anisotropic yield or failure criteria. The derived and applied constitutive equations can be used in the scope of the finite element method or any other numerical approximation method to simulate the macroscopic behavior. However, it may turn out that a derived constitutive equation is not available in a commercial code. Thus, such laws must be implemented based on user-subroutines. Respective approaches for the implementation, that is, integration of the constitutive equations, may be found in the textbooks on nonlinear finite element analysis by Simo and Hughes [12], Crisfield [59, 60], and Belytschko *et al.* [61]. In the second part of this chapter, classical model approaches were introduced. Despite more and more powerful numerical tools and computer hardware, these simple models still remain attractive. Such models allow the derivation of the basic structure of constitutive equations without given exact predictions of the involved material constants. However, such material constants may be obtained from basic mechanical tests such as the uniaxial tensile or pure shear test.

References

1. Chen, W.F. (1982) *Plasticity in Reinforced Concrete*, McGraw-Hill.
2. Chen, W.F. and Saleeb, A.F. (1982) *Constitutive Equations for Engineering Materials. Vol. 1: Elasticity and Modelling*, John Wiley & Sons, Inc.
3. Chen, W.F. and Baladi, G.Y. (1985) *Soil Plasticity*, Elsevier.
4. Hayes, W.C. and Carter, D.R. (1976) *J. Biomed. Res. Symp.*, **7**, 537–544.
5. Reilly, D.T. and Burstein, A.H. (1974) *J. Bone Joint Surg. Am.*, **561**, 1001–1021.
6. Betten, J. (2001) *Kontinuumsmechanik: ein Lehr- und Arbeitsbuch*, Springer-Verlag.
7. Bronstein, I.N. and Semendjajew, K.A. (1988) *Taschenbuch der Mathematik (Erg. Kap.)*, Verlag Harri Deutsch.
8. Lebon, G. (1992) Extended thermodynamics, in *Non-Equilibrium Thermodynamics with Application to Solids* (ed. W. Muschik), Springer-Verlag, pp. 139–204.
9. Altenbach, J. and Altenbach, H. (1994) *Einführung in die Kontinuumsmechanik*, B.G. Teubner.
10. Hahn, H.G. (1985) *Elastizitätslehre*, B.G. Teubner.
11. Hodgskinson, R. and Curry, J.D. (1992) *J. Mater. Sci. Mater. M*, **3**, 377–381.
12. Simo, J.C. and Hughes, T.J.R. (1998) *Computational Inelasticity*, Springer-Verlag.
13. Chen, W.F. and Han, D.J. (1988) *Plasticity for Structural Engineers*, Springer-Verlag.
14. William, K.J. (2001) Constitutive models for engineering materials, in *Encyclopedia of Physical Science and Technology*

(ed. R. Meyers), Academic Press, pp. 603–633.
15. Flügge, W. (1962) *Handbook of Engineering Mechanics*, McGraw-Hill Book Company.
16. Mang, H. and Hofstetter, G. (2000) *Festigkeitslehre*, Springer-Verlag.
17. Golub, G.H. and van Loan, C.F. (1996) *Matrix Computations*, Johns Hopkins.
18. Press, W.H., Flannery, B.P., Teukolsky, S.A., and Vetterling, W.T. (1992) *Numerical Recipes in Fortran*, Cambridge University Press.
19. Armen, H. (1979) *Comput. Struct.*, **10**(1–2), 161–174.
20. Backhaus, G. (1983) *Deformationsgesetze*, Akademie-Verlag.
21. Jirasek, M. and Bazant, Z.P. (2002) *Inelastic Analysis of Structures*, John Wiley & Sons, Inc.
22. Altenbach, H., Altenbach, J., and Zolochevsky, A. (1995) *Erweiterte Deformationsmodelle und Versagenskriterien der Werkstoffmechanik*, Deutscher Verlag für Grundstoffindustrie.
23. Nayak, G.C. and Zienkiewicz, O.C. (1972) *J Struct Div.-ASCE*, **98**(ST4), 949–954.
24. Lubliner, J. (1990) *Plasticity Theory*, Macmillan Publishing Company.
25. Pietruszczak, S., Inglis, D., and Pande, G.N. (1999) *J. Biomech.*, **32**, 1071–1079.
26. Zysset, P. and Rincóm, L. (2006) An alternative fabric-based yield and failure critrion for trabecular bone, in *Mechanics of Biological Tissues* (eds G.A. Holzapfel and R.W. Ogden), Springer-Verlag, pp. 457–470.
27. Zyczkowski, M. (1981) *Combined Loadings in the Theory of Plasticity*, PWN - Polish Scientific Publishers.
28. Betten, J. (2005) *Creep Mechanics*, Springer-Verlag.
29. Kolupaev, V. (2006) *Dreidimensionales Kriechverhalten von Bauteilen aus unverstärkten Thermoplasten*, Papierflieger.
30. Gibson, L.J. and Ashby, M.F. (1997) *Cellular Solids: Structures and Properties*, Cambridge University Press.
31. Gibson, L.J. (1985) *J. Biomech.*, **18**(5), 317–328.
32. Galante, J., Rostoker, W., and Ray, R.D. (1970) *Calcif. Tissue Res.*, **5**, 236–246.
33. Carter, D.R. and Hayes, W.C. (1977) *J. Bone Joint Surg. Am.*, **59**, 954–962.
34. Hertel, H. (1960) *Leichtbau*, Springer-Verlag.
35. Bitzer, T. (1997) *Honeycomb Technology: Materials, Design, Manufacturing, Applications and Testing*, Chapman & Hall.
36. Banhart, J. (2001) *Prog. Mater. Sci.*, **46**(6), 559–632.
37. Hashin, Z. (1962) *J. Appl. Mech. Trans. ASME*, **29**, 143–150.
38. Cohen, L.J. and Ishai, O. (1967) *J. Compos. Mater.*, **1**, 390–403.
39. Feldkamp, L.A., Goldstein, S.A., Parfitt, A.M., Jesion, G., and Kleerekoper, M. (1989) *J. Bone Miner. Res.*, **4**(1), 3–10.
40. Müller, R. and Rüegsegger, P. (1995) *Med. Eng. Phys.*, **17**(2), 126–133.
41. Müller, R. and Rüegsegger, P. (1996) *J. Biomech.*, **29**(8), 1053–1060.
42. Lemaitre, J. (1996) *A Course on Damage Mechanics*, Springer-Verlag.
43. Müller, E.P., Rüegsegger, P., and Seitz, P. (1985) *Phys. Med. Biol.*, **30**, 401–409.
44. Keyak, J.H., Meagher, J.M., Skinner, H.B., and Mote, C.D. (1990) *J. Biomed. Eng.*, **12**, 389–397.
45. van Rietbergen, B., Weinans, H., Huiskes, R., and Odgaard, A. (1995) *J. Biomech.*, **28**(1), 69–81.
46. Marks, I.W. and Gardner, T.N. (1993) *J. Biomed. Eng.*, **14**, 474–476.
47. Gent, A.H. and Thomas, A.G. (1959) *J. Appl. Polym. Sci.*, **1**(1), 107–113.
48. Gent, A.H. and Thomas, A.G. (1963) *Rubber Chem. Technol.*, **36**(3), 597–610.
49. Lederman, J.M. (1971) *J. Appl. Polym. Sci.*, **15**, 693–703.
50. Matonis, V.A. (1964) *SPE J.*, **20**, 1024–1030.
51. Gibson, L.J. and Ashby, M.F. (1982) *Proc. R. Soc. Lond. A*, **382**, 43–59.
52. Kanakkanatt, S.V. (1973) *J. Cell. Plast.*, **9**, 50–53.
53. Whitehouse, W.J. and Dyson, E.D. (1974) *J. Anat.*, **118**, 417–444.
54. Weaver, J.K. and Chalmers, J. (1966) *J. Bone Joint Surg. Am.*, **48**, 289–298.
55. Whitehouse, W.J., Dyson, E.D., and Jackson, C.K. (1971) *J. Anat.*, **108**, 481–496.

56. Beaupre, G.S. and Hayes, W.C. (1985) *J. Biomech. Eng.-Trans. ASME*, **107**, 249–256.
57. Raux, P., Townsend, P.R., Miegel, R., Rose, R.M., and Radin, E.L. (1975) *J. Biomech.*, **8**, 1–7.
58. Ko, W.L. (1965) *J. Cell. Plast.*, **1**, 45–50.
59. Crisfield, M.A. (1991) *Non-Linear Finite Element Analysis of Solids and Structures, vol. 1, Essentials*, John Wiley & Sons, Ltd.
60. Crisfield, M.A. (1997) *Non-Linear Finite Element Analysis of Solids and Structures, vol. 2, Advanced Topics*, John Wiley & Sons, Ltd.
61. Belytschko, T., Liu, W.K., and Moran, B. (2000) *Nonlinear Finite Elements for Continua and Structures*, John Wiley & Sons, Ltd.

7
Mechanical and Magnetic Stimulation on Cells for Bone Regeneration

Humphrey Hak Ping Yiu and Kuo-Kang Liu

7.1
Introduction

Nature experiences the most amazing mechanical stimulations for its maintenance and growth. Everyday, biological tissues or cells are under various mechanical forces; for instances, human skeleton is mechanically stimulated by compression forces during walking; cornea is constantly under mechanical stimulation by eyelid sweeping; even blood vessels are under pulsatile pressure/forces that stem from heartbeat. Hence, it is not surprising that mechanical stimulations have been widely used in biomedical applications, such as tissue regeneration/engineering, regenerative medicine, rehabilitation, and especially for enhancing bone tissue growth.

In bone remodeling, it has been well recognized that mechanical forces have profound and complex influences, which normally include osteoclastic resorption and osteoblastic formation. For example, it has been found that mechanical stimulation is essential for osteoblast cells to differentiate from their precursors. Through analyzing osteoblast cell kinetic, the stage of differentiation and proliferation of cells in response to elastic loading has been characterized [1]. It has been discovered that increasing the mechanical loading (e.g., orthodontic force) has been found to induce a shift in the cell nuclear volume, that is, shift from less-differentiated precursor cells to preosteoblasts [2, 3]. Since bone consists of 75% organic and mineral component and 25% fluid, when it is subjected to external loading, complex mechanical interplay will normally be involved in the fluid–structure interactions within the bone. It is therefore highly challenging to correlate mechanical stimulus with bone regeneration in a quantitative manner. Recent endeavors in the advanced biomechanical measurement instruments open a new vista for studying such "mechanobiological" issues in bone regeneration. Figure 7.1 illustrates how mechanical stimulation influences bone regeneration, which is addressed in detail later in this chapter.

In this chapter, we introduce several techniques that are used for cell stimulation. The principles and their applications are discussed. In the first half, we focus on the more commonly used techniques such as optical tweezers and atomic

Biomechanics of Hard Tissues: Modeling, Testing, and Materials.
Edited by Andreas Öchsner and Waqar Ahmed

ISBN: 978-3-527-32431-6

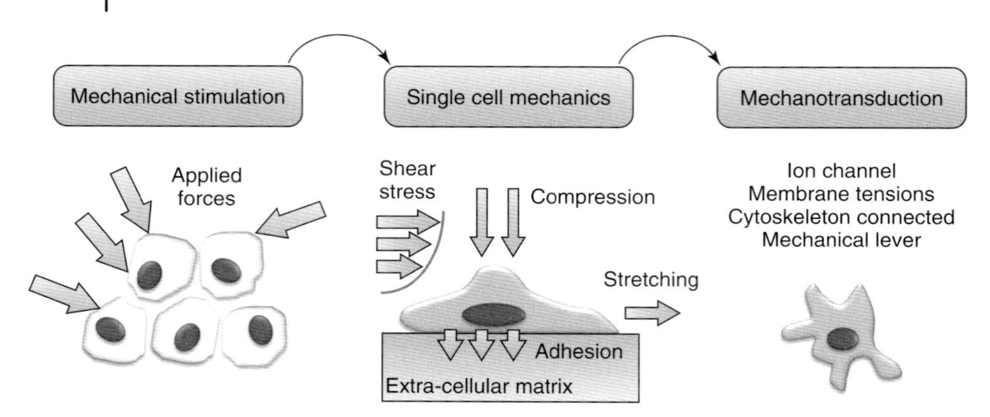

Figure 7.1 Schematic illustration of mechanical stimulation on bone cells.

force microscopy (AFM). Recently, research on mechanical stimulation on cells using magnetic micro- and nanoparticles has increased. This is partly due to the advancement made in nanoparticle technology and the various biological techniques such as electrophysiology and microscopy. In the second half, we discuss the possibility of using magnetic particles in cell stimulation and their potential application in bone regeneration.

7.2 Mechanical Stimulation on Cells

7.2.1 Various Mechanical Stimulations

There has been a long debate regarding the types of forces, that is, tension, compression, or shear force, that play the most important role in bone remodeling and formation. Similar questions have been asked by bioscientists. Which force among strain, stress, and strain rate, plays the most important role in bone growth and remodeling? Although it still needs detailed studies to answer these questions more accurately, animal studies have shown that strain rate is more important than strain amplitude to induce bone formation [4]. Some other recent studies, however, show that both strain and loading frequency are important for bone adaptation [5, 6]. For example, Turner *et al.* showed that the bone formation had increased more than 10-fold when the frequency of loading applied on rat forelimbs had been increased from 1 to 10 Hz [6]. The most common frequency applied for stimulating bone regeneration is around 1–10 Hz [2], which is close to walking pace, while the optimal strain for bone growth is around a few hundreds to thousands microstrain [7]. It is worth pointing out that the *in vivo* strains for osteoblasts are between 0 and 4000 microstrain. Nonetheless, the optimal microstrain and frequency are somewhat cross-related.

7.2.2 Techniques for Applying Mechanical Loading

Applying various mechanical loading on bone cells *in vitro* are by no means trivial since it requires elegant instrumentation designed for accurately imposing specific force types. In recent years, various devices have been developed for mechanically stimulating cells and tissues, as concisely described in an excellent review on this topic [8]. In the review, the current state of the art has been addressed and represented by several typical systems. These systems include the following:

1) Compression loading systems: either using platen displacement or hydrostatic pressure to compress tissues.
2) Longitudinal stretch systems: either stretching an extendable substrate that is placed in grip and which is seeded with cells or alternatively bending a cell-cultured flexure (it is commonly known as *four-point bending*) for generating longitudinal force.
3) Out-of-plane distension systems: pressuring or displacing a cell-seeded circular membrane to be deformed in out-of-plane direction.
4) In-plane substrate distension: using biaxial traction or frictionless platen to deform a cell-cultured substrate to generate in-plane homogeneous stresses.
5) Fluid shear systems: place cultured cells on a cone and plate chamber or in a cylindrical tube and then apply fluid flow to generate flow-induced shear stress on the cells/tissues.

In order to enhance our understanding of how mechanical stimulation is translated in a single cell and consequently affects the cell's function, proliferation and differentiation, single-cell mechanics, mechanotransduction of bone cells, has become increasingly important in exploring this new frontier now [9]. For studying mechanotransduction of single cells, various advanced instruments that are capable of deforming single osteoblastic cells have been developed. They have been applied to study how mechanical forces influence the properties of the single cells. Among these advanced techniques, the most popular modes include AFM [10], optical tweezers [11], and nanoindentation [12, 13]. A schematic illustration of these techniques is shown in Figure 7.2.

The AFM mainly consists of a sharp tip, which is linked with an ultrasensitive cantilever beam, and a photo detector for measuring the deflection/displacement of the beam subjected to external forces generated from the sharp tip interacting with a cell. Optical tweezers is an instrument utilizing a highly focused laser beam shining on a bone cell directly or a particle-attached cell which has higher refractive index compared with the surrounding liquid, and then an optical trap force can be generated on the cell as a mode of mechanical stimulation. Nanoindentation or microneedle indentation is to use an indenter to press a bone cell and simultaneously measure the force and displacement of the deformed cell. Both AFM and nanoindentation are able to apply force in the range of 10^{-9}–10^{-11} N, while optical tweezers work in the range of 1–100 pN (10^{-12} N).

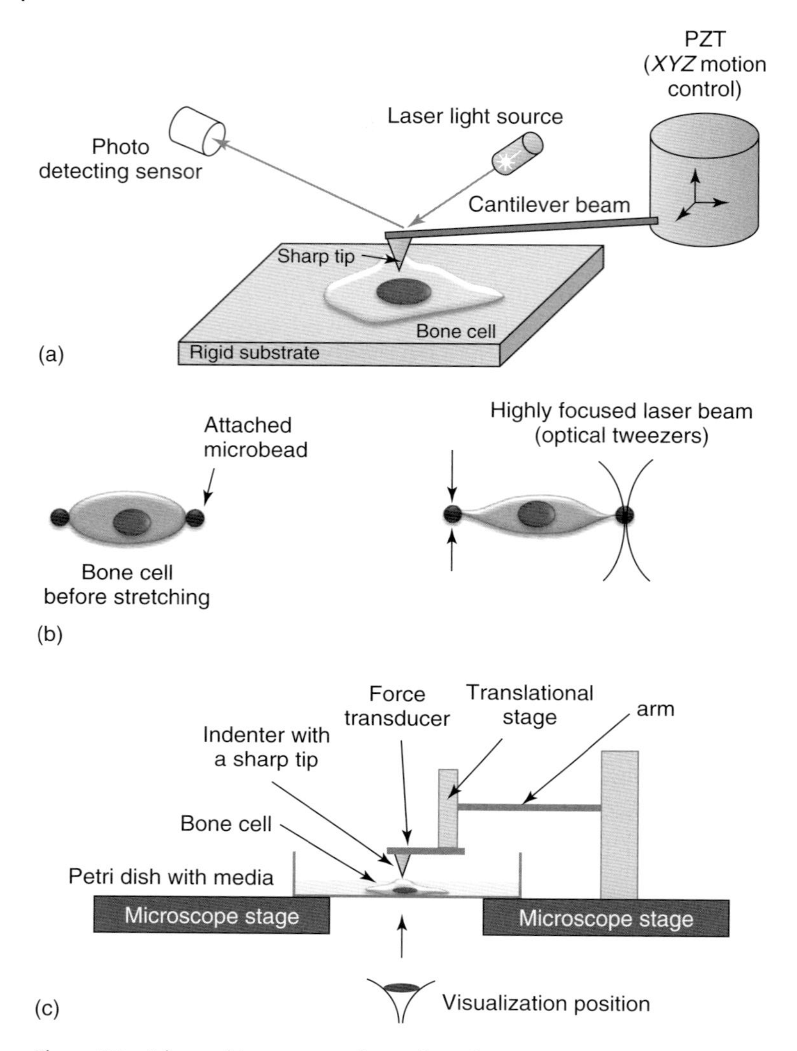

Figure 7.2 Advanced instruments for applying force on single cells: (a) AFM; (b) optical tweezers; and (c) nanoindentation.

7.2.3
Mechanotransduction

Mechanotransduction is to study the mechanisms involved in the translation of mechanical stimulus into biochemical activity when biological cells are subjected to mechanical forces. More specifically, bone cell under mechanical forces will normally induce complex biochemical/biophysical cascades. These complex interplaying mechanisms include strain-related potentials, activation of ion channels, membrane tension, cytoskeleton deformation, and mechanical lever (primary cilium) [14]. For example, osteocytes, which are believed to play an important role as mechanosensing elements in bone tissues, in particular respond strongly to fluid flow by a rapid release of nitric oxide (NO) and prostaglandins [15]. Apart from NO and prostaglandins, others such as Ca^{2+} and adenosine 5′-triphosphate (ATP) are also released as molecular signals in mechanotransduction cascades, which ultimately causes bone formation. Recent studies have shown that several molecular, cellular, and extracellular components and structures play important roles in the mediation of cellular mechanotransduction. These mechanotransduction elements include extracellular matrix (ECM), cell–cell and cell–ECM adhesion, cytoskeleton structures (namely, microfilaments, microtubules, and intermediate filaments), specialized surface processes (e.g., ion channels and surface receptors), and nuclear structures [16].

7.2.4 Mechanical Influences on Stem Cell

Stem cell is a cell from which other types of cells, such as bone, cardiac, and cartilages are developed. Recently, researches on stem cells have become increasingly important for developing new techniques for bone regeneration [17]. Much research effort has been directed to study how biochemical environments may alter the gene expression of stem cells. However, very limited studies have been done to elucidate how mechanical force/stimulation influences the differentiation and proliferation of stem cells in order to form bone cells. Figure 7.3 simply shows the progenic cascade of stem cell.

Mechanical forces affect bone regeneration which can be significantly found in osteogenic differentiation of mesenchymal stem cells (MSCs) [19]. In this study, a short period of cyclic mechanical strains (2000 microstrains for 40 min) was applied to rat MSCs through stretching an MSCs-seeded elastic membrane. The results showed that mechanical strains can promote proliferation of MSCs. Moreover, a more recent study shows that dynamic compression on human mesenchymal stem cells (hMSCs) cultured in a bioreactor for 24 h will lead to a significant increase (about fivefold) in osteopontin, which is a major noncollagenous bone matrix protein [20]. This result also implies that dynamic compression is an important factor for modulating progenitor cell differentiation after mechanical instability caused by a fracture.

In addition to compressive stimulation, it has also been found that marrow stromal osteoblasts exposed to shear stress in the range of 0.01–0.03 Pa in a transverse flow will increase their mineral deposition. When shear stress was in the range of 0.5–1.5 Pa in 2D planar plates, the expansion and differential phenotype of osteoblasts grown in bioreactor systems were enhanced [21, 22]. However, it was also found that an average surface stress of 5×10^{-5} Pa in 3D

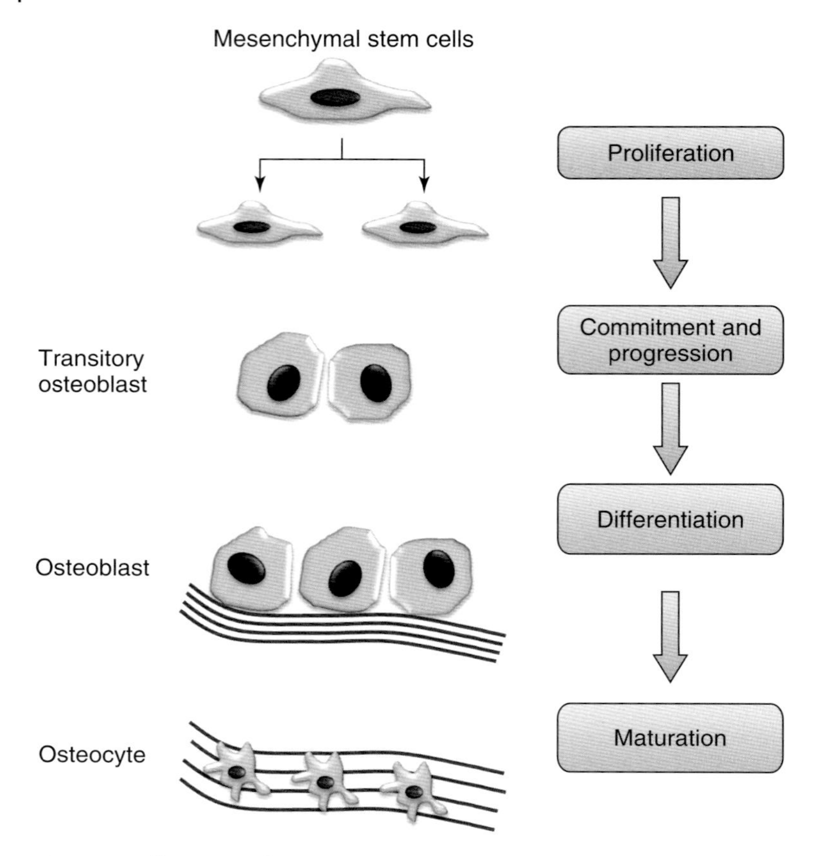

Figure 7.3 Illustration of progenic process of mesenchymal stem cell. Adapted from Caplan and Bruder [18].

constructs corresponds to the increase in cell viability and proliferation, whereas higher shear stresses of 1×10^{-3} Pa were associated with up-regulation of mRNA expressions of Runx2, osteocalcin, and alkaline phosphatase [23, 24]. Moreover, in a perfusion bioreactor, a 1.4 times higher proliferation rate, a higher colony-forming unit fibroblast (CFU-F) formation, and more fibronectin and heat shock protein 47 (HSP-47) secretion were observed for human mesenchymal stem cells (hMSCs) at a lower flow rate corresponding to an initial shear stress of 1×10^{-4} Pa [25]. However, a higher flow rate corresponding to initial interface shear stress of 1×10^{-5} Pa can up-regulate the osteogenic differentiation of hMSCs. Majority of literature has been focused on the effect of shear stress on the monolayer hMSCs culture during *in vitro* 2D or 3D cultivation. More detailed studies are required to understand the way in which hMSCs respond to shear stress in physiological flow.

7.3
Magnetic Stimulation on Cells

Recent advancement in magnetic nanoparticles allows mechanical stimulation forces on cells to be applied through an external magnetic field. These magnetic nanoparticles are designed to target particular receptors on the surfaces of cells, by tagging specific active biomolecules, including antibodies, peptides, and ligands. The idea of using magnetic device for studying cells was introduced over 80 years ago [26, 27]. Later, Crick *et al.* exploited this idea to study the cell rheology [28]. In the 1980s, a "twisting" technique was introduced by Valberg *et al.* [29–31] and this was later used by several research groups to activate mechanosensitive ion channels [32]. Since then, research on magnetic stimulation on cells has been intense. We discuss the latest state of the art in this area of research in the subsequent sections of this chapter.

7.3.1
Magnetic Nanoparticles for Cell Stimulation

Generally, magnetic nanoparticles used for cell biology are mainly iron-oxide-based nanomaterials because of their low toxicity and high magnetization [33]. Indeed, iron oxide micro- or nanoparticles have been clinically approved for several medical applications including as contrast agents for MRI scans [34]. For many biological applications, a layer of protective coating is coated on the surface of these iron oxide particles in order to prevent unwanted corrosion and provide a surface for tethering organic functionalities. This allows binding of biomolecules of interest such as antibodies. Figure 7.4 illustrates the common design of magnetic nanoparticles for cell stimulation.

7.3.1.1 Properties of Magnetic Nanoparticles

The size of a nanoparticle is critical to such an application. Figure 7.5 shows the size relation between a cell and its receptor. The diameter of commercial magnetic nanoparticles ranges from 20 up to 600 nm. Particles with larger dimensions should not be classified as "nanoparticles." However, among these commercial particles, the smaller particles are usually not unfunctionalized with no coating. The size of coated particles starts from around 150 nm while the smallest magnetic particles

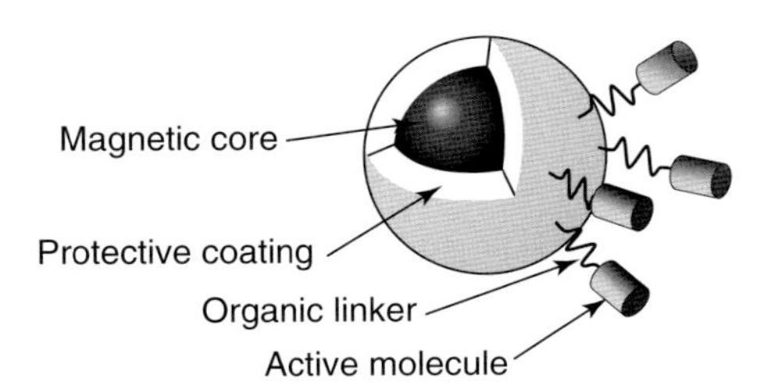

Figure 7.4 A common design of a magnetic nanoparticle for biomedical applications.

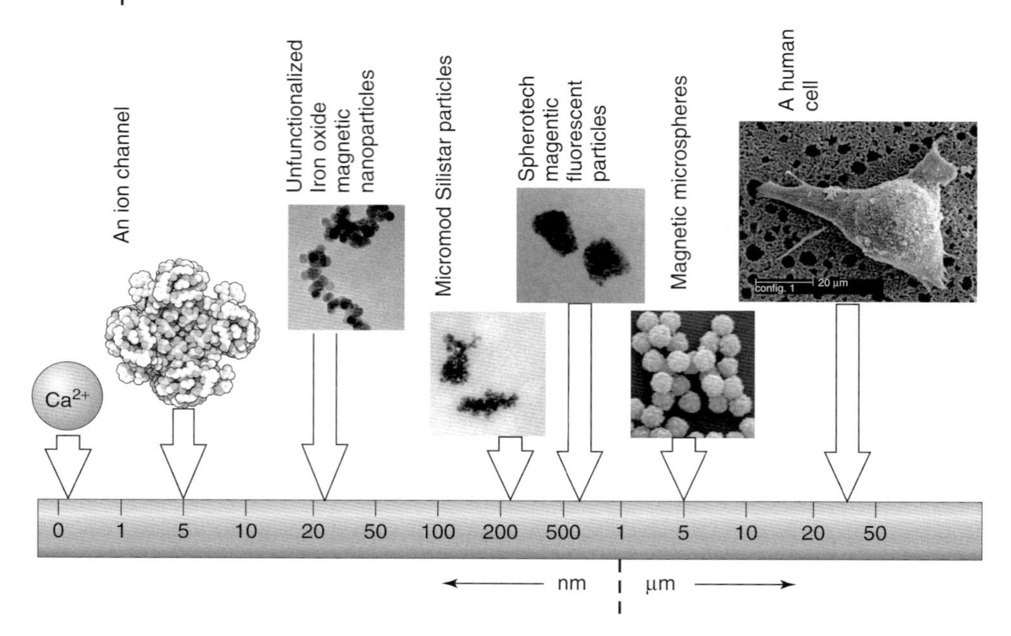

Figure 7.5 A schematic comparison on the size of some magnetic micro-/nanoparticles and a human cell.

with, for example, carboxylic acid functionality are around 250 nm in diameter. However, a number of reports on preparing functionalized magnetic nanoparticles <150 nm are widely available [35].

We should note that an ion channel is only around a few nanometers wide; large magnetic nanoparticles may cause complications. For example, since these particles usually have many functional groups, several ion channels attached to one large particle can occur. On the other hand, small particles, with a small magnetic core, may not generate enough pulling force to stimulate the ion channels. As a result, small particles with a high magnetization would be ideal in this application. This can be achieved by either increasing the iron oxide content or using a core material with a higher specific magnetization value, for example, metallic iron or ferrites ($MFeO_4$, where M = Co, Ni, Mn). Research on these areas has already begun [36].

There is a common misunderstanding on the magnetization properties of these magnetic nanoparticles. The magnetization of any magnetic materials depends on the nature of the materials and the volume of particle domain size. As a result, the magnetization value of a single nanoparticle is usually lower than that of its bulk counterpart [37]. However, most of the common commercial magnetic particles (150 nm to 4 μm) are made of aggregates of small superparamagnetic nanoparticles (10–20 nm). The overall size of these particles may only have a small effect on the overall magnetization value if the iron oxide content remains the same.

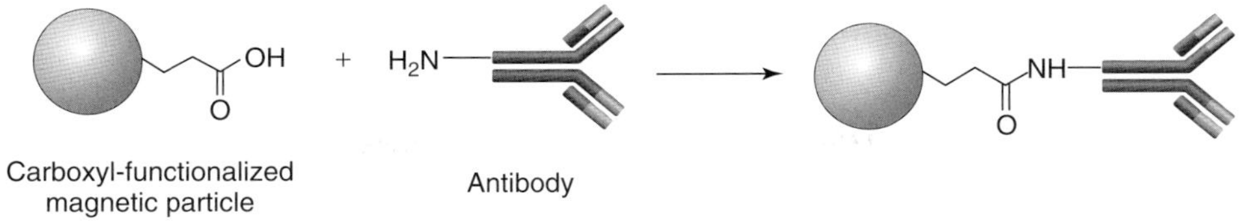

Carboxyl-functionalized magnetic particle

Antibody

Figure 7.6 The amine–carboxylic acid coupling strategy for binding antibody onto magnetic nanoparticles.

In order to improve the biocompatibility of magnetic nanoparticles, most of these particles are coated with materials such as carbohydrate, peptides, synthetic polymers, or inorganic materials (e.g., gold, silica) [38]. Besides, these coating materials also provide a platform for building organic functionalities on the surface of the magnetic nanoparticles.

7.3.1.2 Functionalization of Magnetic Nanoparticles

In order to target the receptors on the cell membrane, specific antibodies are bound on the surface of the nanoparticles. Therefore, nanoparticles with the optimal surface properties are essential for the stimulation of cells. Typically, nanoparticles with amine or carboxylic acid groups are used for binding antibodies, as well as other protein molecules. This is because a simple chemical coupling reaction can be applied by forming an amide bond between the particle and the antibody molecule (see Figure 7.6).

However, using this procedure, antibody molecules are bound in random orientation, and therefore some may not be able to target the receptor. The use of a secondary antibody has been proposed [39]. Although this would ensure that the antibodies are at the right orientation, the overall size of the particles will increase and the whole binding procedure becomes very complicated.

7.3.2 Magnetic Stimulation

In general, there are two main categories for this technique of magnetic stimulation: magnetic pulling and magnetic twisting cytometry (MTC).

7.3.2.1 Magnetic Pulling

In the mid-1990s, Glogauers and coworkers started to study the effect of an external magnetic field, generated by a permanent magnet, on cells loaded with iron oxide magnetic micro- or nanoparticles [40, 41]. In this study, the particles were attached to integrin receptors on the surface of fibroblast cells. To do that, the magnetic particles were coated with collagen. Studies using electromagnets have also been reviewed [42] but this approach is less popular due to the heat generated by the electromagnet. Such heat may cause uncertainty in cell activities or even damages on cells. Whether it is a permanent magnet or electromagnet, the principle is to provide a magnetic field with a high gradient to pull the magnetic particles in a

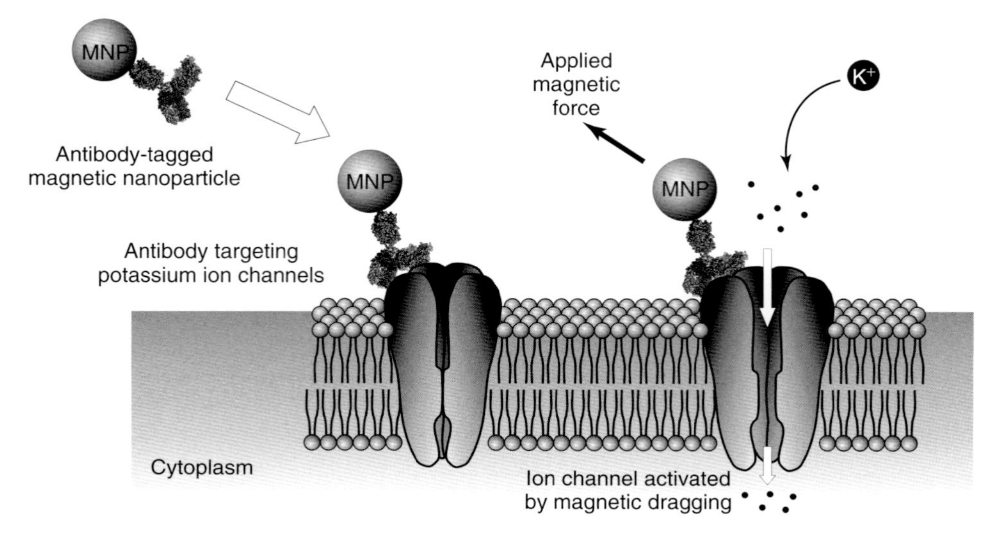

Figure 7.7 The mechanism of an antibody-tagged magnetic nanoparticles targeting and activating the ion channel receptor on the cell membrane.

specific direction. Such a pull force causes tension on the cell membrane and hence activation of the ion channels. In order to control the force of stimulation, several factors need to be considered: (i) the strength of the applied magnetic field; (ii) the magnetic susceptibility of the particles; (iii) the number of particles attached; and (iv) the location of the particles bound on the cell. Magnetic pulling systems with pulling in both vertical [40, 41] and horizontal [43] directions have already been reported.

In addition to stimulation, such magnetic pulling systems have also been used to study the biomechanical property of cells [44–46]. However, these studies are limited by the fact that particles are bound to specific receptors (Figure 7.7). It is difficult to apply the force to the whole cell using magnetic pulling.

7.3.2.2 **Magnetic Twisting**

Magnetic pulling, vertical or horizontal, provides a mechanism to apply forces in one dimension. Wang and coworkers have developed a magnetic twisting technique, MTC, to stimulate the cells with a "two-dimensional" force (see Figure 7.8). In this system, a twisting force was applied by rotating the magnetic particles bound on the surface of cells and the mechanical properties of the cytoskeleton of the cells were studied [32, 47, 48]. Compared with magnetic pulling, this technique provides a highly localized torque to the specific receptors but avoids deformation of the cell membrane. Similar to magnetic pulling, these magnetic particles were tagged with ligands or antibody molecules to target the receptors on the cell

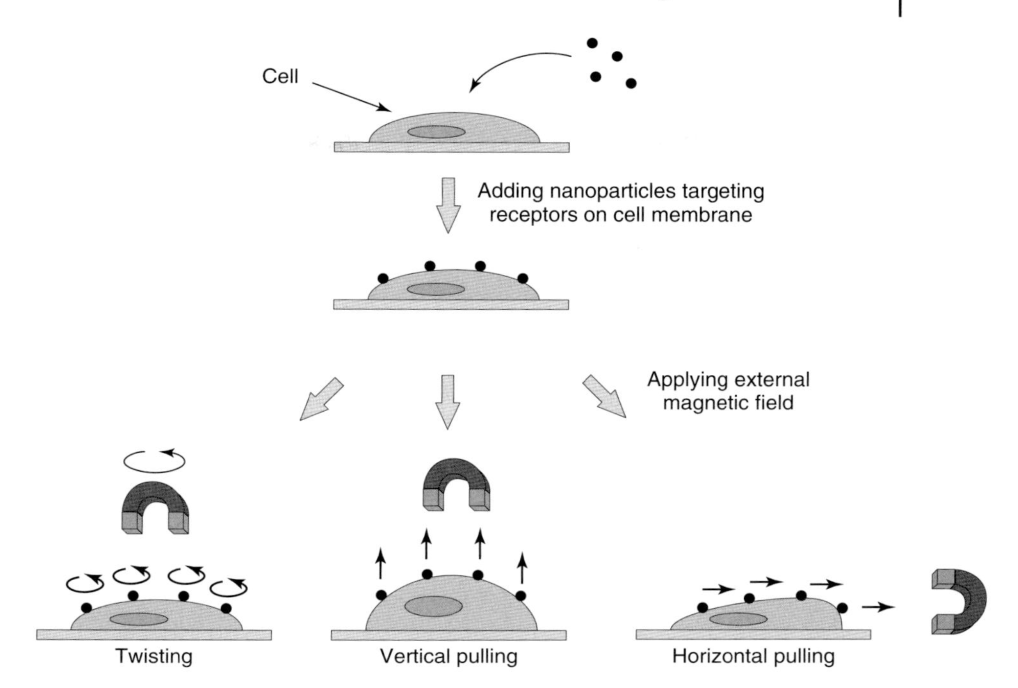

Figure 7.8 An illustration depicts three main techniques in magnetic stimulation of cells: twisting, vertical pulling, and horizontal pulling.

membrane. To create a twisting effect, ferromagnetic particles (particles carry a magnetic moment) are used and the applied magnetic field is applied at an angle to the magnetization vector of the particles. Since this twisting technique offers a localized force without cell deformation, it is an excellent tool to study the function of the receptors and the adhesive protein molecules in mechanotransduction.

7.3.3 Limitation of Using Magnetic Nanoparticles for Cell Stimulation

Since magnetic nanoparticles are biocompatible, internalization of nanoparticles is inevitable. Once internalized, the nanoparticles are no longer attached to the ion channels on the cell membrane, travel through the cytoplasm toward the nucleus, and stimulation is not possible. This internalization process takes place in a matter of hours [49]. Moreover, the speed of internalization generally increases as the particle size decreases [50–52].

One possible solution is to target the receptors inside the cell. This requires a fast internalization process, usually aided by using lipids, and instant attachment of the nanoparticles to the internal receptors. Unfortunately, there is no successful research using such a complicated system.

7.3.4
Magnetic Stimulation and Cell Conditioning for Tissue Regeneration

Magnetic stimulation has been applied to study the property of many cell types. Table 7.1 summarizes the research published in this area. Nonetheless, most of these research works are focused on the mechanotranducers, such as integrins, E-cadherins, E-selectins, and urokinase receptors, because activation of these receptors can trigger many responses from the cell. For example, studies have already shown that magnetic stimulation on integrin receptors on osteoblast cells initiated changes in intracellular pH [49], changes in cytoskeleton, and generation of intracellular calcium transients.

These responses from cells can be very significant in tissue engineering as the changes in cell properties can determine the quality of the engineered tissue in terms of mechanical strength, which is particularly critical to bone regeneration. Therefore, scientists have started to investigate the use of this magnetic stimulating technique to "condition" cells mechanically during the production of engineered tissues. The concept is to grow replacement tissue with patient's own cells *ex vivo* and then implant them back to the patient's body. Although these magnetic stimuli can promote the growth of engineered tissues, so far it has been proved to be difficult to apply these techniques in a complex three-dimensional growth of tissues.

One of the holy grails of tissue regeneration is directing the differentiation of MSCs into bone, muscle, and other connective tissues such as ligaments and tendons. This concept avoids the use of patient's primary organ cells, which may not always be available. The theory relies on applying magnetic stimulation on specific receptors, for example, ion channels, on the stem cell membranes, and regulating the influx of cations, such as calcium and potassium, into cells. Despite the results showing a certain degree of initiation and promotion of bone matrix mineralization *in vitro*, the possibility of using magnetic stimulation on stem cells for bone regeneration is far from well understood.

Table 7.1 Studies of various cell types under magnetic stimulation.

Cell type	Organs to be regenerated	Reference
Osteoblast cells	Bone	[43, 49, 53, 54]
Fibroblast cells	–	[40, 41, 55–57]
Vascular muscle cells	Lung	[58]
Mesenchymal stem cells	–	[59]
Endothelial cells	Blood vessels	[32, 60]
Myoblast cells	Muscles	[61]
Kidney cells	Kidney	[62]
Neurons	Nerve	[63]
Bladder smooth muscle cells	Bladder	[64]
Hepatocytes	Liver	[65]

7.4 Summary

So far, we have discussed the use of various physical means to stimulate bone cells for the regeneration of bone tissues. These include exploiting use of optics (optical tweezers), magnetism (magnetic twisting and pulling), and forces (nanoindentation). Further development on using other methods of stimulation is also likely to emerge.

The major obstacles that the existing methods face are not their precision but the complexity of the mechanical forces our bone experiences everyday. To simulate these forces, a very complex model is needed and one single stimulation method may not be enough.

The formation of bone tissue *in vitro* has already been proved to be successful [66]. To use these engineered bone tissues in repairing damages may be far from straightforward. We need to consider the compatibility of the implant in terms of mechanical strength. This is particularly tricky when we have to consider the interface between the original tissue and the implant. Quite often, such a defect becomes the point of weakness. Nonetheless, the pace of progress in this research area is an encouraging sign that such repairing technique will soon be available to the public.

References

1. Robling, A.G., Castillo, A.B., and Turner, C.H. (2006) *Annu. Rev. Biomed. Eng.*, **8**, 455–498.
2. Roberts, W.E., Mozsary, P.G., and Klingler, E. (1982) *Am. J. Anat.*, **165**, 373–384.
3. Martin, R.B., Burr, D.B., and Sharkey, N.A. (2004) *Skeletal Tissue Mechanics*, Springer, pp. 101–103.
4. Mosely, J.R. and Lanyon, L.E. (1998) *Bone*, **23**, 313–318.
5. Turner, C.H., Forwood, M.R., and Otter, M.W. (1994) *FASEB J.*, **8**, 875–878.
6. Turner, C.H., Owan, I., and Takano, Y. (1995) *Am. J. Physiol. Endocrinol. Metab.*, **269**, E438–E442.
7. Viscontia, L.A., Yenb, E.H.K., and Johnson, R.B. (2004) *Arch. Oral Biol.*, **49**, 485–492.
8. Brown, T.D. (2000) *J. Biomech.*, **33**, 3–14.
9. Liu, K.K. (2006) *J. Phys. D: Appl. Phys.*, **39**, R189–R199.
10. McGarry, J.G., Maguire, P., Campbell, V.A., O'Connell, B.C., Prendergast, P.J., and Jarvis, S.P. (2008) *J. Orthop. Res.*, **26**, 513–521.
11. Walker, L.M., Holm, A., Maxwell, L., Oberg, A., Sundqvist, T., and El Haj, A.J. (1999) *FEBS Lett.*, **459**, 39–42.
12. Pelleda, G., Taib, K., Sheyna, D., Zilbermana, Y., Kumbarc, S., Nairc, L.S., Laurencinc, C.T., Gazita, D., and Ortiz, C. (2007) *J. Biomech.*, **40**, 399–411.
13. Adachi, T. and Sato, K. (2005) Threshold Fiber Strain That Induces Reorganization of Cytoskeletal Actin Structure in Osteoblastic Cells, in *Biomechanics at Micro- and Nanoscale Levels* (ed. H. Wada), World Scientific Publishing, pp. 55–64.
14. Jones, D., Leivseth, G., and Tenbosch, J. (1995) *Biochem. Cell Biol.*, **73**, 525–533.
15. Ajubi, N.E., Klein-nulend, J., Nijweide, P.J., Vriheid-lammer, T., Alblas, M.J., and Burger, E.H. (1996) *Biochem. Biophys. Res. Commun.*, **225**, 62–68.
16. Ingber, D.E. (2006) *FASEB J.*, **20**, 811–827.

17. Simmons, C.A., Matlis, S., Thornton, A.J., Chen, S., Wang, C.Y., and Mooney, D.J. (2003) *J. Biomech.*, **36**, 1087–1096.
18. Caplan, A.I. and Bruder, S.P. (2001) *Trends Mol. Med.*, **7**, 259–263.
19. Qi, M.C., Hu, J., Zou, S.-J., Chen, H.Q., Zhou, H.X., and Han, L.C. (2008) *Int. J. Oral Maxillofac. Surg.*, **37**, 453–458.
20. Tuischer, J., Matziolis, G., Krocker, D., Duda, G., and Perka, C. (2005) *J. Bone Joint Surg.*, **88-B**, 130.
21. Jiang, G.L., White, C.R., Stevens, H.Y., and Frangos, J.A. (2002) *Am. J. Physiol. Endocrinol. Metab.*, **283**, E383–E389.
22. McAllister, T.N., Du, T., and Frangos, J.A. (2000) *Biochem. Biophys. Res. Commun.*, **270**, 643–648.
23. Cartmell, S.H., Porter, B.D., Garcia, A.J., and Guldberg, R.E. (2003) *Tissue Eng.*, **6**, 1197–1203.
24. Porter, B., Zauel, R., Stockman, H., Guldberg, R., and Fyhrie, D. (2005) *J. Biomech.*, **38**, 543–549.
25. Zhao, F., Chella, R., and Ma, T. (2007) *Biotechnol. Bioeng.*, **96**, 584–595.
26. Seifritz, W. (1924) *Br. J. Exp. Biol.*, **2**, 1–11.
27. Heilbrunn, L.V. (1922) *Jahrb. Wiss. Bot.*, **61**, 284.
28. Crick, F.H.C. and Hughes, A.F.W. (1950) *Exp. Cell Res.*, **1**, 37–80.
29. Valberg, P.A. and Albertini, D.F. (1985,) *J. Cell Biol.*, **101**, 130–140.
30. Valberg, P.A. and Feldman, H.A. (1987) *Biophys. J.*, **52**, 551–569.
31. Valberg, P.A. and Butler, J.P. (1987) *Biophys. J.*, **52**, 537–550.
32. Wang, N., Butler, J.P., and Ingber, D.E. (1993) *Science*, **260**, 1124–1127.
33. Xu, Z.P., Zeng, Q.H., Lu, G.Q., and Yu, A.B. (2006) *Chem. Eng. Sci.*, **61**, 1027–1040.
34. Mornet, S., Vasseur, S., Grasset, F., and Duguet, E. (2004) *J. Mater. Chem.*, **14**, 2161–2175.
35. Katz, E. and Willner, I. (2004) *Angew. Chem. Int. Ed.*, **43**, 6042–6108.
36. Sun, S.H., Zeng, H., Robinson, D.B., Raoux, S., Rice, P.M., Wang, S.X., and Li, G.X. (2004) *J. Am. Chem. Soc.*, **126**, 273–279.
37. Sun, X.C., Gutierrez, A., Yacaman, M.J., Dong, X.L., and Jin, S. (2000) *Mater. Sci. Eng. A.*, **286**, 157–160.
38. McBain, S.C., Yiu, H.H.P., and Dobson, J. (2008) *Int. J. Nanomed.*, **3**, 169–180.
39. Ambrosi, A., Castaneda, M.T., Killard, A.J., Smyth, M.R., Alegret, S., and Merkoci, A. (2007) *Anal. Chem.*, **79**, 5232–5240.
40. Glogauer, M., Ferrier, J., and McCulloch, C.A.G. (1995) *Am. J. Physiol.*, **38**, C1093–C1104.
41. Glogauer, M. and Ferrier, J. (1998) *Eur. J. Physiol.*, **435**, 320–327.
42. Goldschmidt, P.L., Devillechabrolle, A., Ait-Arkoub, Z., and Aubin, J.T. (1998) *Clin. Diagn. Lab. Immunol.*, **5**, 513–518.
43. Pommerenke, H., Schreiber, E., Durr, F., Nebe, B., Hahnel, C., Moller, W., and Rychly, J. (1996) *Eur. J. Cell Biol.*, **70**, 157–164.
44. Bausch, A.R., Moller, W., and Sackmann, E. (1999) *Biophys. J.*, **76**, 573–579.
45. Alenghat, F.J., Fabry, J.B., Tsai, K.Y., Goldmann, W.H., and Ingber, D.E. (2000) *Biochem. Biophys. Res. Commun.*, **277**, 93–99.
46. Berrios, J.C., Schroeder, M.A., and Hubmayr, R.D. (2001) *J. Appl. Physiol.*, **91**, 65–73.
47. Wang, N. (1998) *Hypertension*, **32**, 162–165.
48. Wang, N. and Stamenovic, D. (2000) *Am. J. Physiol.*, **279**, 188–194.
49. Bierbaum, S. and Notbohm, H. (1998) *Eur. J. Cell Biol.*, **77**, 60–67.
50. Zauner, W., Farrow, N., and Haines, M.R. (2001) *J. Controlled Release*, **71**, 39–51.
51. Pisken, E., Tuncel, A., Denzili, A., and Ayhan, H. (1994) *J. Biomater. Sci.*, **5**, 451–471.
52. Yamamato, N., Fukai, F., Ohshima, H., Terada, H., and Makino, K. (2002) *Colloids Surf.*, **25**, 157–162.
53. Schmidt, C., Pommerenke, H., Durr, F., Nebe, N., and Rychly, J. (1998) *J. Biol. Chem.*, **273**, 5081–5085.
54. Pommerenke, H., Schmidt, C., Durr, F., Nebe, B., Luthen, F., Muller, P., and Rychly, J. (2002) *J. Bone Miner. Res.*, **17**, 603–611.
55. Koh, S.K., Arora, P.D., and McCulloch, C.A.G. (2001) *J. Biol. Chem.*, **276**, 35967–35977.

56. Wu, Z., Wong, K., Glogauer, M., Ellen, R.P., and McCulloch, C.A.G. (1999) *Biochem. Biophys. Res. Commun.*, **261**, 419–425.
57. D'Addario, M., Arora, P.D., Fan, J., Ganss, B., Ellenm, R.P., and McCulloch, C.A.G. (2001) *J. Biol. Chem.*, **276**, 31969–31977.
58. Goldschmidt, M.E., Kenneth, J., McLeod, W., and Taylor, R. (2001) *Circuit Res.*, **88**, 674–680.
59. Bierbaum, S. and Notbohm, H. (1997) Magnetomechanical Stimulation of Mesenchymal Cells, in *Scientific and Clinical Applications of Magnetic Carriers* (ed. E.A. Hafeli), Plenum Press, New York, pp. 311–322.
60. Ingber, D.E. (2000) *Nat. Cell Biol.*, **2**, 666–668.
61. Yuge, L. and Kataoka, K. (2000) *Cell Dev. Biol. Anim.*, **36**, 383–386.
62. Niggel, J., Sigurdson, W., and Sachs, F. (2000) *J. Membr. Biol.*, **174**, 121–134.
63. Fass, J.N. and Odde, D.J. (2003) *Biophys. J.*, **85**, 623–636.
64. Kushida, N., Kabuyama, Y., Yamaguchi, O., and Homma, Y. (2001) *Am. J. Physiol. Cell Physiol.*, **281**, 1165–1172.
65. Nebe, B., Rychly, J., Knopp, A., and Bohn, W. (1995) *Exp. Cell. Res.*, **218**, 479–484.
66. Bonzani, I.C. (2006) *Curr. Opin. Chem. Biol.*, **10**, 568–575.

8
Joint Replacement Implants

Duncan E. T. Shepherd

8.1
Introduction

Human synovial joints, such as the ankle, fingers, hip, and knee, enable articulation between two bone ends. The bone ends in a joint are covered with articular cartilage that enables the low friction movement of the joints. The joints are outstanding as they can last and be trouble-free for a lifetime. However, the joints of some people can be affected by disease, such as osteoarthritis or rheumatoid arthritis, that causes destruction of the articular cartilage joint surface. In the advanced stages of the disease, this can result in great pain, disability, and a poor quality of life for the person. Joint replacement implants have been one of the great successes in healthcare where the diseased synovial joint is replaced with an artificial joint manufactured from synthetic materials [1]. This chapter deals with joint replacement implants, detailing the materials used in their manufacture and the various designs of implants that exist for the ankle, fingers, hip, knee, and wrist. It is also important to have an understanding of how engineers go about designing joint replacement implants, and later in the chapter the main stages of the design process for joint replacement implants are discussed.

8.2
Biomaterials for Joint Replacement Implants

Materials that are used in the manufacture of joint replacement implants include metals, polymers, ceramics, and elastomers. All these materials must be biocompatible for use in the human body meaning that the materials should not cause an adverse reaction to the body (e.g., nontoxic, noncarcinogenic) and the body should not cause the material to degrade (e.g., corrosion, reduced strength) [2].

Metals are typically used for the stems of implants since they have the necessary strength to prevent fracture, a high enough Young's modulus to prevent appreciable deflection, and good fatigue strength to be able to survive the cyclic loading nature of joint replacement implants in the human body. Stainless steel, cobalt chrome

Biomechanics of Hard Tissues: Modeling, Testing, and Materials.
Edited by Andreas Öchsner and Waqar Ahmed

ISBN: 978-3-527-32431-6

Table 8.1 Typical mechanical properties of metals used for joint replacement implants.

Material	Young's modulus (GPa)	Ultimate tensile strength (MPa)	Fatigue strength @10 million cycles (MPa)
Stainless steel	193	1100	583 [7]
Cobalt chrome molybdenum alloy	200	655	290 [3]
Titanium alloy	114	860	500 [8]

All values from Brown [9], unless indicated otherwise.

molybdenum alloy, and titanium alloy are all used for joint replacement implant stems, and their typical mechanical properties are shown in Table 8.1. Various grades of stainless steel are used with a common one known as *Ortron 90*, while Ti–6Al–4V is the typical titanium alloy [3]. Each metal has a national or international standard associated with it to ensure consistency in composition and mechanical properties. For example, BS 7252-9 [4] is for stainless steel, BS 7252-4 [5] is for cobalt chrome molybdenum alloy, and BS 7252-3 [6] is for titanium alloy.

Metals are also commonly used as bearing materials in joint replacement implants. The metals used for bearing surfaces are stainless steel and cobalt chrome molybdenum alloy. Titanium alloy is not used as a bearing material as it has been shown to have poor wear characteristics on its own. However, there are a range of coatings, such as titanium nitride, that can be applied to titanium alloy to ensure that it has suitable mechanical properties to act as a bearing material [3]. Oxidized zirconium, commercially known as *Oxinium*, is a metal with a ceramic surface that is used as a bearing surface in some joint replacement implants [10].

The ceramics alumina and zirconia have been increasingly used as bearing surfaces in hip joint replacement implants, as they have excellent wear resistance. Alumina has a Young's modulus of 380 GPa and a fracture toughness of 4 MPa $m^{0.5}$. The Young's modulus and fracture toughness of zirconia are 210 GPa and 8 MPa $m^{0.5}$, respectively. The standards for alumina and zirconia are BS 7253-2 [11] and BS 7253-6 [12], respectively. As ceramics have a much lower fracture toughness compared to metals, this did lead to some fractures in the bearing surfaces in the early days of their use. However, improved manufacturing processes for ceramics have generally prevented this problem and led to their increased use [3].

The polymer ultrahigh molecular weight polyethylene is most commonly used as one of the bearing surfaces in joint replacement implants [13]. It has been shown to be a material that has low friction when articulated against a metal or ceramic. Ultrahigh-molecular-weight polyethylene has a typical Young's modulus of 677 MPa, an ultimate tensile strength of 43 MPa, and a typical fatigue strength at 10 million cycles of 18 MPa [14]. Various standards for ultrahigh-molecular-weight polyethylene exist including BS ISO 5834-2 [15]. A newer variation of the material is highly cross-linked polyethylene, where radiation is used to increase the cross-links in the polymer. This material has been shown to have lower wear rates in joint replacement implants compared to conventional ultrahigh-molecular-weight polyethylene, although there has been some concern about reduced mechanical properties [16].

The replacement of joints of the hand and wrist has typically involved the use of implants manufactured from silicone [17]. Silicone is a rubber that has a Young's modulus of 30 MPa and an ultimate tensile strength of 7 MPa [9]. Silicone has excellent fatigue properties over a wide range of strains. However, silicone has been shown to have poor crack growth resistance once a crack has been initiated in the material.

8.3 Joint Replacement Implants for Weight-Bearing Joints

8.3.1 Introduction

Joint replacement implants are commonly used for replacement of diseased synovial joints that are weight-bearing, such as the hip, knee, and ankle. These weight-bearing joints are subjected to large forces (typically two to three times body weight during walking) over 1 million times a year [1]. This section of the chapter discusses the materials that are used to manufacture the various designs of joint replacement implants for the hip, knee, and ankle.

8.3.2 Hip Joint Replacement

The natural human hip consists of the articulation between the femoral head and the acetabulum. Conventional hip replacement surgical procedures involve cutting off the femoral head, broaching the femoral bone to create the shape of the implant stem, and reaming the acetabulum. Surgical navigation and robotic systems are increasing in popularity and have been found to be less invasive and lead to better positioning of implant parts [18].

There is a variety of designs of hip joint replacement implants available, but all consist of femoral and acetabular parts that form a ball and socket joint. The femoral part comprises a stem and a head, as shown in Figure 8.1. The femoral part can be either one-piece or a modular design where different femoral heads can be fitted to the stem. The stems are generally made from stainless steel, cobalt chrome molybdenum alloy, or titanium alloy. Conventionally, metals are also used for the femoral head with stainless steel or cobalt chrome molybdenum alloy used. Depending on the design, the diameter of the femoral head is typically between 22 mm (Charnley hip, DePuy International Ltd, Leeds, UK) and 32 mm (Muller hip, Biomet, Dordrecht, The Netherlands). During manufacture, the femoral head is highly polished, with the surface having an average roughness of around 0.008 μm.

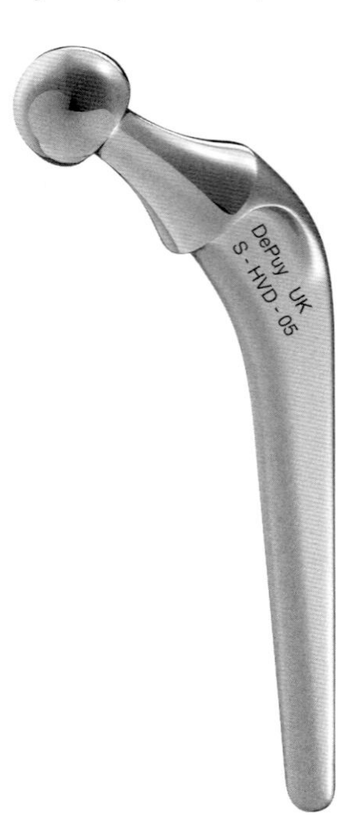

Figure 8.1 Charnley hip stem. (Reproduced with kind permission from DePuy International Ltd, Leeds, UK. © by DePuy International Ltd.)

As an alternative to the use of metals for the femoral head, ceramics, such as alumina and zirconia, can be used and they will be attached to a metal stem [19, 20]. These ceramic heads have lower surface roughness values than metals, typically 0.002 μm. Recent advances in the materials for femoral heads is the use of coatings, which have a very high wear resistance. Smith and Nephew have produced Oxinium, which is a metal with a ceramic surface.

The acetabular part of a hip joint replacement can be a one-piece design (Figure 8.2) that fixes directly into the acetabular bone. This design is manufactured from either ultrahigh-molecular-weight polyethylene or cobalt chrome molybdenum alloy, depending on the bearing coupling the surgeon requires with the femoral head. The ultrahigh-molecular-weight polyethylene acetabular cup has a metal wire attached to its outside so that the part can be seen on X-rays. After manufacture, the ultrahigh-molecular-weight polyethylene bearing surface has an average surface roughness of around 1.29 μm, which is higher than the surface finishes achieved on the surfaces of metals and ceramics. A two-piece design of the acetabular part is also available that consists of a metal shell (that fixes into the acetabular bone) with a liner that inserts into the shell. Liners are made from ultrahigh-molecular-weight polyethylene or a ceramic. Highly cross-linked polyethylene is now in use as a bearing surface for the acetabular part and early clinical results show that the wear rates are far reduced compared with conventional ultrahigh-molecular-weight polyethylene [16].

The range of bearing material combinations available for hip replacement implants consists of metal against polymer, metal against metal, ceramic against

Figure 8.2 One-piece ultrahigh-molecular-weight polyethylene acetabular parts.

polymer, and ceramic against ceramic. These different material combinations result in differences in the lubrication regimes that occur between the bearing surfaces. The ultimate aim of any bearing would be fluid-film lubrication where the two bearing surfaces are completely separated by a thin film of fluid during motion. Boundary lubrication is where there is contact between the bearing surfaces as conditions are not conducive for a film of fluid to be present. The transition zone between boundary and fluid-film lubrication is known as the *mixed lubrication regime*. The metal against polymer-bearing combination for hip replacement implants generally operates with a boundary lubrication regime, which means that there will be contact between the bearing surfaces and the generation of wear debris. The ceramic against polymer combination generally operates with a mixed lubrication regime. Fluid-film lubrication has been shown to occur for metal against metal and ceramic against ceramic material combinations under some conditions. However, different modes of lubrication will operate throughout the walking cycle [21].

The traditional metal against polymer bearing combination has a survivorship rate of around 90% at 15 years. Indeed survivorship rates at 25 years have been shown to be around 80% for some implant designs. The most common clinical problem associated with metal against polymer hip joint replacement implants is loosening of the femoral or acetabular parts. Wear particles can cause osteolysis, where an adverse cellular reaction of the bone to the polymer wear particles causes bone resorption and implant loosening. It has been shown that it is not the total wear volume that causes the adverse cellular reaction, but the number of wear particles in the size range of 0.2–0.8 μm [22]. Wear debris is still generated in metal against metal, ceramic against polymer, and ceramic against ceramic hip implants, but at a lower volume [21].

Loosening of the femoral or acetabular parts of the hip replacement will cause pain in the patient and the motion of the hip may be limited. Revision surgery involves removal of the parts, including removing all bone cement if it was used in the original surgical procedure. Bone loss is a major challenge in revision surgery

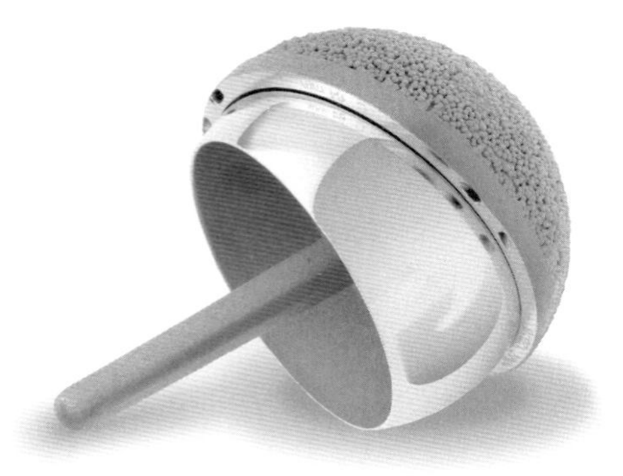

Figure 8.3 The Birmingham hip. (Reproduced with kind permission from Smith & Nephew, Memphis, TN, USA. © by Smith & Nephew.)

and, in many cases, bone graft will be used to replace the missing bone. In most cases, the joint replacement implant will be specifically designed for use in a revision case, which generally involves the use of longer stems for the femoral part [23]. Revision of the acetabular part is a challenging surgical procedure where there is extensive bone loss. A newer approach involves the use of blocks of tantalum, which is a metal in the form of a porous structure that bone can grow into [24].

Hip resurfacing (Figure 8.3) is a newer alternative to the traditional design of hip replacement implants, for use in young and active people. The resurfacing implant has a cobalt chrome molybdenum alloy against cobalt chrome molybdenum alloy bearing articulation, with a large diameter femoral head (typically 50 mm). Hip resurfacing implants have been found to operate under fluid-film lubrication, under some conditions [25]. The femoral part is fixed to the bone with the use of a peg instead of a stem which is used in the traditional femoral part. This means that there is more preservation of the bone. Large diameter femoral heads are also

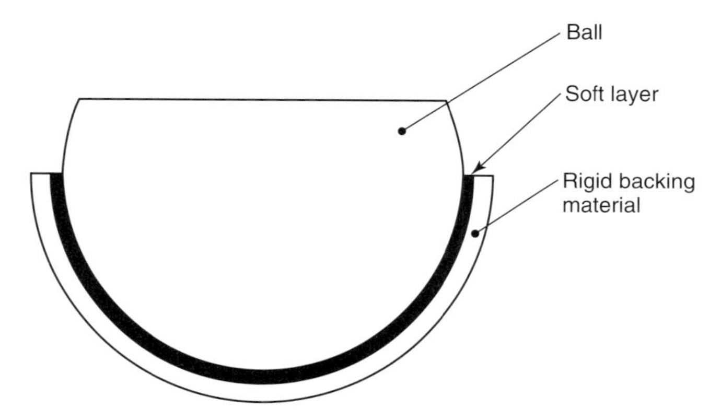

Figure 8.4 Cushion form joint for the hip.

available for attachment to conventional femoral stems where a patient might not be suitable for the peg.

Future developments in hip joint replacement implants are generally investigating the use of different bearing surfaces. This includes the combination of ceramic against metal and the use of diamond-like coatings [20, 21]. An alternative approach that has been investigated in hip simulator studies in the laboratory is to use a soft layer of rubber as one of the bearing surfaces (Figure 8.4). The design rationale behind this is the fact that in natural human synovial joints the articular cartilage is a low modulus material and the use of a rubber layer may benefit from the excellent tribological characteristics seen in natural synovial joints [21]. However, this has yet to reach a clinical investigation.

8.3.3 Knee Joint Replacement

The natural human knee joint is formed by the distal end of the femur, the proximal ends of the tibia and fibula together with the patella. The meniscus covers a large proportion of the surface of the proximal tibia. Four ligaments connect the knee together and provide stability during locomotion: the anterior and posterior cruciate ligaments and the two collateral ligaments. Knee replacement surgery involves using a series of jigs to guide an oscillating saw to cut the bone on the femur and tibia to the appropriate shape for the parts of the implant. In a similar manner to hip replacement surgery, surgical navigation and robotic systems are increasing in popularity [18].

There are a huge variety of designs of knee replacement implants including fixed bearing, mobile bearing, and rotating hinge designs. Various designs also allow for the preservation of both cruciates, preservation of the posterior cruciate, or resection of both cruciates. Designs of total knee replacement implants (Figure 8.5) consist of femoral and tibial parts; some designs may also include a patellar part. The femoral part is generally made from cobalt chrome molybdenum alloy. The tibial part is generally a two-piece design consisting of a tibial tray and an insert. The insert is manufactured from ultrahigh-molecular-weight polyethylene, with the tray made from a metal. With fixed bearing designs (where the tray and insert are rigidly fixed together) the tibial tray is manufactured from titanium alloy. With mobile bearing designs (where the insert is free to move relative to the tray) the tray is manufactured from cobalt chrome molybdenum alloy, since titanium alloy has poor wear characteristics. The design rationale with a mobile bearing is that it will reduce wear and allow for more natural motion [20, 26, 27].

Metal against polymer knee joint replacement implants operate with a boundary/mixed lubrication regime and therefore wear will occur [28]. Knee replacement implants typically have a 95% survival rate at 10 years, with osteolysis being a major cause of revision surgery. With fixed bearing designs, wear has occurred on the top side and bottom side of the polymer insert. Mobile bearing designs have been shown to have much reduced wear compared to fixed bearing designs [29].

Figure 8.5 PFC sigma fixed bearing knee system cruciate retaining. (Reproduced with kind permission from DePuy International Ltd, Leeds, UK. © by DePuy International Ltd.)

An alternative to the use of a total knee replacement implant is the use of a unicondylar (or unicompartmental) implant. This is used where damage to the joint surface of the natural knee is confined to either the medial or lateral side of the knee. The materials used are the same as for the total knee system, with fixed and mobile bearings also available [30].

Alternative materials in knee replacement implants include the use of titanium nitride coatings or Oxinium for the femoral part. Highly cross-linked polyethylene has been used for the tibial part [31]. Future development may involve the use of cushion form joints, which have so far been confined to laboratory simulator studies [32, 33].

8.3.4 Ankle Joint Replacement

The ankle joint consists of the proximal articulating surfaces of the talus and the distal ends of the tibia and fibula. Arthritis in the ankle is less frequent than that of the hip and knee and therefore the use of ankle joint replacement implants is less. However, the smaller number of ankle joint replacements performed can also be attributed to the designs not being as successful as those for the hip and knee. The surgical procedure for ankle joint replacement is also more challenging as the

Figure 8.6 BOX® total ankle replacement. (Reproduced with kind permission from Finsbury Orthopaedics, Leatherhead, Surrey, UK. © by Finsbury Orthopaedics.)

ankle joint is not as accessible as the hip or knee. Most surgical procedures for ankle joint replacement involve distraction where the surfaces of the tibia and talus are separated to enable access. Jigs are then used to guide a saw to cut the bone from the tibia and the talus to the shape of the implant.

The designs of ankle joint replacement implants have generally comprised three parts: a metal tibial part, a metal talar part, and a polymer insert. Early designs had a fixed bearing design where the polymer insert was rigidly attached to the tibial part. Newer designs now have mobile bearings, where the polymer part is free to articulate against the metal parts attached to the tibia and talus. An example of a mobile bearing ankle joint replacement is shown in Figure 8.6. The articulating bearing surfaces of ankle joint replacement implants generally consist of stainless steel or cobalt chrome molybdenum alloy against ultrahigh-molecular-weight polyethylene [34–36]. This metal against polymer material combination is likely to operate with a boundary lubrication regime in a similar manner to metal against polymer hip and knee joint replacement implants.

Owing to the smaller number of ankle joint replacements that are implanted compared to hips and knees, follow-up studies are less frequent. However, osteolysis has been noted in some studies and *in vitro* simulator studies have shown wear debris to be generated. Revision surgery involves replacing the implant, or, in many cases, performing an ankle fusion [37, 38].

8.3.5
Methods of Fixation for Weight-Bearing Joint Replacement Implants

Secure fixation of joint replacement implants within the bone is essential to ensure clinical success. The methods of fixation available are bone cement, bone ingrowth, and mechanical [26].

Bone cement has been used for the fixation of implant stems within bone for over 40 years. Bone cement is based on polymethylmethacrylate and is made by mixing together polymethylmethacrylate powder, liquid methylmethacrylate, benzoyl peroxide, and barium sulfate (radiopaque additive so the cement can be seen on X-rays) to form a paste. Additionally, an antibacterial agent may also be added. The cement is mixed in the operating theater by either hand or machine, and may sometimes be mixed in a vacuum. Mixing these components together causes an exothermic reaction and the temperature of the cement is typically in the range 67–124 °C. Once the cement has been mixed, it is injected or applied manually into the bone cavity. It is important that pressure is applied to the cement so that it can be forced into the interstices of the bone, which will increase the strength at the bone/cement interface. In some cases, such as the acetabular component of a hip replacement, holes may be drilled into the bone so that the cement can enter deeper into the bone [39].

Implant stems for use in cement are generally designed with a tapered shape so that as the stem is pushed into the bone cement it will cause pressurization of the cement. Once the implant is in place, the cement will cure and harden in around 6–15 min, depending on the type of cement used. Bone cement is not an adhesive, but acts as a filler between the bone and the implant stem and therefore relies on mechanical interlock. The main advantage of using bone cement is that after it has cured, there is immediate fixation of the implant within the bone and the patient can weight-bear soon after surgery. Disadvantages can include bone necrosis from the temperature of the exothermic reaction, blood contamination, and fatigue failure.

Fixation of implants can also be achieved through bone ingrowth. The implant stem is manufactured so that it is roughened or a porous surface is applied. The bone does not attach itself to the implant, but instead grows into the roughened or porous surface. There is a variety of techniques for producing a porous surface such as sintered titanium alloy beads or the use of hydroxyapatite, which is applied by a plasma spray [40, 41]. Bone ingrowth techniques do require the patient to have good bone stock and the ultimate fixation is not immediate as time is required for the bone to grow into the porous coating.

Mechanical fixation includes the use of an interference fit and screws. An interference, or press fit, involves a component being pushed into the bone. This can be used for hip and knee joint replacement implants, but generally there is a roughened or porous surface for bone ingrowth as well. Some implants rely completely on a mechanical fixation. Screw fixation is used in the Hintegra total ankle prosthesis (Newdeal, Lyon, France) and is also used for additional fixation in hip replacement surgery where the acetabular part is revised.

8.4 Joint Replacement Implants for Joints of the Hand and Wrist

8.4.1 Introduction

Joint replacement in the hand and wrist is not as common as replacement of the hip and knee. Some of this can be attributed to the implants not matching the clinical success of hip and knee joint replacement implants. However, rheumatoid arthritis which commonly affects the joints of the hand and wrist causes extensive damage to the surrounding soft tissues and bone, leading to severe deformities. This is an added complication when it comes to the design of a successful replacement implant for these joints. In this section, the designs of replacement implants for the joints of the hand and wrist are discussed.

8.4.2
Finger Joint Replacement

The finger joints in the human hand include the metacarpophalangeal (MCP), proximal interphalangeal (PIP), and distal interphalangeal (DIP) joints and are essential for hand function. The MCP is the finger joint most commonly replaced with an implant [42]. For the MCP joint, the surgical procedure involves cutting to remove the bone ends of the joint and broaching the shaft of the bone to create a cavity for the implant stems. There is a variety of designs of finger joints available and they can be broadly divided into single-piece silicone joints and articulating bearing surface joints [43, 44].

The most common design of joint replacement implant for the fingers is a single-piece silicone spacer. The Swanson finger implant (Wright Medical Technology, Arlington, Tennessee, USA) has been in use since the 1960s and comprises two stems that are joined to a central barrel (Figure 8.7). Since its introduction, the implant has been manufactured from different grades of silicone (including conventional and high-performance silicone); it is currently manufactured from Flexspan®.

The stems of the Swanson implant are not actually fixed within the bone; during finger movement, the implant pistons in and out of the bone cavity. The Swanson implant has been found to have a poor survivorship, compared with joint replacement implants for the hip and knee. Failure of the Swanson implant has typically involved fractures occurring between the barrel and the distal stem. Studies have shown that 34% of implants had fractured at 48 months follow-up [45]. The crack is believed to be initiated from contact with the sharp bone edges or from abrasion against the bone during finger movement as the implant pistons in and out of the bone. To protect the Swanson implant from the bone, titanium

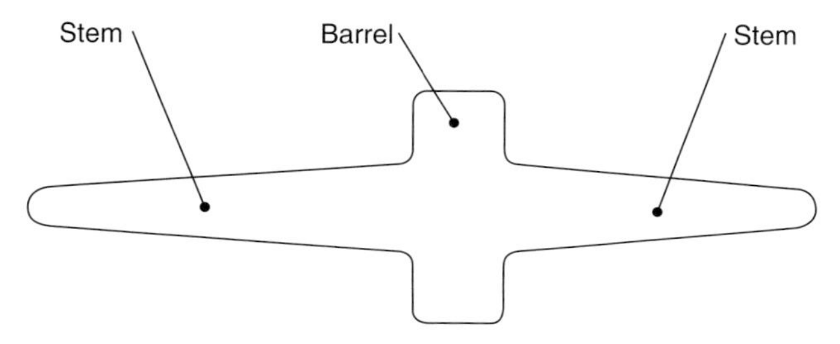

Figure 8.7 Swanson finger implant.

grommets, which fit over the stems of the implant, were introduced to act as a shield between the silicone and the bone. However, fractures have still been observed. The Avanta Soft Skeletal Implant (Avanta Orthopaedics, San Diego, California, USA) was introduced as the Sutter implant. It is a similar design to the Swanson and the problems of fracture have also been seen clinically. At 48 months follow-up, 26% of Sutter implants had fractured in one study [45].

The Swanson implant has been subjected to *in vitro* mechanical testing using a finger joint simulator and the implants have also been found to develop a crack at the junction between the barrel and the distal stem. The crack was found to grow as the implant was subjected to flexion/extension in the simulator until fracture occurred after about 3 million cycles of flexion/extension [46]. Similar results have also been found when the Sutter implant was mechanically tested using a finger joint simulator, with fracture occurring at the junction of the distal stem and the hinge between 5 million and 10 million cycles of flexion/extension [47].

Newer designs of single-piece silicone implants that have been designed for the finger joint include the Preflex (Avanta Orthopaedics, San Diego, CA, USA) and Neuflex (DePuy, Leeds, UK) [44]. The Neuflex is shown in Figure 8.8. These newer designs have the neutral position of the implant with one stem at an angle of 30° to the other. The design philosophy for this type of implant is that the natural resting position of the hand has the MCP joint at about 30°. If the implant is straight,

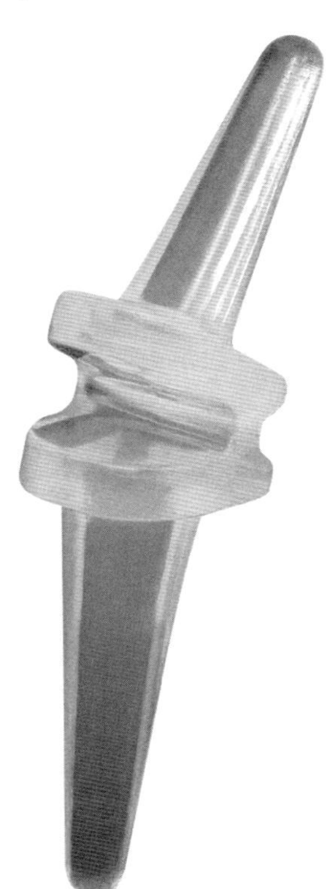

Figure 8.8 Neuflex finger joint. (Reproduced with kind permission from DePuy International Ltd, Leeds, UK. © by DePuy International Ltd.)

Figure 8.9 Total Metacarpophalangeal Replacement, TMPR™. (Reproduced with kind permission from Finsbury Orthopaedics, Leatherhead, Surrey, UK. © by Finsbury Orthopaedics.)

as with the Swanson, the neutral position of the hand at 30° will induce a higher strain in the material. *In vitro* testing of the Neuflex using a finger joint simulator has shown that it lasts for much longer than the Swanson and Sutter implants, with between 9 million and 20 million cycles of flexion/extension completed before fracture was seen along the central hinge [48].

An alternative design of finger joint replacement implant to the use of silicone implants is the use of articulating bearing surfaces. For the MCP joint, these designs comprise metacarpal and phalangeal parts that form a ball and socket joint. However, many surgeons believe that this type of implant is inappropriate for patients with rheumatoid arthritis, as the soft tissues are damaged by the disease and there is inadequate support for the implant [44]. Examples of articulating bearing surface implants include the Total Metacarpophalangeal Replacement, TMPR (Finsbury Orthopaedics, Leatherhead, Surrey, UK) which has a cobalt chrome molybdenum alloy against ultrahigh-molecular-weight polyethylene articulation (Figure 8.9) and the SR™ (Small Bone Innovations, Morrisville, PA, USA). The TMPR™ design has a mechanical fixation with a series of ultrahigh-molecular-weight polyethylene fins that wedge against the inside of the bone.

Other designs of articulating finger joint replacements include the Pyrocarbon Total Joint (Ascension Orthopaedics, Austin, TX, USA) which has an articulation of pyrocarbon against itself. Current designs of articulating finger joint replacements have been shown to operate with a boundary lubrication (metal against polymer) or mixed lubrication regime (pyrocarbon against itself) [49].

8.4.3
Wrist Joint Replacement

The human wrist joint is made up of the radius, the ulna, and the eight carpal bones. Wrist joint replacement involves cutting away the diseased bone and using a broach to create a cavity within the radius and carpal bones for insertion of the implant stems. A single-piece silicone Swanson implant is available for the wrist, with the design along similar lines to the Swanson finger implant, with stems joined to a central barrel (Figure 8.10). Titanium grommets are also available to shield the silicone from the edges of the bone. The Swanson wrist implant also suffers from the same problems as the Swanson finger implant, with fractures occurring at the junction between the barrel and the distal stem. Fracture rates of up to 65% have been reported at follow-up studies of 6 years [50, 51].

Articulating surface implants, with a cobalt chrome molybdenum alloy against ultrahigh-molecular-weight polyethylene articulation, are also available for wrist replacement implants (Figure 8.11). Designs include the Biaxial (DePuy International Ltd, Leeds, UK), Universal 2 (Integra, Plainsboro, NJ, USA), and Re-motion (Small Bone innovations, Morrisville, PA, USA) [51]. These cobalt chrome molybdenum alloys against ultrahigh-molecular-weight polyethylene articulating implants operate with a boundary lubrication regime [52] and therefore wear is likely. However, owing to the early failure of these wrist implants, wear problems have only been detailed in a few studies. Revision surgery is more often required for loosening, dislocation, and subsidence of the stems into the soft rheumatoid bone. In many cases, a wrist fusion is performed rather than attempting to implant another wrist replacement.

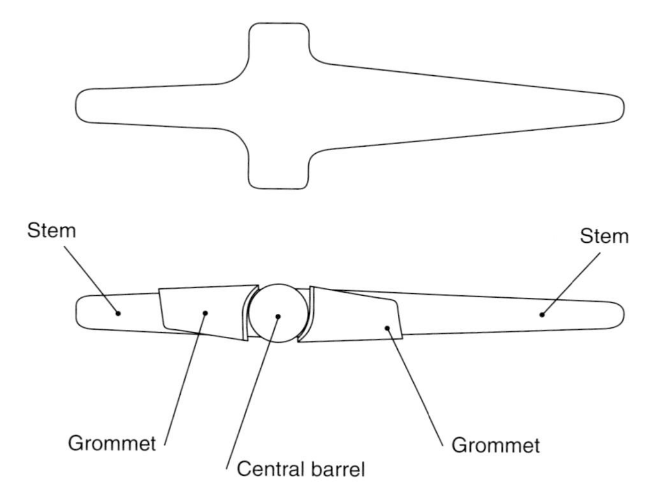

Figure 8.10 Swanson wrist implant. Lower part of the figure shows grommets fitted.

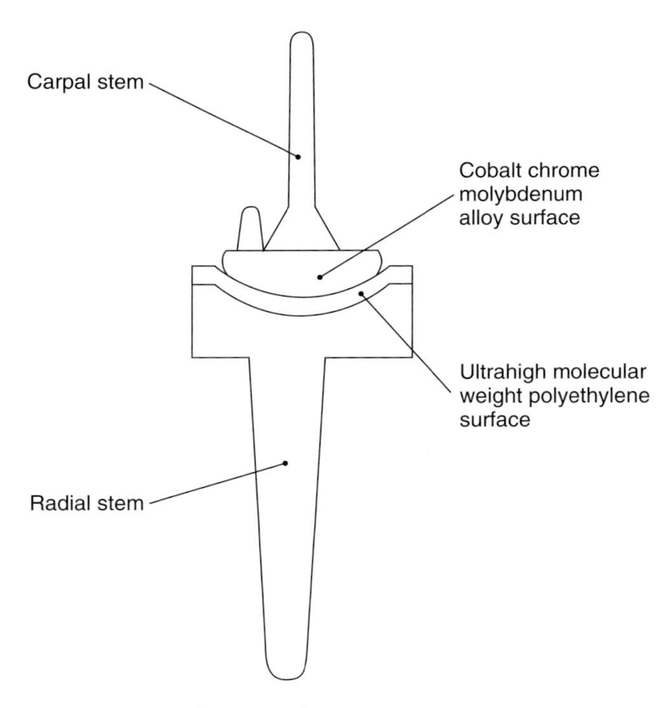

Figure 8.11 Biaxial wrist implant.

8.5 Design of Joint Replacement Implants

8.5.1 Introduction

The design process for joint replacement implants is highly regulated to ensure the safety of patients. In Europe, the Medical Device Directive [53] was developed, while in the United States of America, the Food and Drug Administration (FDA) has the responsibility for the regulation of implants [54]. Companies producing joint replacement implants are required to have a quality management system in place to ensure that the designs of implants are managed in a systematic and repeatable manner. Medical device companies are required to maintain a design history file and document the whole design process of an implant, from start to finish, to ensure that all decisions are traceable. Once a joint replacement implant is in use, the medical device company has responsibility for the lifetime of the device and they must have a postmarket surveillance system in place to ensure that any problems are rectified, or, in extreme cases, the implant is withdrawn from the market. An overview of the design process for joint replacement implants is shown in Figure 8.12. At each key stage, it is important that a design review is carried out

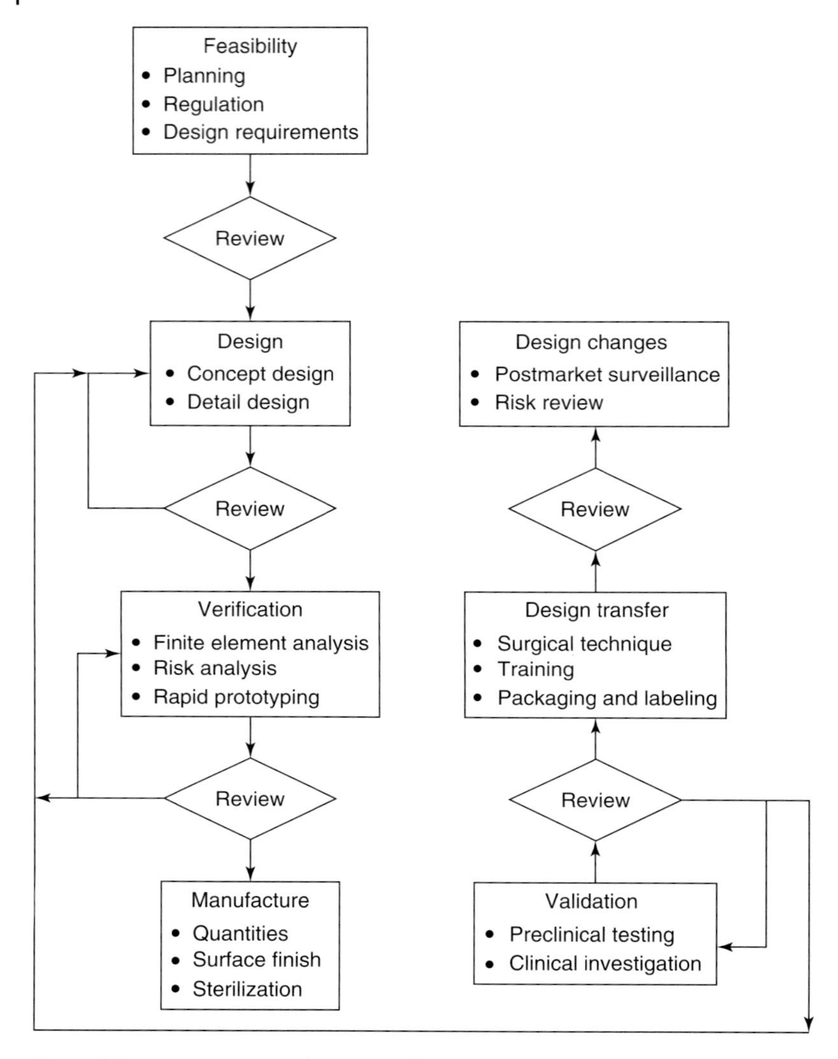

Figure 8.12 Design process for joint replacement implants.

by the design team to ensure that all aspects are considered. This section of the chapter discusses the main stages of the design process, namely

- feasibility
- design
- verification
- manufacture
- validation
- design transfer
- design changes.

8.5.2
Feasibility

Design ideas for new or improved joint replacement implants will come from orthopedic surgeons, sales people, or engineers. In the initial stages of the project the feasibility will be assessed. This will include outlining the clinical indications, assessing the market potential of a new joint replacement implant, and reviewing the intellectual property to assess if there is scope to protect the design through patents.

Before a joint replacement implant can be designed, it is essential that the design requirements for the implant are produced. The design requirements capture everything that is required of the implant to ensure that it fulfills the required performance once implanted into a patient. Various standards exist to help determine the design requirements for implants ranging from a general standard BS EN ISO 14630 [55] to standards that are specific to certain joint replacement implants, such as the hip BS EN 12563 [56] and knee BS EN 12564 [57]. Besides considering the requirements of the joint replacement implant, it is vital to consider the requirements of the packaging and labeling. In addition, the requirements of any surgical instruments must be considered early in the design process. It is not always the design of an implant on its own that dictates good clinical success; high-quality instruments that are easy to use and enable accurate placement of the implants are vital. The design requirements for a joint replacement implant will include details on

- intended performance
- design attributes
- materials
- design evaluation
- manufacture
- testing
- instruments required
- sterilization
- packaging
- information to be supplied by the manufacturer.

8.5.3
Design

At the concept design stage, design solutions that meet the design requirements are generated. It is important that design concepts are not dismissed in the initial stages. An open mind is important to try and consider all possible solutions. The concept design stage may involve

- sketches of design solutions;
- computer-aided design models; and

- analytical calculations, for example, to calculate the likely stress or deflection in the implant.

A number of techniques, such as brainstorming and TRIZ (theory of inventive problem solving) exist to help with the development of design concepts. The developed range of concepts for a joint replacement implant can then be rated by the design team to determine the most suitable concept to develop into a final design. This may involve rating each concept against known criteria, such as cost or ease of manufacture, or against a known "gold standard" joint replacement implant [58]. During the decision process to determine which concept to develop into the final design, it is important to not just rely on the opinion of the engineers. It is vital that orthopedic surgeons and sales personnel are involved. Surgeons have to implant the joint replacement devices, while the sales people have to actually sell them.

Once the concept to be developed has been decided, the process moves on to the detail of the design stage where the concept design is worked up into a full design for a joint replacement implant. This stage will include

- producing models using computer-aided design software – this helps to visualize the implants and to see how the instruments will interact with it;
- specification of appropriate materials;
- drafting engineering drawings – this helps to determine if any parts of the design can be simplified;
- consultation with manufacturers – this helps to determine if the device can be manufactured at a reasonable cost and whether any special manufacturing processes are likely to be required.

8.5.4 Verification

During the design stage, it is essential to verify that the proposed design of joint replacement implant meets the design requirements. Verification is essentially asking the question: "are we building the thing right?" [59]. Verification of designs of joint replacement implants can include

- finite element analysis
- rapid prototyping
- risk analysis.

Finite element analysis is a computational technique that can be used to simulate the loading conditions in the human body to verify if a joint replacement implant has sufficient strength and stiffness to withstand the expected loading conditions [60]. The technique can also be used to predict the wear of joint replacement implants [61]. However, finite element analysis must not be used on its own; the models must be validated against known results, whether they be analytical solutions or mechanical testing.

Rapid prototyping is a very effective technique for verification as it aids communication between design engineers and orthopedic surgeons. Models of joint replacement implants created using computer-aided design software can be prototyped in plastic, using techniques such as three-dimensional printing, within a few hours. The plastic models can then be used to size against a skeleton and to check for functionality. The interactions of surgical instruments with the prototype implant can also be investigated.

Risk analysis is an important stage in the design process to consider all the hazards associated with a joint replacement implant (e.g., wear, lack of strength, incorrect materials, etc.) and to reduce the risk as far as possible. A common technique used in the risk analysis of joint replacement implants is failure mode and effect analysis, which is a bottom-up approach where each part, and then the assembly (including packaging, sterilization, and labeling), is considered [62]. In the risk analysis of possible failure modes, the effect of each failure mode and the possible cause of the failure are identified. The occurrence, severity, and detection of each failure are then rated on a scale from 1 to 10. A risk priority number is then calculated by multiplying the three ratings together. The design team must then decide if a possible failure mode is acceptable or if ways need to be found to reduce it. Table 8.2 shows the results of a partial risk analysis for a Swanson finger implant that is described in Section 8.4.2.

Once the final design has been produced, manufacture of the final implants for preclinical testing is undertaken.

Table 8.2 Results of a risk analysis for the Swanson finger implant.

Possible hazard or failure mode	Effect of hazard or failure	Cause of hazard or failure	O	S	D	RPN
Fractures	Device does not function as intended	Fatigue failure	10	6	5	300
		Crack initiated from sharp edges of bone	10	6	5	300
		Crack initiated from abrasion with bone	10	6	5	300
		Damaged during implantation	2	6	5	60
		Incorrect implantation technique	2	6	5	60
Wear particles of silicone	Silicone synovitis	Implant rubbing against bone	8	6	7	336

Occurrence (O), severity (S), detect (D), risk priority number (RPN).

8.5.5
Manufacture

Before the design of a joint replacement implant is transferred to production, the appropriate manufacturing processes need to be chosen to ensure that the implants are manufactured in a repeatable and reliable manner. The choice of manufacturing processes for joint replacement implants is wide and will depend on the chosen materials, shape of the implant, and the required surface finishes. For example, the femoral stem of a hip joint replacement implant could be forged or cast, followed by machining. Femoral heads could be machined from bar or cast. An appropriate sterilization process will also need to be finalized.

8.5.6
Validation

Validation of joint replacement implants involves the implants being subjected to actual or simulated conditions of use. Validation is essentially asking the question "have we built the right thing?" to ensure that the implant meets the design requirements [59]. Validation can include

- preclinical testing of implants
- a clinical investigation.

Preclinical testing of joint replacement implants will involve mechanical testing to validate the design under simulated conditions of use. Mechanical testing enables the predicted mechanical conditions in the human body to be simulated in the laboratory. This will ensure that the joint replacement implant will have sufficient strength, stiffness, and wear resistance, but will give no indication of how the implant will behave in the body from a biological point of view.

There are a large number of standards available to guide the preclinical mechanical testing of joint replacement implants. For a total hip joint replacement it would be necessary to

- determine the endurance properties of the stem to BS 7251-12 [63] by using a materials testing machine to apply a sinusoidally varying force (between 300 and 2300 N) for up to 5 million cycles;
- to determine the wear of the bearing surfaces to BS ISO 14242-1 [64] by using a hip simulator to subject a hip replacement implant to loads of up to 3000 N and motions similar to those encountered in the body. The amount of wear is determined at 0.5 million and 1 million cycles and then at least every 1 million cycles up to 5 million cycles using either the gravimetric or dimensional methods to determine the wear rate.

Similarly, a knee simulator would be used to investigate the wear of the bearing surfaces of a knee joint replacement to BS ISO 14243-3 [65] and the endurance properties of the tibial tray would be determined to BS ISO 14879-1 [66].

While standards exist for the testing of hip and knee replacement implants, none are available for the ankle, finger, or wrist joint replacement implants, although various methods of testing have been described. Ankle joint replacement implants have been tested using a modified knee simulator [67]. A test protocol for finger implants has been proposed following the development of a finger joint simulator [68]. Details on the development of a wrist simulator have been described [69].

Once preclinical testing has been completed, manufacturers of joint replacement implants are required to make a decision as to whether a clinical investigation is required. This is guided by the standard BS EN ISO 14155-1 [70]. A critical review of the literature is required to ascertain similar implants and how they have performed in patients. Analogy may be used to justify not undertaking a clinical investigation. If the manufacturer does decide that a clinical investigation is required, a clinical investigation plan must be produced in accordance with BS EN ISO 14155-2 [71]. The investigation will need approval of the Ethics Committee and the manufacturer will be required to decide on the length of the investigation, the number of patients to be involved, and the type of data to be collected.

8.5.7 Design Transfer

Once the design and testing of a joint replacement implant has been completed, it can be transferred to production. At this stage, it is necessary to have all the documents and training associated with the use of the implant finalized. Design transfer will include

- generation of instructions for use;
- planning the training of orthopedic surgeons who will implant the devices;
- finalization of the packaging and labeling.

8.5.8 Design Changes

After a joint replacement implant is on the market and being implanted into patients, the medical device company must have a postmarket surveillance system in place to ensure that any adverse reports about the implant are dealt with. This may include some design changes being made to the implant based on the feedback from orthopedic surgeons. Any changes to a joint replacement implant needs to be fully documented and any design changes fully investigated. This may include repeating verification and validation processes, depending on the type of design changes made.

8.6 Conclusions

This chapter has described the design process for joint replacement implants and the various biomaterials that are used in their manufacture. The biomaterials used in the manufacture of implants for the ankle, fingers, hip, knee, and wrist are metals, polymers, ceramics, and elastomers. Future developments in joint replacement implants will include improved designs and new biomaterials to produce longer lasting implants.

References

1. Shepherd, D.E.T. and Azangwe, G. (2007) *Appl. Bionics Biomech.*, **4**, 179–185.
2. Williams, D.F. (2008) *Biomaterials*, **29**, 2941–2953.
3. Dearnley, P.A. (1999) *Proc. Inst. Mech. Eng. H*, **213**, 107–135.
4. BS 7252-9 (1993) *Metallic Materials for Surgical Implants. Specification for High-Nitrogen Stainless Steel*, British Standards Institute, London.
5. BS 7252-4 (1997) *Metallic Materials for Surgical Implants. Specification for Cobalt-Chromium-Molybdenum Casting Alloy*, British Standards Institute, London.
6. BS 7252-3 (1997) *Metallic Materials for Surgical Implants. Specification for Wrought Titanium 6-Aluminium 4-Vanadium Alloy*, British Standards Institute, London.
7. Smethurst, E. (1981) *Biomaterials*, **2**, 116–119.
8. Li, S.J., Cui, T.C., Hao, Y.L., and Yang, R. (2008) *Acta Biomater.*, **4**, 305–317.
9. Brown, S.A. (2006) Synthetic biomaterials for spinal applications, in *Spine Technology Handbook* (eds S.M. Kurtz and A.A. Edidin), Elsevier, London, pp. 11–33.
10. Good, V., Widding, K., Hunter, G., and Heuer, D. (2005) *Mater. Des.*, **26**, 618–622.
11. BS 7253-2 (1997) *Non-metallic Materials for Surgical Implants. Specification for Ceramic Materials Based on High Purity Alumina*, British Standards Institute, London.
12. BS 7253-6 (1997) *Non-metallic Materials for Surgical Implants. Specification for Ceramic Materials Based on Yttria-Stabilized Tetragonal Zirconia (Y-TZP)*, British Standards Institute, London.
13. Li, S. and Burstein, A.H. (1994) *J. Bone Joint Surg. Am.*, **76A**, 1080–1090.
14. Sauer, W.L., Weaver, K.D., and Beals, N.B. (1996) *Biomaterials*, **17**, 1929–1935.
15. BS ISO 5834-2 (1998) *Implants for Surgery. Ultra-high Molecular Weight Polyethylene. Moulded Forms*, British Standards Institute, London.
16. Glyn-Jones, S., Saac, S., Hauptfleisch, J., McLardy-Smith, P., Murray, D.W., and Singh, H. (2008) *J. Arthroplasty*, **23**, 337–343.
17. Yoda, R. (1998) *J. Biomater. Sci. Polym. Edn.*, **9**, 561–626.
18. Davies, B.L., Baena, F.M.R.Y., Barrett, A.R.W., Gomes, M.P.S.F., Harris, S.J., Jakopec, M., and Cobb, J.P. (2007) *Proc. Inst. Mech. Eng. H*, **221**, 71–80.
19. Sanfilippo, J.A. and Austin, M.S. (2006) *Expert Rev. Med. Dev.*, **3**, 769–776.
20. Lee, K. and Goodman, S.B. (2008) *Expert Rev. Med. Dev.*, **5**, 383–393.
21. Dowson, D. (2001) *Proc. Inst. Mech. Eng. H*, **215**, 335–358.
22. Ingham, E. and Fisher, J. (2000) *Proc. Inst. Mech. Eng. H*, **214**, 21–37.
23. Brubaker, S.M., Brown, T.E., Manaswi, A., Mihalko, W.M., Cui, Q., and Saleh, K.J. (2007) *J. Arthroplasty*, **22** (Suppl. 3), 52–56.

24. Levine, B.R., Sporer, S., Poggie, R.A., Della Valle, C.J., and Jacobs, J.J. (2006) *Biomaterials*, **27**, 4671–4681.

25. Vassitiou, K., Elfick, A.P.D., Scholes, S.C., and Unsworth, A. (2006) *Proc. Inst. Mech. Eng. H*, **220**, 269–277.

26. Park, S.H., Llinás, A., Goel, V.K., and Keller, J.C. (2000) Hard tissue replacements, in *The Biomedical Engineering Handbook*, Chapter 44 (ed. J.D. Bronzino), CRC Press, Boca Raton, pp. 1–35.

27. Walker, P.S. and Sathasivam, S. (2000) *Proc. Inst. Mech. Eng. H*, **214**, 101–119.

28. Stewart, T., Jin, Z.M., and Fisher, J. (1997) *Proc. Inst. Mech. Eng. H*, **211**, 451–465.

29. Gupta, S.K., Chu, A., Ranawat, A.S., Slamin, J., and Ranawat, C.S. (2007) *J. Arthroplasty*, **22**, 787–799.

30. Tanavalee, A., Choi, Y.J., and Tria, A.J. (2005) *Orthopedics*, **28**, 1423–1433.

31. Muratoglu, O.K., Rubash, H.E., Bragdon, C.R., Burroughs, B.R., Huang, A., and Harris, W.H. (2007) *J. Arthroplasty*, **22**, 435–444.

32. Auger, D.D., Dowson, D., and Fisher, J. (1995) *Proc. Inst. Mech. Eng. H*, **209**, 73–81.

33. Scholes, S.C., Unsworth, A., and Jones, E. (2007) *Phys. Med. Biol.*, **52**, 197–212.

34. Vickerstaff, J.A., Miles, A.W., and Cunningham, J.L. (2007) *Med. Eng. Phys.*, **29**, 1056–1064.

35. Lewis, G. (2004) *Clin. Orthop. Relat. Res.*, **424**, 89–97.

36. Guyer, A.J. and Richardson, E.G. (2008) *Foot Ankle Int.*, **29**, 256–264.

37. Schutte, B.G. and Louwerens, J.W.K. (2008) *Foot Ankle Int.*, **29**, 124–127.

38. Bell, C.J. and Fisher, J. (2007) *J. Biomed. Mater. Res. Part B: Appl. Biomater.*, **81B**, 162–167.

39. Lewis, G. (1997) *J. Biomed. Mater. Res. (Appl. Biomater.)*, **38**, 155–182.

40. Kienapfel, H., Sprey, C., Wilke, A., and Griss, P. (1999) *J. Arthroplasty*, **14**, 355–368.

41. Sun, L.M., Berndt, C.C., Gross, K.A., and Kucuk, A. (2001) *J. Biomed. Mater. Res. (Appl. Biomater.)*, **58**, 570–592.

42. Pylios, T. and Shepherd, D.E.T. (2007) *J. Mech. Med. Biol.*, **7**, 163–174.

43. Beevers, D.J. and Seedhom, B.B. (1993) *Proc. Inst. Mech. Eng. H*, **207**, 195–206.

44. Joyce, T.J. (2004) *Expert Rev. Med. Dev.*, **1**, 193–204.

45. Parkkila, T.J., Belt, E.A., Hakala, M., Kautiainen, H.J., and Leppilahti, J. (2006) *Scand. J. Plast. Reconstr. Surg. Hand Surg.*, **40**, 49–53.

46. Stokoe, S.M., Unsworth, A., Viva, C., and Haslock, I. (1990) *Proc. Inst. Mech. Eng. H*, **204**, 233–240.

47. Joyce, T.J., Milner, R.H., and Unsworth, A. (2003) *J. Hand Surg. [Br]*, **28**, 86–91.

48. Joyce, T.J. and Unsworth, A. (2005) *Proc. Inst. Mech. Eng. H*, **219**, 105–110.

49. Joyce, T.J. (2007) *Med. Eng. Phys.*, **29**, 87–92.

50. Costi, J., Krishnan, J., and Pearcy, M. (1998) *J. Rheumatol.*, **25**, 451–458.

51. Shepherd, D.E.T. and Johnstone, A.J. (2002) *Med. Eng. Phys.*, **24**, 641–650.

52. Pylios, T. and Shepherd, D.E.T. (2004) *J. Biomech.*, **37**, 405–411.

53. Council Directive 93/42/EEC of 14 June (1993) *Off. J. Eur. Communities*, **L169**, 1–43.

54. McAllister, P. and Jeswiet, J. (2003) *Proc. Inst. Mech. Eng. H*, **217**, 459–467.

55. BS EN ISO 14630 (2005) *Non-active Surgical Implants – General Requirements*, British Standards Institute, London.

56. BS EN 12563 (1999) *Non-active Surgical Implants – Joint Replacement Implants – Specific Requirements for Hip Joint Replacement Implants*, British Standards Institute, London.

57. BS EN 12564 (1999) *Non-active Surgical Implants – Joint Replacement Implants – Specific Requirements for Knee Joint Replacement Implants*, British Standards Institute, London.

58. King, A.M. and Sivaloganathan, S. (1999) *J. Eng. Des.*, **10**, 329–349.

59. Alexander, K. and Clarkson, P.J. (2000) *J. Med. Eng. Technol.*, **24**, 53–62.

60. Bennett, D. and Goswami, T. (2008) *Mater. Des.*, **29**, 45–60.

61. Fialho, J.C., Fernandes, P.R., Eca, L., and Folgado, J. (2007) *J. Biomech.*, **40**, 2358–2366.
62. Shepherd, D.E.T. (2002) *Proc. Inst. Mech. Eng. H*, **216**, 23–29.
63. BS 7251-12 (1995) *Orthopaedic Joint Prostheses. Specification for Endurance of Stemmed Femoral Components with Application of Torsion*, British Standards Institute, London.
64. BS ISO 14242-1 (2002) *Implants for Surgery. Wear of Total Hip Joint Prostheses. Loading and Displacement Parameters for Wear-Testing Machines and Corresponding Environmental Conditions for Test*, British Standards Institute, London.
65. BS ISO 14243-3 (2004) *Implants for Surgery. Wear of Total Knee-Joint Prostheses. Loading and Displacement Parameters for Wear-Testing Machines with Displacement Control and Corresponding Environmental Conditions for Test*, British Standards Institute, London.
66. BS ISO 14879-1 (2000) *Implants for Surgery. Total Knee-Joint Prostheses. Determination of Endurance Properties of Knee Tibial Trays*, British Standards Institute, London.
67. Affatato, S., Leardini, A., Leardini, W., Giannini, S., and Viceconti, M. (2007) *J. Biomech.*, **40**, 1871–1876.
68. Joyce, T.J. and Unsworth, A. (2002) *Proc. Inst. Mech. Eng. H*, **216**, 105–110.
69. Leonard, L., Sirkett, D.M., Langdon, I.J., Mullineux, G., Tilley, D.G., Keogh, P.S., Cunningham, J.L., Cole, M.O.T., Prest, P.H., Giddins, G.E.B., and Miles, A.W. (2002) *Proc. Inst. Mech. Eng. B*, **216**, 1297–1302.
70. BS EN ISO 14155-1 (2003) *Clinical Investigation of Medical Devices for Human Subjects. General Requirements*, British Standards Institute, London.
71. BS EN ISO 14155-2 (2003) *Clinical Investigation of Medical Devices for Human Subjects. Clinical Investigation Plans*, British Standards Institute, London.

9
Interstitial Fluid Movement in Cortical Bone Tissue

Stephen C. Cowin

9.1
Introduction

Blood and interstitial fluid have many functions in a bone. They transport nutrients to, and carry waste from, the bone cells (osteocytes) buried in the bony matrix. They are involved in the transport of minerals to the bone tissue for storage and the retrieval of those minerals when the body needs them. Interstitial flow is considered to have a role in bone's mechanosensory system. Bone deformation causes the interstitial flow over the cell processes of the osteocyte creating a drag on the fibers that connect the cell; the drag force created by the flowing interstitial fluid is sensed by the cell [1–4]. A full physiological understanding of this mechanosensory system will provide insight into the following three important clinical problems: (i) how to maintain the long-term stability of bone implants, (ii) the physiological mechanism underlying osteoporosis, and (iii) how to maintain bones in long-duration space flights and long-term bed rest.

Since one purpose of this work is to describe how these fluid systems work, consideration is limited to cortical bone in the mid-diaphysis of a long bone. Although most of what is described is also applicable to the bone tissue at other anatomical sites, the discussion is more concise and direct if this limitation is stipulated.

The majority of the motive force for the blood flow is from the heart, but the contraction of muscles attached to the bone and the mechanical loading of bone also contribute to this motive force. The majority of the motive force for the interstitial fluid flow is due to the mechanical loading of bone, but the contraction of muscles attached to bone and the heart also supply some of its motive force. The influence of the mechanical loading of a whole bone on the fluid system's maintenance of the bone tissue is critical. The fluid flow resulting from the mechanical loading is modeled by the theory of poroelasticity. This theory models the interaction of deformation and fluid flow in a fluid-saturated porous medium. The theory was proposed by Biot [5, 6] as a theoretical extension of soil consolidation models developed to calculate the settlement of structures placed on fluid-saturated porous soils. The theory has been widely applied to geotechnical problems beyond soil

Biomechanics of Hard Tissues: Modeling, Testing, and Materials.
Edited by Andreas Öchsner and Waqar Ahmed

ISBN: 978-3-527-32431-6

consolidation, most notably problems in rock mechanics. Certain porous rocks, marbles, and granites have material properties that are similar to those of bones [7].

The structure of this chapter is first to describe the vascular system in a bone and then describe the interstitial fluid movement in the bone as well as the factors that drive these flows and cause changes in the flow patterns associated with diseases, surgery, and whole body movement. Thus, the sections that follow immediately describe the arterial system, the microvascular network of marrow, the microvascular network of cortical bone, and the venous drainage of bone. The connections between the vascular system and the interstitial fluid system are then described in Section 9.6 on bone lymphatics and blood vessel trans-vessel-wall transport. Attention then turns to the spaces in bone tissue occupied by these two fluid systems, the vascular porosity (PV) and the lacunar–canalicular porosity (PLC), and the interfaces between the systems. The remainder of the chapter considers different aspects of interstitial fluid flow.

9.2 Arterial Supply

9.2.1 Overview of the Arterial System in Bone

All elements of a bone, including the marrow, perichondrium, epiphysis, metaphyses, and diaphyses, are richly supplied by vasculature. Mature long bones in all species have three sources of blood supply: (i) the multiple metaphyseal–epiphyseal vessel complex at the ends of the bones, (ii) the "nutrient" artery entering the diaphyses (Figure 9.1), and (iii) the periosteal vessels (Figures 9.1 and 9.2). After entering the diaphyses, the nutrient artery divides into ascending and descending branches, which have further, radially orientated, branches streaming to the bone cortex (Figure 9.1). Usually a single nutrient artery enters the diaphyses of a long bone, though many human long bones such as the femur, tibia, and humerus often have two. When the nutrient artery enters a bone, the vessel has a thick wall consisting of several cell layers, but within the medulla it rapidly becomes a thin-walled vessel with two cell layers and minimal supporting connective tissue [8]. After reaching the medullary cavity the nutrient artery divides into ascending and descending branches, which proceed toward the metaphyseal bony ends (Figure 9.1). These branches approach the epiphyseal ends of the bone, subdividing repeatedly along the way into branches, which pursue a helical course in the juxta-endosteal medullary bone. The terminal branches of the main ascending and descending branches supply the ends of the long bone and anastomose freely with the metaphyseal vessels. The vessels divide and subdivide to feed into a complex network of sinusoids (Figures 9.1 and 9.3). In the immature bone, the open cartilaginous epiphyseal growth plate separates the epiphyseal and metaphyseal vessel complexes.

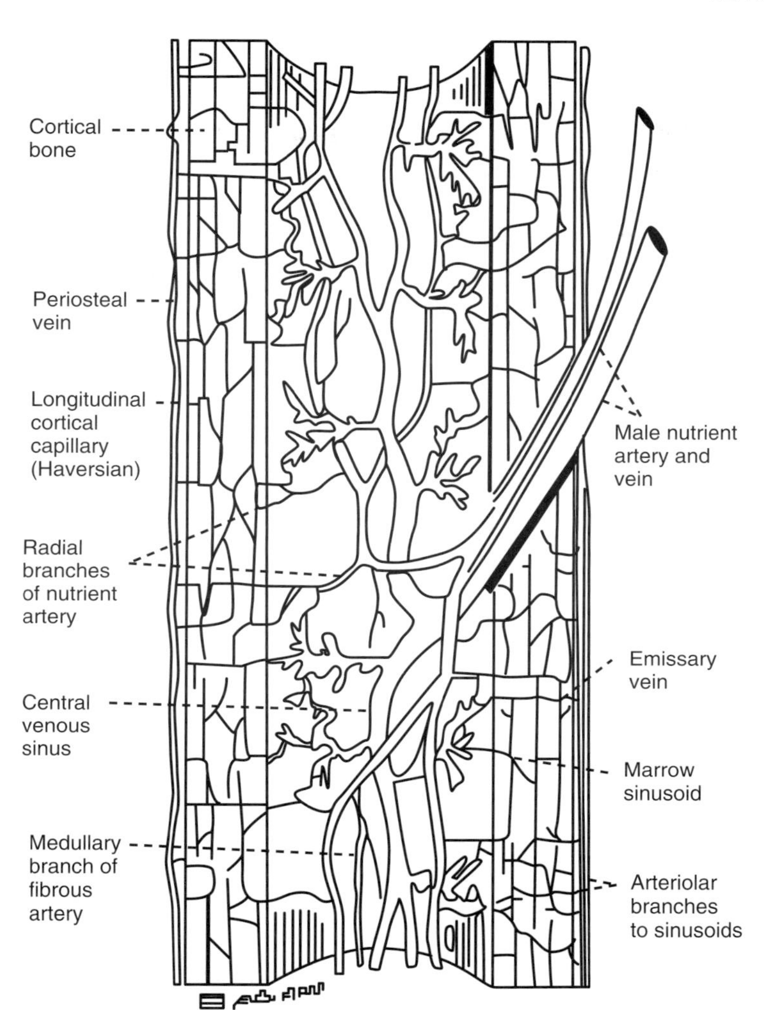

Figure 9.1 Schematic diagram showing the vascular arrangement in the long bone diaphysis. (Modified from Williams *et al.* [9].)

The blood supply to cortical bone may come from either the medullary canal (younger animals) or the periosteum (older humans) (Figure 9.4). The transcortical blood supply transits in the Volkmann canals and the longitudinal blood supply transits in Haversian systems or osteons. Haversian arteries run longitudinally in osteons (Haversian systems), oriented roughly about 15° to the long axis of a bone (Figure 9.5). Human cortical bone is largely Haversian at a rather young age compared to other animals. The thin-walled vessels in the cortical canals of Haversian and Volkmann canals are contained in hard unyielding canals in the cortical bone and serve to connect the arterioles (the afferent system) with the

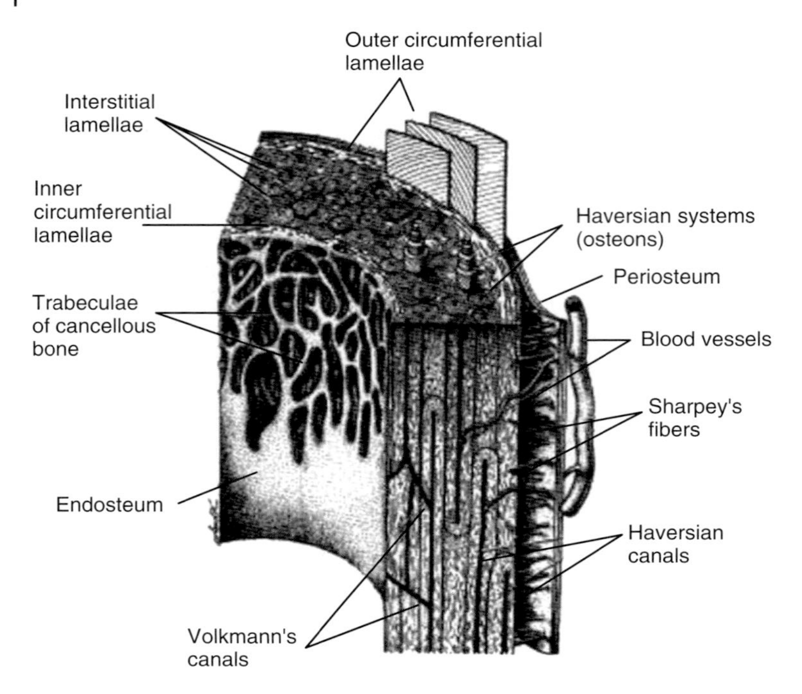

Figure 9.2 A detailed view of the structure of a typical long bone. (From figure 2.1 of Martin *et al.* [10].)

venules (the efferent system), but unlike true capillaries, they apparently are not able to change diameter in response to physiologic needs [11]. Diffusion from Haversian vessels to the bone cells buried in the bony matrix is insufficient to maintain their nutrition; convection driven by the interstitial fluid pressure gradients is necessary for the viability of these cells. Canaliculi serve to connect osteocytic processes [12]. Increased distance from the vascular source (the Haversian artery) probably accounts for the finding that the interstitial bone is more susceptible to ischemia than is the Haversian bone [13].

9.2.2 Dynamics of the Arterial System

In considering the hemodynamics of any tissue, the important elements to be considered are fluid and tissue pressures, fluid viscosity, vessel diameter, and the capillary bed. Blood vessels in a bone are richly supplied with nerves and are intimately connected to vasomotor nerve endings; these nerves presumably exert a precise control over blood flow in the bone [14]. It is known that in most soft tissues, the arteriolar mechanism reduces the blood pressure from 90 mmHg or more in arteries to about 35 mmHg at the arterial end of capillaries. Arterial vessels will close unless the transmural pressure is positive, that is, to say, unless the blood

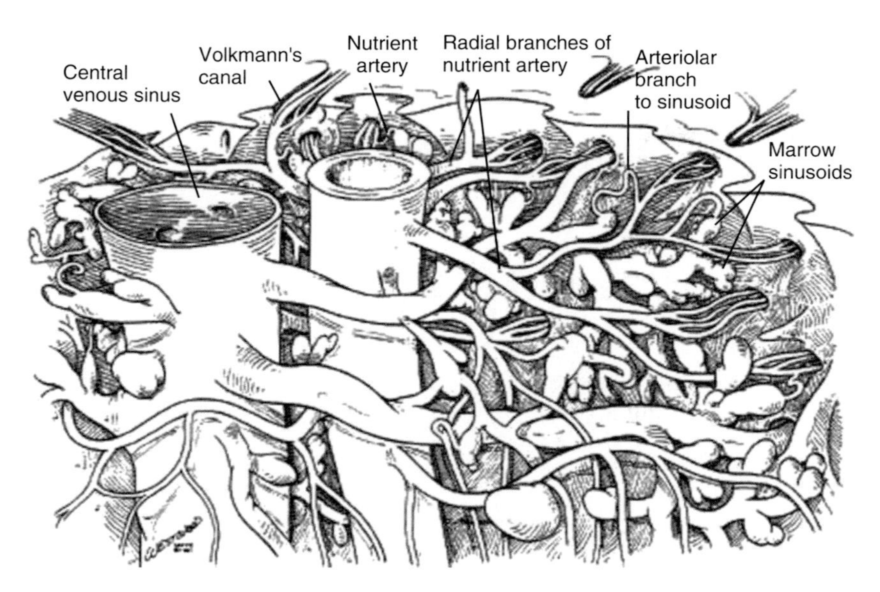

Figure 9.3 The relationship between the marrow and cortical bone circulations. The radial branches of the nutrient artery form a leash of arterioles that penetrate the endosteal surface to supply the bone capillary bed. Small arterioles from these radial branches supply the marrow sinusoids adjacent to the bone. (Modified from Williams *et al.* [9].)

pressure in the capillaries does not fall below that in the extravascular space, the interstitial fluid pressure. Note that transmural pressure (blood pressure minus the interstitial fluid pressure) must initially exceed osmotic pressure if filtration is to occur. Absorption of tissue fluid depends upon the transmural pressure being less than the osmotic pressure of the blood at the end of the sinusoid. Osmotic pressure is generally held to be about 20 mmHg. It follows that the pressure in the collecting sinuses of the diaphyseal marrow may be of the order of 55 mmHg. Note that 1 mmHg is 133.3 Pa or that 3 mmHg is approximately 400 Pa, 60 mmHg is approximately 8 kPa. Bone fluids are interesting in that they exhibit metabolically produced differential diffusion gradients [15, 16]. They are sometimes limited in range, but well documented. Thus, many ions, such as potassium, calcium, and phosphorus, exist in very different concentrations between the blood and bone [17].

9.2.3
Transcortical Arterial Hemodynamics

Bridgeman and Brookes [18] have shown that aged bone cortex is supplied predominantly from the periosteum in contrast to the medullary supply in young human and animal bones, based on cross sections through the mid-diaphyses. They argue that this change is attributed to increasingly severe medullary ischemia with age, brought on by arteriosclerosis of the marrow vessels. They note that

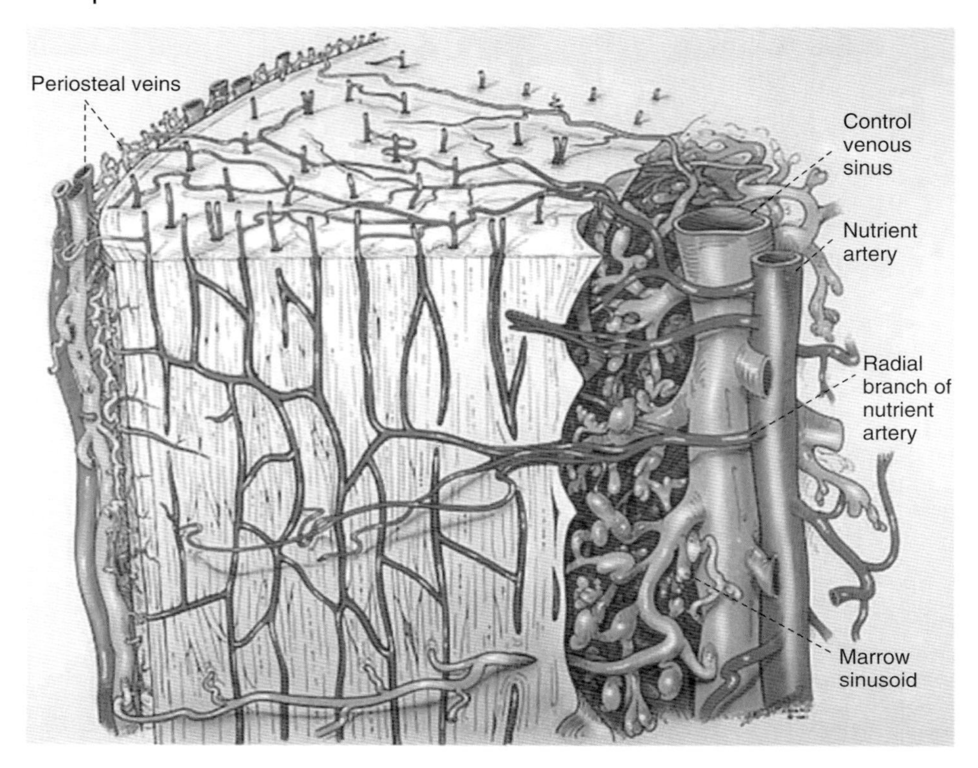

Figure 9.4 The capillary network within the cortical bone. The major arterial supply to the diaphysis is from the nutrient artery. There is an abundant capillary bed throughout the bone tissue that drains outward to the periosteal veins. (Modified from Williams *et al.* [9].)

an examination of the findings reported by investigators of animal bone blood supply in the past 40 years shows a large measure of agreement. Long standing controversy seems to be based on a failure to recognize that marrow ischemia accompanies natural senescence affecting transcortical hemodynamics and entraining an increasing periosteal supply for bone survival in old age The change over from a medullary to a periosteal blood supply to bone cortex is the consequence of medullary ischemia and reduced marrow arterial pressure, brought about by medullary arteriosclerosis.

9.2.4
The Arterial System in Small Animals may be Different from that in Humans

The marrow and cortical vascular networks in the rodent are thought to be in series while they are in parallel in the human. From perfusion studies on small mammals (guinea pig, rat, and rabbit), it has been concluded that the blood flow in

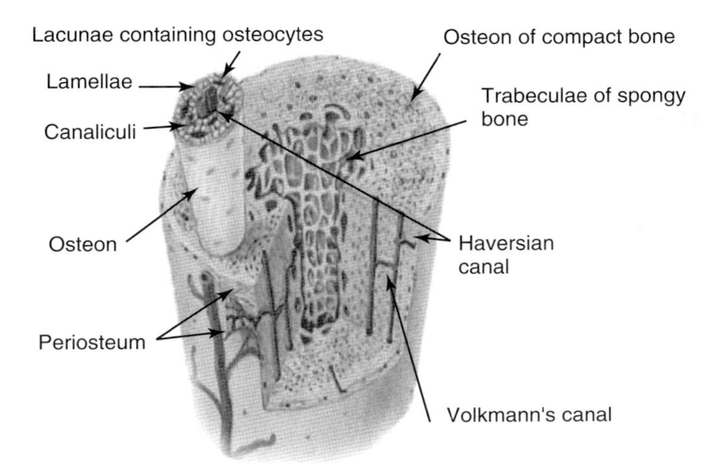

Figure 9.5 The osteon at the top of this figure is entirely PLC porosity except for its central lumen, called the *osteonal canal* or *Haversian canal*, which is part of the PV porosity. The PV porosity consists of the volume of all the tunnels in bones that contain blood vessels and includes all the osteonal canals and all the Volkmann canals, less the volume of the tunnels occupied by the blood vessels.

long bones is such that the major blood supply to the bone marrow is transcortical [19]. This means that the marrow and cortical vascular networks in the rodent are in series. Anatomic and perfusion studies in humans suggest that the circulations of the cortex and marrow are arranged in parallel from a longitudinally running nutrient artery [20]. It was shown that the marrow sinusoids near the endosteal surface of the bone typically receive a small arteriolar branch off the major conduit vessel as it enters the bone cortex.

9.3
Microvascular Network of the Medullary Canal

The vascularization of the marrow is illustrated in Figure 9.3. The radial branches of the nutrient artery form a leash of arterioles that penetrate the endosteal surface to supply the bone capillary bed. Small arterioles from these radial branches supply the marrow sinusoids adjacent to the bone. In the adult dog, the marrow consists of adipose tissue (yellow marrow), which provides support for the lateral branches of the nutrient artery as they run toward the endosteal surface of the bone. However, in the immature animal, much of the marrow cavity is filled with active hemopoietic tissue (red marrow). The type of capillary varies between red and yellow marrow. Although it is easy to distinguish these types of marrow macroscopically, when seen microscopically, there is no clear-cut separation. The appearance can range from highly cellular to completely fatty. In active red marrow, the small vessels are thin-walled sinusoids, so called because they are many times the size of ordinary capillaries. Despite the thin walls of these vessels, Trueta and Harrison [21] were not able to demonstrate open fenestrations between the endothelial cells. However, it should be noted that Zamboni and Pease [22], using electron microscopy, considered the vessels in red marrow to consist of flattened reticulum cells with many fenestrations and no basement membrane. This would mean that there is minimal hindrance at the sinusoid wall for molecular exchange. In the fatty marrow, the capillaries are closed and continuous like those of other tissues such as muscle [21]. This is supported by *in vivo* observations in the rabbit [23] that the vessels varied according to the functional state of the marrow. It was estimated that the sinusoid was up to seven times the size of the marrow capillaries, which have a diameter of 8 μm.

9.4
Microvascular Network of Cortical Bone

Throughout the cortex of long bones, there is a capillary network housed in small passages (Figures 9.4 and 9.5). In immature bones, these are arranged rather haphazardly, but as the bone remodels and matures, a more distinct pattern emerges. In the mature dog and the human, there are two basic systems, the Haversian canals, which run longitudinally, and the Volkmann canals, which run radially (Figures 9.2 and 9.5). The two systems are intimately anastomosed to each other. The vessels within the Haversian canals of the human tibia have been examined by microscopy of decalcified sections [24]. The majority of the vessels were observed to be a single layer of endothelial cells. Occasionally, near the endosteal surface of the cortex, small arterioles with a muscular coat were seen, usually accompanied by a larger vein.

A comprehensive examination of the cortical bone of mature and immature dogs by electron microscopy has been reported by Cooper *et al.* [25]. This revealed considerable detail of the capillaries in bones. The Haversian canals ranged in size from 5 to 70 μm and contained either one or two vessels that had the ultrastructure of capillaries. On the transverse section, they were lined by one or more endothelial cells, which were surrounded by a continuous basement membrane 400–600 Å thick. The junctions of the endothelial cells varied from simple juxtapositioning to a complex interlocking. These investigators found no smooth muscle cells in the walls of the vessels in the Haversian canals. This picture is supported by electron microscopy studies [26] that showed that the cortical capillaries of the growing rat were similar to those found in the skeletal muscle, although a basement membrane surrounding the capillaries could not be demonstrated. Thus, it appears that the capillaries of the bone are a closed tube formed from a single layer of endothelial cells. It has been suggested that transendothelial passage of substances involves two separate pathways: one through the intercellular clefts for hydrophilic substances and another across the endothelial cells themselves for lipophilic substances. If the intercellular capillary clefts are present, they are probably filled with material that makes their permeability low. This is suggested by the work of Cooper *et al.* [25], who observed spaces of 175 Å between adjacent endothelial cells that were filled with an amorphous material seen by electron microscopy.

9.5 Venous Drainage of Bone

The venous complexes draining a long bone parallel those of the arteries. Many workers have commented [21, 24] on the extreme thinness of their walls. In the marrow, the venous sinusoids drain into a large, single-cell-walled, central venous sinus, which in turn drains into the "nutrient" veins of the diaphyses. In the adult dog, this thin-walled "nutrient" vein accounts for only 10% of the drainage from the diaphyses [27]. The multiple, penetrating, venous radicles in the metaphyses and epiphysis are also thin walled and run a more tortuous course than the arteries [21]. The major share of the venous blood leaving long bones has been shown by phlebography to travel by this route [27]. The abundantly anastomosing periosteal network of veins is considered by some workers to drain the diaphyseal bone cortex completely under normal conditions [28]. Many of the veins leaving the long bone pass through muscles, in particular, the calf muscle in the case of the lower limb. The alternate contraction and release of the muscles containing the veins is effectively a pump returning the blood toward the heart and away from the bone and decreasing the intermedullary pressure. The intermedullary pressure can be reduced by exercise of the muscles of the calf. This arrangement in the case of the calf muscles is illustrated in Figure 9.6. It is clear that the long bone as a whole has multiple venous pathways, the relative importance of which can vary with time and circumstance.

Impaired venous circulation (venous stasis) has been shown to stimulate periosteal bone formation or increase bone mass in the young dog [29], the young goat [30], and in a disuse, hind limb suspended, rat model [31]. Venous stasis was induced in the experimental animals by applying tourniquets or vein ligation that lasted from 10 days (with additional 30 days for recovery [32]) up to 42 days [29] before the bones were examined. There are many other studies demonstrating similar effects [32–35]. A hypothesis for the underlying mechanism of the periosteal bone formation induced by venous stasis has been presented [36].

9.6 Bone Lymphatics and Blood Vessel Trans-Wall Transport

The purpose of this section is to indicate where the interstitial fluid can flux from a blood vessel and where it may flux back into a blood vessel. Interstitial fluid can flux from an arterial blood vessel with nutritional solutes that are used to nourish the cells in the bone. Interstitial fluid may flux back into venous blood vessels with wastes as these vessels leave the bone; this becomes a possibility because of evidence (see below) of a lack of lymphatics in the periosteum.

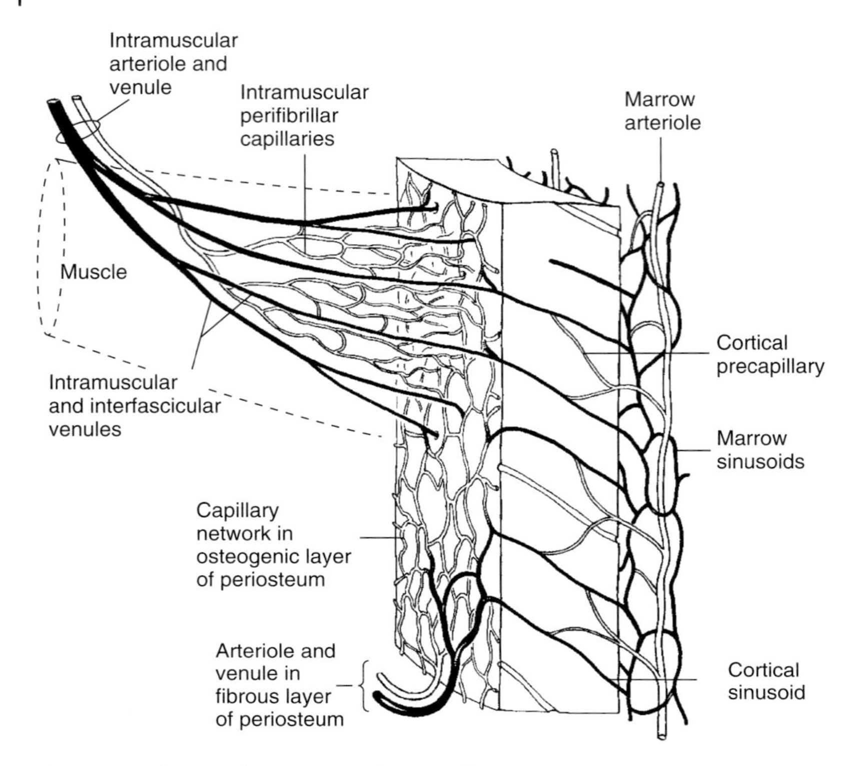

Figure 9.6 The vascular connection between the bone marrow, cortex, periosteum, and attached muscle. (From figure 9.37 of Brooks and Revell [28].)

The PV occupied by interstitial fluid is the space outside the blood vessels and nerves in the Volkmann and Haversian canals (Figures 9.2 and 9.7). This bone interstitial fluid freely exchanges with the vascular fluids because of the thin capillary walls of the endothelium, the absence of a muscle layer, and the sparse basement membrane. There are both outward filtration due to a pressure gradient and inward reabsorption due to the osmotic pressure. The function of these flows is to deliver nutrients to, and remove wastes from, the bone interstitial fluid. Lymph is the fluid that is formed when interstitial fluid enters the conduits of the lymphatic system. The lymphatic system has three interrelated functions. It is responsible for the removal of interstitial fluid from tissues. It absorbs and transports fatty acids and fats as chyle to the circulatory system. The last function of the lymphatic system is the transport of antigen-presenting cells, such as dendritic cells, to the lymph nodes where an immune response is stimulated. The lymph, unlike blood, is not pumped through the body; it is moved mostly by the contractions of skeletal muscles.

The existence of lymphatic vessels in bones remains unclear. On the basis of physiologic evidence, some sort of lymph circulation must be present. Large

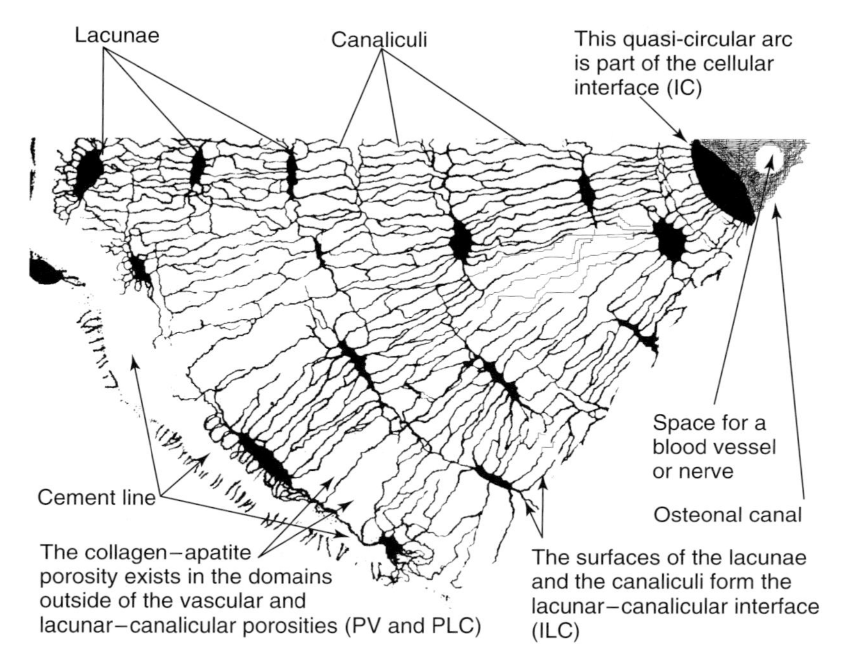

Figure 9.7 A transverse cross section of a pie-shaped section of an osteon. The osteonal canal is on the upper right, and the cement line is to the left. The osteonal canal is part of the vascular porosity, the lacunae and the canaliculi are part of the lacunar–canalicular porosity, and the material in the space that is neither vascular porosity nor lacunar–canalicular porosity contains the collagen–apatite porosity. The three interfaces, the cement line, the cellular interface, and the lacunar–canalicular interface, are indicated separately. The radius of an osteon is usually about 100–150 mm, and the long axis of a lacuna is about 15 mm. Using this information, it should be possible to establish the approximate scale of the printed version of this illustration.

molecules, such as albumin (mol wt 68 000) and horseradish peroxidase (mol wt 40 000), have been shown to leak out of bone capillaries into the interstitial fluid [37, 38], and they must have a pathway to return to the general circulation. Kolodny [39] demonstrated that 2 weeks after India ink was injected into the medullary cavity of long bones, carbon particles were found in the regional lymph nodes.

However, attempts to demonstrate discrete lymphatic vessels within the marrow and bone tissue have been consistently unsuccessful. It has been shown with injection studies using thorotrast [40] that this substance leaks from the capillaries of cortical bone into the perivascular fluid and that eventually it can be seen in the periosteal lymphatic vessels. A similar finding has been observed in the cortical bone after the use of India ink [9]. The indirect conclusion seems to be that, although there are no demonstrable lymphatic channels in bone tissue, the perivascular fluid as a whole circulates toward the periphery of the bone, carrying with it substances such as large proteins and carbon particles to be taken up by a mechanism at or near the periosteum.

Anderson [41] noted that high arterial pressure in the bone marrow probably correlates with an absence of lymphatics in the bone marrow and cortex. We note that lymphatic circulation is unlikely to play a role in bone fluid transport in a normal bone, because lymphatic vessels are absent in the bone. A chapter on the physiology of blood circulation in a book entitled *Blood Vessels and Lymphatics in Organ Systems* [42] contains no description of the bone lymphatic system. It has been shown using immunohistochemistry that lymphatics were not present in the normal bone [43, 44].

9.7 The Levels of Bone Porosity and their Bone Interfaces

There are three levels of bone porosity containing blood or interstitial fluid within the cortical bone and within the trabeculae of the cancellous bone. A section of a long bone indicating the vascular structure is shown in Figure 9.1 and more detailed views of the local bone structure are shown in Figures 9.2 and 9.5. The three levels of bone porosity include the PV associated with the Volkmann canals (Figures 9.2 and 9.5) and the Haversian or osteonal canals (Figures 9.2 and 9.5), which are of the order of 20 μm in radii; the PLC associated with the fluid space surrounding the osteocytes and their processes (Figure 9.7), which is of the order 0.1 μm in radii; and the collagen-hydroxyapatite porosity (PCA) associated with the spaces between the crystallites of the mineral hydroxyapatite (order: 0.01 μm radius). The total volume of the bone fluid PV is about one-half or less than that of the PLC [45–47].

9.7.1 The Vascular Porosity (PV)

The PV occupied by bone fluid is the space outside the blood vessels and nerves in the Volkmann and Haversian canals. The typical pore size (20 μm in radii) of the PV channels is not the blood vessel pore size; rather it is the size of the tubular tunnels (Haversian systems or osteons and Volkmann canals) containing the blood vessels, the arterioles, and the venules, with the actual dimensions of these vessels subtracted from the volume of the tubular tunnels.

The PV is a low-pressure reservoir that can interchange fluid with the PLC. This is the case because the lineal dimension associated with the bone fluid PV is 2 orders of magnitude larger than the lineal dimension associated with the PLC, and the PV is typically at blood pressure, which is low in bones. The total volume of the PV is, however, considerably less than that of PLC [45–47].

The measurement of the permeability of the PV has not been accomplished with sufficient accuracy to date, primarily because of the topological intertwining of the PV with the PLC. This difficulty and others are discussed in some detail in [48], where it is noted that experiments reported [49] bovine cortical bone permeabilities on the order of 10^{-14} m^2. Factors are present in [48] to suggest that the actual permeability could be much larger than this value. The PV permeabilities reported in the past are thought to be significant underestimates because these previous values represent PLC and PV lumped measurements rather than a PV measurement alone. These lumped measurements also compromise the measurement of the permeability of the PLC as pointed out [50] in the discussion of the reported PLC permeabilities [51, 52]. Estimates for the ratio of the permeabilities of the PV to those of the PLC are of the order of 10^{10}. These ratio estimates and the approximate ratio of the diameters as 167 are measures of the significant size difference, pore pressure difference, and relaxation times in these two distinct pore size porosities.

9.7.2 The Lacunar–Canalicular Porosity (PLC)

The PLC consists of the fluid spaces surrounding the osteocytes and their processes (Figure 9.7), less the volume of the soft tissue structures in these fluid spaces. The pore size estimate of the effective radii is of the order 0.1–0.2 μm. There are four reported estimates of the permeability of the PLC. A theory-based estimate of the intrinsic permeability [53] of the PLC is $k = 1.47 \times 10^{-20}\ \mathrm{m}^2$ (this corresponds to a hydraulic permeability of $k = k/m = 1.47 \times 10^{-17}\ \mathrm{m}^4\ \mathrm{N}^{-1}\ \mathrm{s}^{-1}$). An experiment-based estimate [54] is $k = 2.2 \times 10^{-22}\ \mathrm{m}^2$ (this corresponds to a hydraulic permeability of $k = k/m = 2.2 \times 10^{-19}\ \mathrm{m}^4\ \mathrm{N}^{-1}\ \mathrm{s}^{-1}$). More recently a nanoindentation technique was combined with a poroelastic analysis to provide an estimate of $k = 4.1 \times 10^{-24}\ \mathrm{m}^2$ (this corresponds to a hydraulic permeability of $k = k/m = 4.1 \times 10^{-21}\ \mathrm{m}^4\ \mathrm{N}^{-1}\ \mathrm{s}^{-1}$) [55]. The assumption of an incompressible fluid in the analysis of experimental data renders the estimate [55] smaller than it should be. In the incompressible model, all of the hydrostatic stress is transferred from the solid matrix material to the pore fluid pressure; in the compressible model, only a fraction of the hydrostatic stress is transferred from the solid matrix material to the pore fluid pressure. The reason for this is that the compressible model applied to the situation in which the solid matrix material is much stiffer than the bulk modulus of the fluid and the solid shields the fluid from a fraction of its hydrostatic stress (see Table 9.1). A measurement of the PLC permeability reported in [56] based on the analytical model of Gailani and Cowin [57] yielded values on the order of 10^{-24}–$10^{-25}\ \mathrm{m}^2$. Each of the latter estimates above is 2 orders of magnitude smaller than the previous one for the permeability of the PLC.

9.7.3 The Collagen–Hydroxyapatite Porosity (PCA)

The collagen–hydroxyapatite porosity (PCA) has the smallest pore size (approximately 10 nm diameter) [7, 58]. The interstitial pore fluid in the collagen–apatite porosity has been shown [59] to be bound to the solid structure, and it is not of interest in the present considerations of interstitial flow for that reason. This portion of the interstitial fluid is considered to be part of the collagen–hydroxyapatite structure.

Table 9.1 The material properties reported from different sources for the elastic moduli, Poissons ratios, and so on. The superscript d reflects the drained properties and the superscript m reflects the matrix material properties. The water compressibility (K_f) is 2.3 GPa, its viscosity (μ) is 0.001 Pa s, and the amplitude of ε_o, the applied strain, is taken as 0.0005. The transversely isotropic elastic constant data is from Cowin and Mehrabadi [93]. The modal value for land mammal long bones for R_i/R_o was taken as 0.5, varying effectively from 0 (solid) to 0.73 [94]. Human bones are thought to be in the same range.

Material parameters	PLC (*L*)	PV (*V*)
$E_1^d = E_1^d$	15.17 GPa	12.7 GPa
E_3^d	15.96 GPa	14.9 GPa
$\nu_{12}^d = \nu_{21}^d$	0.316	0.27
$\nu_{31}^d = \nu_{32}^d$	0.308	0.285
$\nu_{13}^d = \nu_{23}^d$	0.282	0.27
$E_1^m = E_1^m$	18.6 GPa	18.6 GPa
E_3^m	22.32 GPa	22.32 GPa
$\nu_{12}^m = \nu_{21}^m$	0.322	0.322
$\nu_{31}^m = \nu_{32}^m$	0.312	0.312
$\nu_{13}^m = \nu_{23}^m$	0.255	0.255
Outer radius (osteon, whole bone)	$r_o = 160\ \mu m$	$R_o = 0.03$ m
Inner radius (osteon, whole bone)	$r_i = 40\ \mu m$	$R_i = 0.015$ m
Porosity ϕ	0.05 or 0.1 or 0.15	0.05
Permeability K_{rr}	$2.2 \times 10^{-22}\ m^2$ to $2.2 \times 10^{-24}\ m^2$	$6.35 \times 10^{-14}\ m^2$ to $6.35 \times 10^{-8}\ m^2$

9.7.4 Cancellous Bone Porosity

The cancellous bone porosity is the porosity external to, and surrounding, the trabeculae. It is the bone porosity with the largest pores (up to 1 mm). The porosity is well connected to the marrow cavity and it contains marrow, fat, and blood vessels. The magnitude of the porosity varies with anatomical location; it is smaller near the load-bearing surfaces and increases to its greatest magnitude as the medullary canal is approached. The permeability associated with this porosity is surveyed in [60]; the concentration in the present work is upon the porosity associated with the interstitial bone fluid contained in the bone matrix, and not upon that associated with the bone marrow.

9.7.5 The Interfaces between the Levels of Bone Porosity

There are two external boundaries to the several porosity domains of interest in bones: the periosteum and endosteum. These are, mechanically and biologically, very different structures. The endosteum is mechanically insignificant while the periosteum is like a pretensioned, stiff, relatively impermeable fiber stocking attached to the exterior surface of the bone. It has been reported that the periosteum acts as a barrier to bone fluid flow [61]. The endosteum is simply a monolayer of cells. Multiple layers of cells at various stages of differentiation lie under the periosteum.

There are two interfaces between the three levels of bone porosity within the cortical bone and within the trabeculae of the cancellous bone: the PV/PLC interface and the lacunar–canalicular/collagen–hydroxyapatite porosity interface. Topologically, the entire lacunar–canalicular/collagen–hydroxyapatite porosity interface is completely contained within the PV/PLC interface.

The first important point concerning these interfaces, and the porosities described above, is that they change rapidly after birth, being quite porous at birth and subsequently reducing their porosity as the bone tissue becomes fully mineralized [61, 62]. Experimental permeability studies clearly show time-dependent changes in the interstitial pathways as the bone matures. At the earliest times, the unmineralized collagen-proteoglycan bone matrix is porous to large solutes. A study [63] with ferritin (10 nm in diameter) in a two-day-old chick embryo shows a continuous halo around primary osteons 5 min after the injection of this tracer. The halo passes right through the lacunar–canalicular system suggesting that, before complete mineralization, pores of a larger size can exist throughout the bone matrix. It was later demonstrated [64, 65] that such halos were very likely an artifact of histological processing and could be eliminated by shortened fixation methods [4]. These studies found that ferritin was confined exclusively to the vascular canals and blood vessels and did not enter the PLC. The porosity in puppies is 3.5 times higher than that in dogs [61]. In this work, only the adult or fully mineralized situation is described. The PV/PLC interface, which separates the mineralized tissue from the vascular channels, is considered first.

The region interior to the PV/PLC is called the "*milieu intérieur*," and the existence of a "bone membrane" that would coincide with what we call the *vascular/lacunar–canalicular porosity* interface was suggested [66]. This interface is a continuous layer of bone lining cells [67]; all the surfaces of the Haversian canals and the Volkmann canals are a part of this interface as is the endosteum. There is a report on tight junctions occurring in the bone lining cells on the interface [62]. In [15], it was noted that the bone fluid of the PV (serum) and the bone fluid of the PLC (extracellular fluid) were nearly equivalent in composition (pH, Ca^{++}, Na^{+}, etc.), but it was argued that there must be some diffusion barrier, some ion gradient or ion pump, between the two fluid compartments, a view revised later [17]. It has been analytically demonstrated that there are high transient pressure gradients across the interface that serve to move the bone water across the interface [53]. The bone lining cells with tight junctions do not form a significant barrier to the transport of bone water across this interface. During each cycle of bone loading, the bone fluid of the PV (serum) briefly mix with the bone fluid of the PLC (extracellular fluid). As a first approximation it appears reasonable to assume that the permeability of this interface is equal to the permeability of the PLC.

The lacunar–canalicular/collagen–hydroxyapatite porosity interface is considered next. Evidence suggests that this interface is generally impermeable. Again, the evidence is from tracer studies. This conclusion is supported by the studies on the alveolar bone of five-day-old rats using the small tracer microperoxidase (MP) (2 nm) [68]. These studies clearly showed that the MP only penetrated the unmineralized matrix surrounding the lacunae and the borders of the canaliculi (see figure 9.13 of that study) and was absent from the mineralized matrix. Using more mature rats, another study confirmed the failure of the small (2 nm) MP tracer to penetrate the mineralized matrix tissue from the bone fluid compartments [69]. Further confirmation comes from studies that observed that the tracers of ruthenium red (MW 860, 1.13 nm in the largest dimension) and procion red (MW 300–400) did not penetrate the bone mineral porosity, but were present in the PLC [70] (S.D. Doty, 1997, Private communication).

An important physiological consideration arises from the fact that the bone serves as a reservoir for calcium and phosphorus, and these mineral reserves should be connected to the circulation. Clearly, these minerals must cross the PV/PLC interface, but should they cross the lacunar–canalicular/collagen–hydroxyapatite porosity interface? That is, to ask, can the necessary minerals be supplied by the bone lining cells from the bone matrix they are situated upon, or should the osteocytes be involved in this process? Estimates show that sufficient mineral can be supplied by the bone lining cells, consistent with the suggestions above that the lacunar–canalicular/collagen–hydroxyapatite porosity interface is generally impermeable and that the permeability of the PV/PLC interface is equal to the permeability of the PLC. However, the possibility that the interface permeability between the PLC and the collagen–hydroxyapatite porosity might be changed by physiological demands is worthy of consideration. The exact method of mineral retrieval and redeposition lies at the root of many studies [62, 70–73].

9.8 Interstitial Fluid Flow

9.8.1 The Different Fluid Pressures in Long Bones (Blood Pressure, Interstitial Fluid Pressure, and Intramedullary Pressure)

Since the blood is encased in very thin-walled blood vessels that are contained within the PV, the interstitial fluid pressure is less than the blood pressure. The difference between the blood pressure and the interstitial fluid pressure is the transmural pressure. The PV is a vast low-pressure reservoir for interstitial fluid that can interchange that fluid with the PLC. This is the case because the lineal dimension associated with the bone fluid PV is 2 orders of magnitude larger than the lineal dimension associated with the PLC, and the interstitial fluid pressure in the PV is typically lower than the blood pressure within the blood vessels.

The intramedullary pressure in the normal bone is the pressure of blood in a local pool of hemorrhage from ruptured intraosseous vessels obtained by drilling into the marrow cavity through the cortex to insert a steel cannula through which the marrow cavity pressure is measured. This is pointed out in [74] and supported by Shim *et al.* [75]. Therefore, the measurable marrow cavity pressure varies to some extent by the size and type of vessels ruptured as well as by the vasomotor action in the marrow cavity under a condition of anticoagulation. The differences in the intramedullary pressure from region to region in a given bone, from bone to bone, and from animal to animal in the same and different species are noted in [75]. If the femoral vein was occluded, the intramedullary pressure was elevated and nutrient venous outflow increased – an indication of venous congestion of bone. If the nutrient or femoral artery was occluded, there was an immediate fall in the intramedullary pressure and a profound decrease in nutrient venous outflow. The intramedullary pressure can be increased by mechanical loading of the bone and by venous ligature, and the intramedullary pressure can be reduced by exercise of the muscles of the calf (Figure 9.6).

The pore fluid pressures in these two pore size bone porosities are distinct and vary very differently with time under mechanical loading of the whole bone. Under physiologically possible rapid rise-time loadings of bone, the pore fluid pressure may rise considerably in the PLC [53]. The decay time for this pore pressure rise is much larger in the PLC than it is in the PV [48, 53]. The PV is a low-pore-fluid-pressure domain because the PV permeability is sufficiently large to permit a rapid decay of a pressure pulse. This must be the case because the PV contains thin-walled blood vessels carrying blood with a pressure of 40–60 mm of Hg; a pore-fluid pressure significantly greater than 40–60 mm of Hg will collapse these blood vessels and a prolonged increase in the pore-fluid-pressure significantly above 40–60 mm of Hg would deprive the tissue of oxygen and nutrients.

9.8.2 Interstitial Flow and Mechanosensation

Since bone fluid in the porosity with the largest lineal dimension, the PV, is always at a low pressure, the middle porosity – the PLC – appears to be the most important porosity for the consideration of mechanical and mechanosensory effects in the bone. A detailed theoretical model of the contents of the PLC is given in [1] and [2]. The PLC is the primary porosity scale associated with the relaxation of the excess pore pressure due to mechanical loading. It is the porosity associated with the osteocytes that is the prime candidate for the mechanosensory cell in bone.

In addition to mechanosensation, a function of these flows is to deliver nutrients to, and remove wastes from, the osteocytes housed in the lacunae buried in bone matrix (Figure 9.7). An osteocyte left *in vitro* without nutrient exchange for 4 h will die [76]. This observation makes sense given the estimate that osteonecrosis *in vivo* is significant if the bone is ischemic for 6 h or more [77].

Since the interstitial pore fluid pressure in the porosity with the largest lineal dimension, the PV, is always low; the middle porosity – the PLC – appears to be the most important porosity for the consideration of mechanical and mechanosensory effects in the bone. The interstitial pore fluid pressure in the PLC can be, transiently, much higher. The PLC is the primary porosity scale associated with the relaxation of the excess pore pressure due to mechanical loading. This relaxation of the interstitial pore fluid pressure was illustrated by Wang *et al.* [78]. In this work, the interstitial pore fluid pressure distributions across a bone are calculated using an idealized bone microstructural model consisting of six abutting square osteons with circular osteonal canals (Figure 9.8). This idealized model is shown in Figure 9.8(a,b); it has a length of 1200 μm and a width of 200 μm. The interstitial pore fluid pressure profiles are given for different conditions of loading and of permeability of the cement line that forms the outer boundary of the osteon. The completely free flow across the osteonal cement line represents 100% coupling of the osteon with its neighboring osteons, and 0% osteonal coupling is the case in which there is no flow across the cement line. In Figure 9.8(c,d) the interstitial pore fluid pressure profiles for a bone model specimen with 40 μm osteonal canals were subjected to an external loading applied at 1.5 Hz for 100% coupling (Figure 9.8c) and for 0% osteonal coupling (Figure 9.8d). In Figure 9.8(c,d), the interstitial pore fluid pressure profiles are plotted along lines whose x distance is expressed as a multiple of the osteonal canal diameter d; the interstitial pore fluid pressure profiles along the line $x = 0$ are the profiles along a line passing through the canal centers; $x = d/4$ are interstitial pore fluid pressure profiles along a line halfway between the canal centers and the cement line; $x = d/2$ are interstitial pore fluid pressure profiles along a line passing through the cement lines. In Figure 9.8(e), the local interstitial pore fluid pressure gradients for 0% coupling and 100% coupling are compared in the case $x = 0$. In Figure 9.8(f), the effects of the different sized osteonal canals ($d = 0, 40,$ or 60 μm) on the pressure profiles and the transcortical interstitial pore fluid pressure difference are illustrated for 100% osteonal coupling with the loading applied at 1.5 Hz. In Figure 9.8(g), the effects of two different

Figure 9.8 Dimensionless pressure distributions from one surface of the bone specimen ($y'' = -600$ μm) to the other surface ($y = 600$ μm) for different conditions. (a) and (b) the spacing of the osteonal lumen across the test section. (c) and (d) Pressure profiles for a specimen with 40 μm osteonal canals with the external loading applied at 1.5 Hz for 100% osteonal coupling (c) and for 0% osteonal coupling (d). $x = 0$: profile along a line passing through the canal centers; $x = d/4$: profile along a line halfway between the canal centers and the cement line; $x = d/2$: profile along a line passing through the cement lines. (e) Comparison of the local pressure gradients for 0% coupling and 100% coupling ($x = 0$). (f) Effects of the size of the osteonal canals ($dc = 0$, 40, or 60 μm) on the pressure profiles and the transcortical pressure difference (Δ_p) for 100% osteonal coupling with the loading applied at 1.5 Hz. The transcortical pressure difference is the pressure difference between the points marked "∇", "x" or "O" on the external surfaces. (g) Comparison of the local pressure gradients and transcortical pressure difference between loading applied at 1.5 and 15 Hz.

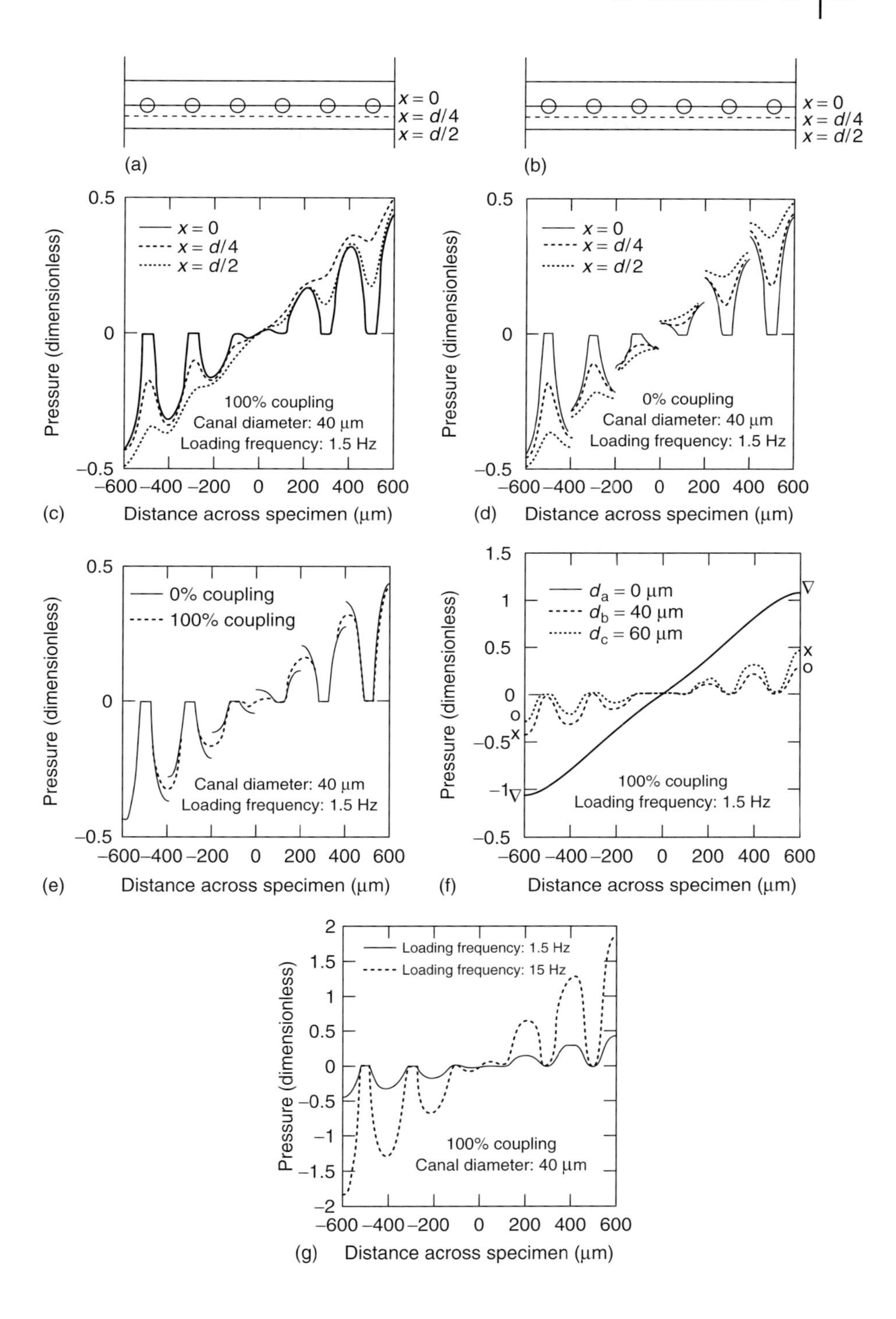

x = 0
x = d/4
x = d/2
(a)
(b)
(c)
(d)
(e)
(f)
(g)
Pressure (dimensionless)
Distance across specimen (μm)
100% coupling
0% coupling
Canal diameter: 40 μm
Loading frequency: 1.5 Hz
Loading frequency: 15 Hz
d_a = 0 μm
d_b = 40 μm
d_c = 60 μm

loading frequencies, 1.5 and 15 Hz, are illustrated by plotting the transcortical interstitial pore fluid pressure differences at these frequencies. This panel also illustrates the effect of frequency on the interstitial pore fluid pressure gradients. The PLC is the porosity associated with the osteocytes that are the prime candidates for the mechanosensory cell in bones because of the fluid movement induced by the interstitial pore fluid pressure gradients [1–4]. The bone fluid in the smallest porosity, the collagen–hydroxyapatite porosity, is considered to be immovable under normal conditions, because it is bound to the collagen–hydroxyapatite structure.

Over the last 40 years, many researchers have used tracers to document bone fluid transport [37, 45, 61, 63, 68, 70, 79, 80] (S.D. Doty, 1997, Private communication); see [70] for a summary of the tracers employed. An excellent recent summary of these efforts is given by Fritton and Weinbaum [4]. These tracers show that the normal bone fluid flow is from the marrow cavity to the periosteal lymphatic vessels through the Volkmann and Haversian canals. The flow passes from the Haversian canal into the PLC to the cement line of the osteon.

9.8.3
Electrokinetic Effects in Bone

Electrodes placed on two different bone surfaces will measure a difference in voltage when wet bones are deformed. These voltages are called strain-generated potentials (SGPs). SGPs in wet bone are now recognized as dominantly electrokinetic phenomena explained by an extension of poroelasticity. From an experimental viewpoint, SGPs are a significant technique for the investigation of the poroelastic behavior of the bone. The source of SGPs stems from the fact that the extracellular bone matrix is negatively charged due to negative fixed charges on carbohydrates and proteins; thus, a fluid electrolyte bounded by the extracellular matrix will have a diffuse double layer of positive charges. When the fluid moves, the excess positive charge is convected, thereby developing streaming currents and streaming potentials. The fluid motion is caused by the pore fluid pressure gradients induced by the deformation of the extracellular matrix due to whole bone mechanical loading. Pollack and coworkers [81–83] have laid an important foundation for explaining the origin of SGPs. The foundation is based on poroelasticity and begins with the fluid movement in the bone channels convecting the charge accumulated in the diffuse double layer of positive charges. The charge distribution in the channel is determined from the linear Poisson–Boltzmann equation. The electrical potential attenuates exponentially with distance into the fluid, perpendicular to the charged surface, divided by λ, where λ is the Debye length characterizing the diffuse double layer. The typical Debye length λ for normal physiological saline is 1 nm or less; hence, the decay of the potential with respect to distance from the surface is very rapid. The total streaming current vector per unit area $\mathbf{j}$ passing through all the channels of a material because of the pressure-driven axial flow can be obtained by multiplying the charge density by the local velocity field in the channel and integrating this result over the cross section of the channel.

The total streaming current vector **j** is a flux vector similar to the fluid mass flow rate vector **q**. Linear irreversible thermodynamics [81–84] provides a structure for relating these fluxes. For small departures from equilibrium in isotropic materials, the total streaming current vector **j** and the fluid mass flow rate vector **q** are related by

$$\mathbf{q} = -\mathbf{K}^{(qp)} \cdot \nabla p + \mathbf{K}^{(qV)} \cdot \nabla V \text{ and } \mathbf{j} = \mathbf{K}^{(jp)} \cdot \nabla p - \mathbf{K}^{(jV)} \cdot \nabla V \tag{9.1}$$

where the superscripted **K** matrix coefficients are material properties. These equations state that the total streaming current vector **j** and fluid mass flow rate vector **q** are linearly dependent upon the voltage gradient as well as the pressure gradient. The first equation of Eq. (9.1) when $\nabla V = 0$ is Darcy's law (i.e., $\mathbf{q} = -\mathbf{K}^{(qp)} \cdot \nabla p$). The second equation of Eq. (9.1) when $\nabla p = 0$ ($\mathbf{j} = -\mathbf{K}^{(jV)} \cdot \nabla V$) is Ohm's law in its field version. The Onsager reciprocity theorem relates the cross-flux coefficients, $\mathbf{K}^{(qV)} = \mathbf{K}^{(jp)}$.

The significant and useful connection between pressure and voltage arises from the recognition that the convective and the conduction currents are equal and opposite so that there is no net current flow, $\mathbf{j} = 0$; thus, from the second equation of Eq. (9.1):

$$\nabla p = [\mathbf{K}^{(jp)}]^{-1} \cdot \mathbf{K}^{(jV)} \nabla V \tag{9.2}$$

Integration of this result yields the fact that the pressure must be proportional to the voltage plus a function of time. This means that when a potential difference in a bone is measured in connection with an electrokinetic event, the potential difference is proportional to a pore pressure difference between the same two points used to measure the potential difference. Since electrodes are much smaller than pore pressure probes, this theory provides a useful tool for probing the poroelastic response of the bone. The formulation of the model presented by Salzstein and Pollack [83] was extended to the Biot poroelastic formalism, and it removed the incompressibility assumption [1, 2]. The revised model has been applied to study the mechanosensory system in the bone [1, 2, 85–87].

The anatomical site in the bone tissue that contains the fluid source of the experimentally observed SGPs was not agreed upon, but it was argued by Cowin *et al.* [2] that it should be the PLC. That argument is summarized here. Earlier, it had been concluded that the site of SGP creation was the collagen–hydroxyapatite porosity of the bone mineral, because small pores of approximately 16 nm radius were consistent with their experimental data if a poroelastic–electrokinetic model with unobstructed and connected circular pores was assumed [83]. In [2], using the model of Weinbaum *et al.* [1], it is shown that the published data [83, 88, 89] are also consistent with the argument that the larger pore space (100 nm) of the PLC is the anatomical source site of the SGPs if the hydraulic drag and electrokinetic contribution associated with the passage of bone fluid through the surface matrix (glycocalyx) of the osteocytic process are accounted for. The mathematical models [1, 83] are similar in that they combine poroelastic and electrokinetic theories to describe the phase and magnitude of the SGP. The two theories differ in the description of the interstitial fluid flow and streaming currents at the microstructural level and in the anatomical structures that determine the flow. In [1], the resistance to fluid flow and the source of the SGP reside in the PLC, that is, to say, in the fluid annulus that surrounds the osteocytic processes, that is, the space between the cell membrane of the osteocytic process and the walls of the canaliculi–the space containing the glycocalyx or fiber matrix. In [2], the presence of the glycocalyx increases the SGPs and the hydraulic resistance to the strain-driven flow. The increased SGP matches the phase and amplitude of the measured SGPs. In the model [83], this fluid resistance and SGP are explained by assuming that an open, continuous small pore structure ($\approx$ 16 nm radius) exists in the mineralized matrix.

Experimental evidence indicating that the collagen–hydroxyapatite porosity of the bone mineral is unlikely to serve as the primary source of the SGP is obtained from several sources, including the estimates of the pore size in the collagen–hydroxyapatite porosity [58] and the impermeability of the lacunar–canalicular/collagen–hydroxyapatite porosity interface described in Section 9.7. It is thought that this impermeability is inconsistent with the suggestion of Mak *et al.* [90] that both the PLC and the collagen–hydroxyapatite porosity are sources of the experimentally observed SGPs. It was noted by Mak *et al.* [90] that, since there were many assumptions associated with the physical constants in their model, their study should be considered as a parametric study of their model. For example, the authors assume a value for the interface permeability between the PLC and the collagen–hydroxyapatite porosity that appears quite high in view of the tracer studies summarized in Section 9.7.

9.8.4
The Poroelastic Model for the Cortical Bone

Poroelasticity is a well-developed theory for the interaction of fluid and solid phases of a fluid-saturated porous medium. It is widely used in geomechanics, and it has been applied to bones by many authors in the last 40 years. A review of the literature related to the application of poroelasticity to the bone fluid is presented in [7]. This work also describes the specific physical and modeling considerations that establish poroelasticity as an effective and useful model for deformation-driven bone fluid movement in the bone tissue. The application of poroelasticity to bone differs from its application to soft tissues in two important ways. First, the deformations of the bone are small while those of soft tissues are generally large. Second, the bulk compressibility of the mineralized bone matrix is about seven times stiffer than that of the fluid in the pores, while the bulk compressibilities of the soft tissue matrix and the pore water are almost the same. Poroelasticity and electrokinetics can be used to explain SGPs in a wet bone. It is noted that SGPs can be used as an effective tool in the experimental study of local bone fluid flow, and that the knowledge of this technique will contribute to the answers for a number of questions concerning bone mineralization and the bone mechanosensory system.

A poroelastic model for the interstitial fluid flow space in bone tissue, with a reasonably accurate anatomical model for the architecture of its pore space structure, is presented in [91]. In order to characterize the special type of porous material's pore structure considered, the phrase "hierarchical" was used as an adjective to modify "poroelasticity." Alternatively, it could be described as a set of nested porosities like a set of Russian nested dolls or babushka (matryoshki); babushka is a set of dolls of decreasing sizes placed one inside another; each doll but the smallest may be opened to reveal another doll of the same sort inside. The idea of a smaller structure within a larger, similarly shaped, structure is the idea that is to be transferred from a set of babushka to sets of different pore structures in a porous material. The body fluids in tissues reside in such nested, topologically similar, pore structures with different pore sizes in the bone, and other tissue types. Examples of these porosities in the bone tissue are the PV, the PLC, and PCA.

The animal vascular tree is an example of a pore structure with two such nested systems that are connected. In a microcirculatory bed, blood flows from arteries to arterioles, then to capillaries, and then to venules and into the veins; in each of these pore structures, the pore size is relatively uniform, but it monotonically varies between the levels of porosity characterized by their pore size. The arterial system consists of the capillaries nested within the arterioles that are nested within the arteries. The venous system consists of the capillaries nested within the venules that are nested within the veins. The capillary plexuses of the two nested systems are connected.

Body fluids in tissues reside in such nested, topologically similar, pore structures with different pore sizes in the bone, tendon, meniscus, and possibly other tissue types. The nesting or ordering criterion is the porosity or pore size. The nested porosities are connected; so the pore fluid may easily flow through each and across the boundaries between the two nearest neighbor porosities, but any particular pore size porosity may only interchange its pore fluid with the next larger pore size porosity and the next smaller pore size porosity. The flow of interstitial fluid in tissues like bones, tendons, meniscus, and possibly other tissue types is similar to the blood flow in the vascular system in the sense that the different pore size porosities are nested, but unsimilar in three important aspects: (i) there are only two levels of pore size porosity important for bone fluid flow, (ii) there is no flow out of, or in from, the smaller pore size porosity into any even smaller pore size porosity, but only into, or in from, a larger pore size porosity, and (iii) the fluid flow direction reverses in the normal physiological function of these tissues.

With the exception of only the animal vascular tree, the applications of poroelasticity to fluid movement in biological tissues have simply transferred the models of the pore structure employed in geomechanics to tissues. Existing theories of the poroelasticity of materials with multiple connected porosities with different characteristic sizes and therefore different permeabilities do not address the case of nested porosities. These existing theories are appropriate for their intended use, fractured porous geological structures, but they are not appropriate for the biological tissues of interest; the nested porosities in biological tissues are hierarchical based on the average diameters of their fluid transport channels while the multiple porosity poroelasticity theories for fractured geological structures are democratic; their fluid transport channels of a particular size may exchange fluid with transport channels of any pore size. The primary objective of this work is to provide a model of a poroelastic pore structure that is appropriate for bone tissue; it is a model that is easily extended to other tissues such as the tendon and the meniscus. Concerning bone, the principal focus is on the modeling of the mechanical and blood pressure load-driven movements of the interstitial bone fluid flow.

The absence of the assumption of incompressible constituents is a significant difference between the version of poroelasticity theory employed in [91] and the poroelasticity theory used for previous published solutions involving soft tissues. The assumption of incompressible constituents, while appropriate for soft tissues, is inaccurate for hard tissues. The solution for the unconfined compression of an annular, transversely isotropic, poroelastic hollow cylinder with compressible constituents was recently presented [57]. On the basis of this solution, a protocol has been devised for an experimental test procedure to determine tissue permeabilities for the smallest nested bone porosity, the osteonal lumen wall, and the osteonal cement line. This protocol will extend to bone tissue an experimental technique that has been very effective in determining soft tissue poroelastic properties [56].

As noted above, current theoretical and experimental evidence suggests that the bone cells in the lacunae (pores) of the PLC are the principal mechanosensory cells of the bone, and that they are activated by the induced drag from fluid flowing through the PLC [1, 2]. The movement of bone fluid from the region of the bone vasculature through the canaliculi and the lacunae of the surrounding mineralized tissue accomplishes three important tasks. First, it transports nutrients to the cells in the lacunae buried in the mineralized matrix. Second, it carries away the cell waste. Third, the bone fluid exerts a force on the cell process – a force that is large enough for the cell to sense. This is thought to be the basic mechanotransduction mechanism in the bone – the way in which the bone senses the mechanical load to which it is subjected. Understanding bone mechanotransduction is fundamental to the understanding of how to treat osteoporosis, how to cope with microgravity in long-term manned space flight, and how to design prostheses that are implanted in bone tissues to function for longer periods.

These considerations suggest that the PV and PLC function almost independently, the prime mechanical influences for the two porosities being very different as are the timescales of their response. The mechanical loading of the whole bone moves the bone fluid in the PLC. When the bone is compressed, the bone fluid is driven from the PLC into the low-pressure PV, and when the bone is in tension, or the compression is reduced, the bone fluid is sucked from the PV into the PLC. These drainage and imbibing processes occur on a pressure timescale that is much larger than the short pressure adjustment relaxation time for the PV and therefore have minimum influence on the pressure in the PV. The change in interstitial pore fluid pressure in the PV due to inflow or outflow of bone fluid from the PLC is insignificant because of the time period of pressure adjustment, which is much shorter (estimates of these time periods are in Zhang *et al.* [53]) than the pressure adjustment time period for the PLC. While the bone fluid in the PLC is significantly affected for a longer time by the mechanical loading of the whole bone, the bone fluid pressure in the PV is hardly affected because the PV relaxes the pressure pulse very rapidly by diffusion.

9.8.5
Interchange of Interstitial Fluid between the Vascular and Lacunar–Canalicular Porosities

Using the hierarchical scheme described in the previous subsection, a model was formulated in [91] for the transport of bone interstitial fluid between the PV and PLC porosity levels in the osteonal cortical bone. A section of this bone is illustrated in Figure 9.5. The osteon at the top of this figure is entirely PLC porosity except for its central lumen, called the *osteonal canal* or *Haversian canal*, which is part of the PV porosity. The PV porosity consists of the volume of all the tunnels in bones that contain blood vessels and includes all the osteonal canals and all the Volkmann canals, less the volume of the tunnels occupied by the blood vessels.

The PV and the PLC are both modeled as poroelastic hollow circular cylinders. The poroelastic hollow circular cylinder model of the PLC connects through its inner cylindrical wall to the PV; the hollow part of this cylinder is actually part of the PV. The inner surface of the cylinder representing the PLC is the surface across which the two porosities exchange pore fluids. The PLC is assumed to permit flow across its inner radial boundary, but not across its outer radial boundary. While other assumptions are possible, an earlier study [78] showed that this is a reasonable assumption. The PV is assumed to permit flow across both of its radial boundaries. The PLC hollow cylinder is the osteon of Figure 9.5. The PV hollow cylinder is the entire bone of Figure 9.5 with central lumen of the whole bone, the medullary canal, constituting the hollow part of the PV model. Both of these models are continuum models and the transport connection between is the outflow–influx across the osteonal or Volkmann inner wall between the PLC and the PV [92]. In the domain between the inner surface and outer surface of the PV cylinder, there are areal sources–sinks that permit interchange of fluid between the two continuum models representing the PV and PLC.

In this model, the fluid movement will be driven by two force systems. The whole bone and its surrounding soft tissue structures are assumed to be cyclically strained in the long bone direction by the axial strain $\varepsilon(t) = \varepsilon_o e^{i\omega t}$. This straining occurs as a result of environmental loading and muscle stimulation. At the endosteum, the wall of the central lumen of the whole bone, the medullary canal, the pore fluid pressure is assumed to be the same as the blood pressure, $p^{BPo}e^{i\Omega t}$ where p^{BPo}= 60 mmHg. We assume that the periosteum is impermeable; it has been identified as a barrier to the interstitial fluid flow [61]. We note that lymphatic circulation is unlikely to play a role in bone fluid transport in the normal bone, because lymphatic vessels are absent in the bone (see Section 9.6 above).

As a first step in describing the results in [91], the generic poroelastic hollow circular cylinder model of this section is specialized to the PLC. As a second step,

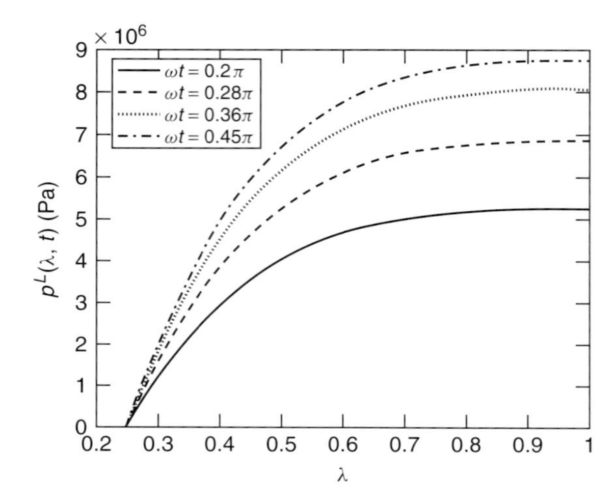

Figure 9.9 A plot of the pore pressure in the PLC, $p^L(\lambda, t)$ as a function of the dimensionless radial coordinate of the model osteon λ, at various temporal points in the cyclic oscillation; this plot is for the special case when the pore pressure in the PV vanishes. The values of parameters used in the plot are listed in Table 9.1. (From Cowin *et al.* [91].)

the volume of the fluid leaving the PLC and moving to the PV in one cycle is described. As a third step, the generic poroelastic hollow circular cylinder model of this section is specialized to the PV, accounting for the influx from/outflow to the PLC.

The pore pressure in the PLC as a function of the dimensionless radial coordinate of the model osteon λ, at various temporal points in the cyclic oscillation, is shown in Figure 9.9; this plot is for the special case when the pore pressure in the PV vanishes. A plot of $V_{\text{total due to } \varepsilon_0}(\omega)$, the fluid volume exchanged between the PLC and the PV in the case of zero PV pressure, against the driving frequency ω for values of ϕ, the PLC porosity, equal to 0.05, 0.1, and 0.15 is shown in Figure 9.10. Since the nonzero PV pressure reduces $V_{\text{total due to } \varepsilon_0}(\omega)$, the case of zero PV pressure yields the maximum value of V_{total}.

Expressions relating the pore pressure field in the PV to the two driving forces, the cyclic mechanical straining with amplitude ε_o and a frequency ω and the blood pressure with amplitude p^{BPo} and a frequency Ω, are plotted in Figures 9.11 and 9.12. A plot of the pore pressure in the PV, $p^V(\beta, t)$, is shown in Figure 9.11 as a function of the dimensionless radial coordinate of the model bone β at various temporal points in the cyclic oscillation. Figure 9.12 is again a plot of the same equation for the pore pressure in the PV, but with the PV permeability reduced by 2 orders of magnitude, $6.35 \times 10^{-8}\ \text{m}^2$, from that employed in Figure 9.11, $6.35 \times 10^{-10}\ \text{m}^2$. Notice from Figure 9.11 that the peak pressure is near 15 MPa for a PV permeability of $6.35 \times 10^{-10}\ \text{m}^2$, and from Figure 9.12, it is found to be about 9.5 kPa for a PV permeability of $6.35 \times 10^{-8}\ \text{m}^2$. Since 8 kPa is about 60 mmHg, the

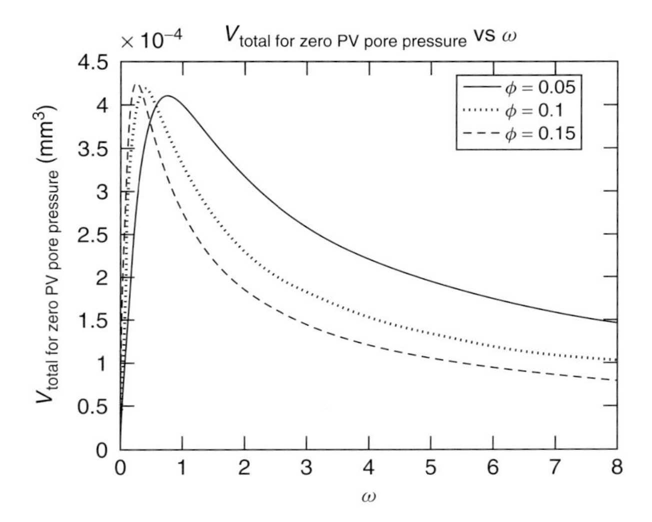

Figure 9.10 A plot of V_{total}, the fluid volume exchanged between the PLC and the PV is the case of zero PV pressure, against the driving frequency ω for values of ϕ^L, the PLC porosity, equal to 0.05, 0.1, and 0.15. Since the nonzero PV pressure reduces V_{total}, the case of zero PV pressure yields the maximum value of V_{total}. The values of parameters used in the plot are listed in Table 9.1. (From Cowin *et al.* [91].)

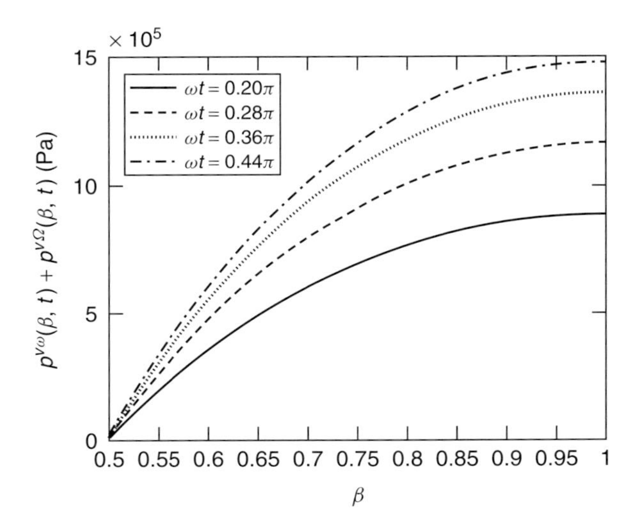

Figure 9.11 A plot of the pore pressure in the PV, $p^V(\beta, t)$ given by Zhang *et al.* [87] using a PV permeability on the order of 10^{-10} m^2, as a function of the dimensionless radial coordinate of the model bone β at various temporal points in the cyclic oscillation. The values of parameters used in the plot are listed in Table 9.1. (From Cowin *et al.* [91].)

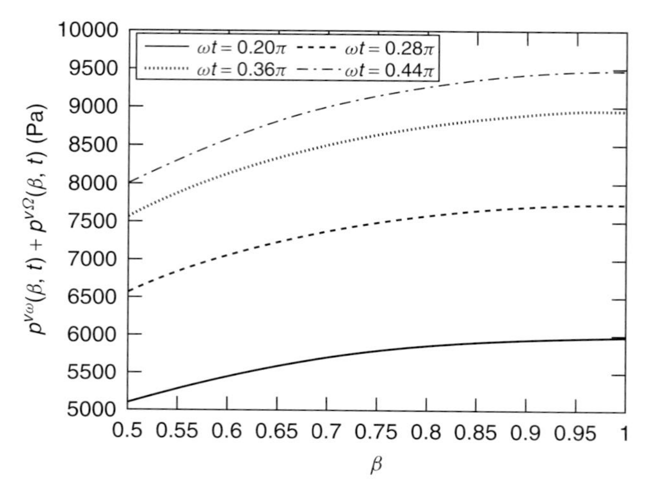

Figure 9.12 A plot of the pore pressure in the PV, $p^V(\beta, t)$ using a PV permeability on the order of 10^{-8} m^2, as a function of the dimensionless radial coordinate of the model bone β at various temporal points in the cyclic oscillation. The values of parameters used in the plot are listed in Table 9.1. (From Cowin *et al.* [91].)

peak blood pressure, a PV pore pressure in the range of 8 kPa is a more reasonable result for several reasons. First, as the comparison of Figures 9.11 and 9.12 shows, the gradient of the PV pore pressure across the bone cortex is smaller in the case of the lower PV permeability; the bone pressure is almost uniform across the cortex. Since the gradient is small at the endosteal surface, there is less flow in and out of the medullary canal; there is no known physiological advantage for interstitial flow in and out of the medullary canal. Last, the pore pressure in the PV cannot exceed a pore pressure that would collapse the blood vessels present for any significant length of time. Figure 9.13 is a plot of the pore pressure due to blood pressure in the PV, $p^{V\Omega}(\beta, t)$, as a function of the dimensionless radial coordinate of the model bone β, at various temporal points in the cyclic oscillation. Figure 9.13a is at a scale that obscures the small spatial gradient in the blood pressure; Figure 9.13b expands the scale for a single time point in the oscillation so that the small gradient is apparent. These considerations concerning the plots in Figures 9.11 and 9.12 suggest a PV permeability lower than 10^{-9} m^2 and perhaps a little greater than 10^{-8} m^2.

The interpretation in [91] of results presented in a recent paper [95] reinforces the point made at the end of the previous paragraph concerning a PV permeability lower than 10^{-9} m^2 and perhaps a little greater than 10^{-8} m^2. In [95], a finite element model whose geometry was generated from a quantitatively computed tomography scan of a section of human tibia was employed. The fluid velocities and the pore fluid pressure in the PLC and in the PLC + PV were calculated for

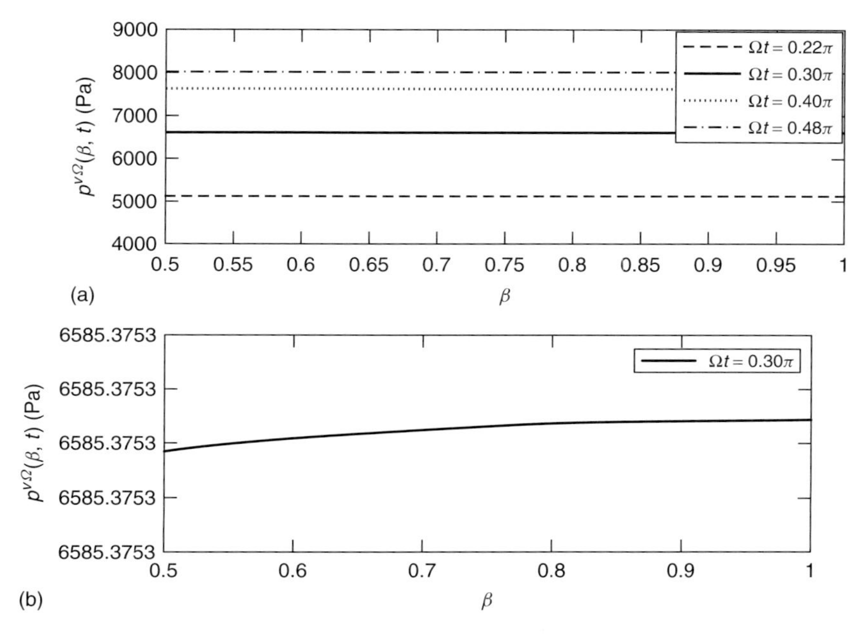

Figure 9.13 A plot of the pore pressure due to blood pressure in the PV, $p^{V\Omega}(\beta, t)$, as a function of the dimensionless radial coordinate of the model bone *b* at various temporal points in the cyclic oscillation. Plot (a) is at a scale that obscures the small spatial gradient in the blood pressure; Plot (b) expands the scale for a single time point in the oscillation so that the small gradient is apparent. Matching the temporally oscillating blood pressure is a boundary condition at the medullary canal ($\beta = 0.5$). The decay in the oscillating blood pressure across the cortex of the whole bone from its value at the medullary canal ($\beta = 0.5$) to the periosteum ($\beta = 1$) is small, as can be seen from the above discussion. Almost no change is seen in these curves when the PV permeability is reduced from the order of 10^{-12} m^2, to the order of 10^{-10} m^2, or to the order of 10^{-8} m^2. The values of parameters used in the plots are listed in Table 9.1. (From Cowin *et al.* [91].)

a specified applied loading consisting of combined axial loading and bending. Figure 4b in that paper shows PV pore pressures in the megapascal range. We think that this pore pressure is too high because of likely constriction of the blood vessels in the PV and the significant volume of interstitial fluid that is drained into the medullary canal. The model in [95] is likely correct, and the high PV pore pressures in the megapascal range are due to the authors' use of a reported PV permeability of 10^{-14} m^2, which we now think to be much too small for the PV. As work began on this chapter, we thought that a PV permeability of 10^{-14} m^2 was possible. The original version of the figure that is now reproduced as Figure 9.11 employed a PV permeability of 10^{-14} m^2, and the results in [91] are similar to those in [95] in predicting an unrealistically high pore pressure in the PV. It is the analysis of the last paragraph that changed our minds. A subsequent literature evaluation of the references reporting the low PV permeabilities revealed that they were very rough estimates.

9.8.6 Implications for the Determination of the Permeabilities

The measurement of the permeability of the PV has not been accomplished with sufficient accuracy to date, primarily because of the topological intertwining of the PV with the PLC. The pore fluid pressures in both the PLC and the PV depend upon the permeability of their own porosity as well as upon the permeability of the other porosities. The PV permeabilities reported in the past are thought to be significant underestimates, because these previous values represent a PLC and PV lumped permeability measurement rather than a PV or PLC measurement alone. The fact that, in general, the pore pressure in each porosity depends upon the permeability of the other porosities means that special strategies should be designed for the experimental determination of each porosity's permeability. Traditional permeability measurement techniques based on Darcy's technique of measuring the volume of flow per unit area per unit time across a porous layer and dividing by the pore pressure gradient across the layer requires modification for porous media structured hierarchically.

The two porosities of the bone tissue occupy the same three-dimensional volume of the bone tissue. In the cross section of the bone shown in Figure 9.5, the PV channels are the central lumens of the osteons and the region immediately surrounding the osteon is part of the PLC. Thus, it is impossible to obtain a reasonably sized specimen of bone tissue that contains some PV porosity, without containing any PLC porosity. This topological fact makes it difficult to design an experiment to measure the permeability of the PV. The analytical basis for a number of different approaches to this measurement technique for the measurement of the permeability of the PV was provided by Cowin *et al.* [91]. From this discussion, it is clear that previous measurements of the permeability of the PV were not accurate as they were based on models that did not include the interchange of interstitial fluid with the PLC. This is one reason why we have not honored the reported measurements of the permeability of the PV that are present in the literature – permeabilities as low as 10^{-14} m^2. As noted in the previous section, the results of this study suggest a PV permeability lower than 10^{-9} m^2 and perhaps a little greater than 10^{-8} m^2. The second reason is that permeabilities as low as 10^{-14} m^2 predict PV pore pressures that are too high as the results in [91] and [95] show.

We now turn to the question of measurement of the permeability of the PLC. The entire content of a single osteon that is considered as a hollow circular cylinder is composed of the PLC porosity and no PV porosity; this may be used to measure the permeability of the PLC. This has been accomplished [57] by isolating single osteons, imposing a compressive stress on them, and measuring the time-dependent axial strain. The PLC permeability determined was on the order of 10^{-24} m^2. The analysis of these experimental data employed the poroelastic model of this situation analyzed in [56].

References

1. Weinbaum, S., Cowin, S.C., and Zeng, Yu. (1994) *J. Biomech.*, **27**, 339.
2. Cowin, S., Weinbaum, S., and Zeng, Yu. (1995) *J. Biomech.*, **28**, 1281.
3. Sharma, U., Mikos, A.G., and Cowin, S.C. (2007) Mechanosensory mechanisms in bone, in *Textbook of Tissue Engineering*, 3rd edn (eds R. Lanza, R. Langer, and J.P. Vacanti), Elsevier/Academic Press, pp. 919–934.
4. Fritton, S.P. and Weinbaum, S. (2009) *Annu. Rev. Fluid Mech.*, **41**, 347–374.
5. Biot, M.A. (1935) *Ann. Soc. Sci. Brux.*, **B55**, 110.
6. Biot, M.A. (1941) *J. Appl. Phys.*, **12**, 155.
7. Cowin, S.C. (1999) *J. Biomech.*, **32**, 218.
8. Yoffey, J.M., Hudson, G., and Osmond, D.G. (1965) *J. Anat.*, **99**, 841.
9. Williams, E.A., Fitzgerald, R.H., and Kelly, P.J. (1984) Microcirculation of bone, in *The Physiology and Pharmacology of the Microcirculation*, vol. 2, Academic Press.
10. Martin, R.B., Burr, D.B., and Sharkey, N.S. (1998) *Skeletal Tissue Mechanics*, Springer.
11. Rhinelander, F.W. and Wilson, J.W. (1982) in *Bone in Clinical Orthopaedics* (ed. G. Summer-Smith), W. B. Saunders, Philadelphia, p. 81.
12. Doty, S.B. (1981) *Calcif. Tissue Int.*, **33**, 509.
13. Kornblum, S.S. (1962) The microradiographic morphology of bone from ischaemic limbs. Graduate thesis. University of Minnesota.
14. Herskovits, M.S., Singh, I.J., and Sandhu, H.S. (1991) Innervation of bone, in *Bone: A Treatise*, Bone Matrix, vol. 3 (ed. B.K. Hall), Telford Press.
15. Neuman, W.F. and Neuman, M.W. (1958) *The Chemical Dynamics of Bone*, University of Chicago Press, Chicago.
16. Martin, G.R., Firschein, H.E., Mulryan, B.J., and Neuman, W.F. (1958) *J. Am. Chem. Soc.*, **80**, 6201.
17. Neuman, M.W. (1982) *Calcif. Tissue Int.*, **34**, 117.
18. Bridgeman, G. and Brookes, M. (1996) *J. Anat.*, **188**, 611.
19. De Bruyn, P.P.H., Breen, P.C., and Thomas, T.B. (1970) *Anat. Rec.*, **168**, 55.
20. LopezCurto, J.A., Bassingthwaighte, J.B., and Kelly, P.J. (1980) *J. Bone Joint Surg.*, **62**, 1362.
21. Trueta, J. and Harrison, M.H.M. (1953) *J. Bone Joint Surg.*, **35B**, 442.
22. Zamboni, L. and Pease, D.C. (1961) *J. Ultrastruct Res.*, **5**, 65.
23. Branemark, P.-I. (1961) *Angiology*, **12**, 293.
24. Nelson, G.G., Kelly, P.J., Lowell, F.A., Peterson, L.F.A., and Janes, J.M. (1960) *J.Bone Joint Surg.*, **42A**, 625.
25. Cooper, R.R., Milgram, J.W., and Robinson, R.A. (1966) *J. Bone Joint Surg.*, **48A**, 1239.
26. Hughes, S. and Blount, M. (1979) *Ann. R. Coll. Surg. Engl.*, **61**, 312.
27. Cuthbertson, E.M., Siris, E., and Gilfillan, R.S. (1965) *J. Bone Joint Surg., Am.*, **47**, 965.
28. Brooks, M. and Revell, W.J. (1998) *Bloods Supply of Bone*, Springer, London.
29. Kelly, P.J. and Bronk, J.T. (1990) *Microvasc. Res.*, **39**, 364.
30. Welch, R.D., Johnston, C.E. II, Waldron, M.J., and Poteet, B. (1993) *J. Bone Joint Surg., Am.*, **75**, 53.
31. Bergula, A.P., Huang, W., and Frangos, J.A. (1999) *Bone*, **24**, 171.
32. Lilly, A.D. and Kelly, P.J. (1970) *J. Bone Joint Surg., Am.*, **52**, 515.
33. Arnoldi, C.C., Lemperg, R., and Linderholm, H. (1971) *Acta Orthop. Scand.*, **42**, 454.
34. Green, N.E. and Griffin, P.P. (1982) *J. Bone Joint Surg., Am.*, **64**, 666.
35. Liu, S.L. and Ho, T.C. (1991) *J. Bone Joint Surg., Am*, **73**, 194.

36. Wang, L., Fritton, S.P., Weinbaum, S., and Cowin, S.C. (2003) *J. Biomech.*, **36**, 1439.
37. Doty, S.D. and Schofield, B.H. (1972) in *Calcium, Parathyroid Hormone and the Calcitonins* (eds R.V. Talmage and P.L. Munson), Excerpta Medica, Amsterdam, p. 353.
38. Owen, M. and Triffitt, J.T. (1976) *J. Physiol.*, **257**, 293.
39. Kolodny, A. (1925) *Arch. Surg. (Chicago)*, **11**, 690.
40. Seliger, W.G. (1970) *Anat. Rec.*, **166**, 247.
41. Anderson, D.W. (1960) *J. Bone Joint Surg.*, **42A**, 716.
42. Abramson, D.I. and Dobrin, P.B. (eds) (1984) *Blood Vessels and Lymphatics in Organ Systems*, Academic Press, Orlando.
43. Vittas, D. and Hainau, B. (1989) *Lymphology*, **22**, 173–177.
44. Edwards, J.R., Williams, K., Kindblom, L.G., Meis-Kindblom, J.M., Hogendoorn, P.C.W., Hughes, D., Forsyth, R.G., Jackson, D., and Athanasou, N.A. (2008) *Hum. Pathol.*, **39**, 49–55.
45. Morris., M.A., Lopez-Curato, J.A., Hughes, S.P.F., An, K.N., Bassingthwaighte, J.B., and Kelly, P.J. (1982) *Microvasc. Res.*, **23**, 188–200.
46. Ciani, C., Ramirez Marin, P.A., Doty, S.B., and Fritton, S.P. (2007) High-resolution measurement of cortical bone porosity in normal and osteopenic rats. Transactions of the 53rd Meeting of the Orthopaedic Research Society, vol. 32, p. 263.
47. Ciani, C., Doty, S.B., and Fritton, S.P. (2008) Osteopenia increases osteocyte lacunar-canalicular porosity in rat metaphyseal bone. Transactions of the 54th Meeting of the Orthopaedic Research Society, vol. 33, p. 344.
48. (a) Johnson, M.W. (1984) *Cal. Tissue Int.*, **36**, S72–S76; (b) Cowin, S.C. and Lanyon, L.E. (1984) *Functional Adaptation in Bone Tissue,* Supplement to vol. 36, Calcified Tissue International, American Society of Mechanical Engineers.
49. Rouhana, S.W., Johnson, M.W., Chakkalakal, D.A., and Harper, R.A. (1981) Permeability of the osteocyte lacuno-canalicular compact bone. Proceedings of the Joint ASME-ASCE Conference Biomechanics Symposium AMD, vol. 43, pp. 169–172.
50. Beno, T., Yoon, Y.J., Cowin, S.C., and Fritton, S.P. (2006) *J. Biomech.*, **39**, 2378–2387.
51. Dillaman, R.M., Roer, R.D., and Gay, D.M. (1991) *J. Biomech.*, **24**, 163–177.
52. Steck, R., Niederer, P., and Knothe Tate, M.L. (2003) *J. Theor. Biol.*, **220**, 249–259.
53. Zhang, D., Weinbaum, S., and Cowin, S.C. (1998) *J. Biomech. Eng.*, **120**, 697–703.
54. Smit, T.H., Huyghe, J.M., and Cowin, S.C. (2002) *J. Biomech.*, **35**, 829–836.
55. Oyen, M.L. (2008) *J. Mater. Res.*, **23**, 1307–1314.
56. Gailani, G., Benalla, M., Mahamud, R., Cowin, S.C., and Cardoso, L. (2009) *J. Biomech. Eng.* **131**, 101007.
57. Gailani, G.B. and Cowin, S.C. (2008) *Mech. Mater.*, **40**, 507–523.
58. Holmes, J.M., Davies, D.H., Meath, W.J., and Beebe, R.A. (1964) *Biochemistry*, **3**, 2019–2024.
59. Wehrli, F.W. and Fernández-Seara, M.A. (2005) *Ann. Biomed. Eng.*, **33**, 79–86.
60. Arramon, Y.P. and Cowin, S.C. (1997) *FORMA*, **12**, 209–221.
61. Li, G., Bronk, J.T., An, K.N., and Kelly, P.J. (1987) *Microcirc. Res.*, **34**, 302.
62. Soares, A.M.V., Arana-Chavez, V.E., Reid, A.R., and Katchburian, B. (1992) *J. Anat.*, **181**, 343–356.
63. Dillaman, R.M. (1984) *Anat. Rec.*, **209**, 445.
64. Wang, L., Ciani, C., Doty, S.B., and Fritton, S.P. (2004) *Bone*, **34**, 499–509.
65. Ciani, C., Doty, S.B., and Fritton, S.P. (2005) *Bone*, **37**, 379–387.
66. Neuman, W.F. (1969) *Fed. Proc. Fed. Am. Soc. Exp. Biol.*, **28**, 1846–1850.
67. Miller, S.C. and Jee, W.S.S. (1992) in *Bone*, Bone Metabolism and Mineralization, vol. 4 (ed. B.K. Hall), CRC Press, Boca Raton, pp. 1–19.
68. Tanaka, T. and Sakano, A. (1985) *J. Dent. Res.*, **64**, 870–876.
69. Asasaka, N., Kondo, T., Goto, T., Kido, M.A., Nagata, E., and Tanaka, T. (1992) *Arch. Oral Biol.*, **37**, 363–368.

70. Knothe Tate, M.L., Niederer, P., and Knothe, U. (1998b) *Bone*, **22**, 107–117.

71. Williams, S.M. and McCarthy, I.D. (1986) *J. Biomed. Eng.*, **8**, 235–243.

72. McCarthy, I.D. and Yang, L. (1992) *J. Biomech.*, **25**, 441–450.

73. Martin, B. (1994) *J. Orthop. Res.*, **12**, 375–383.

74. Azuma, H. (1964) *Angiology*, **15**, 396.

75. Shim, S., Hawk, H., and Yu, W. (1972) *Surg. Gynecol. Obstet.*, **135**, 353.

76. James, J. and SteijnMyagkaya, G.L. (1986) *J. Bone Joint Surg.*, **68B**, 620.

77. Catto, M. (1976) in *Aseptic Necrosis of Bone NY* (ed. J.K. Davidson), Elsevier, American, p. 2.

78. Wang, L., Fritton, S.P., Cowin, S.C., and Weinbaum, S. (1999) *J. Biomech.*, **32**, 663.

79. Seliger, W.G. (1969) *Anat. Rec.*, **166**, 247.

80. Simonet, W.T., Bronk, J.T., Pinto, M.R., Williams, E.A., Meadows, T.H., and Kelly, P.J. (1988) *Mayo Clin. Proc.*, **63**, 154.

81. Pollack, S.R., Salzstein, R., and Pienkowski, D. (1984) *Calcif. Tissue Int.*, **36**, S77–S81.

82. Salzstein, R.A., Pollack, S.R., Mak, A.F.T., and Petrov, N. (1987) *J. Biomech.*, **20**, 261–270.

83. Salzstein, R.A. and Pollack, S.R. (1987) *J. Biomech.*, **20**, 271–280.

84. De Groot, S.R. and Mazur, P. (1969) *Non-Equilibrium Thermodynamics*, North Holland, Amsterdam.

85. Zeng, Yu., Cowin, S.C., and Weinbaum, S. (1994) *Ann. Biomed. Eng.*, **22**, 280–292.

86. Zhang, D., Cowin, S.C., and Weinbaum, S. (1997) *Ann. Biomed. Eng.*, **25**, 357–374.

87. Zhang, D., Weinbaum, S., and Cowin, S.C. (1998a) *Ann. Biomed. Eng.*, **26**, 644–659.

88. Scott, G.C. and Korostoff, E. (1990) *J. Biomech.*, **23**, 127–143.

89. Otter, M.W., Palmieri, V.R., Wu, D.D., Seiz, K.G., MacGinitie, L.A., and Cochran, G.V.B. (1992) *J. Orthop. Res.*, **10**, 710–719.

90. Mak, A.F.T., Huang, D.T., Zhang, J.D., and Tong, P. (1997) *J. Biomech.*, **30**, 11–18.

91. Cowin, S.C., Gailani, G., and Benalla, M. (2009) *Philos. Trans. R. Soc. A*, **367**, 3401–3444.

92. Fritton, S.P., Wang, L., Weinbaum, S., and Cowin, S.C. (2001) Interaction of mechanical loading, blood flow, and interstitial fluid flow in osteonal bone. Proceedings of the Bioengineering Conference, BED, vol. 50, pp. 341–342.

93. Cowin, S.C. and Mehrabadi, M.M. (2007) *J. Mech. Phys. Solids*, **55**, 161–193.

94. Currey, J.D. and Alexander, R.McN. (1985) *J. Zool.*, **206**, 453–468.

95. Goulet, G.C., Hamilton, N., Cooper, D., Coombe, D., Tranc, D., Martinuzzia, R., and Zernicke, R.F. (2008) *J. Biomech.*, **41**, 2169–2175.

10
Bone Implant Design Using Optimization Methods

Paulo R. Fernandes, Rui B. Ruben, and Joao Folgado

10.1
Introduction

Total hip arthroplasty (THA) is a successful treatment for joint and bone diseases as well as in fracture repair [1]. It involves the replacement of the natural joint by an artificial one, composed of a femoral stem and an acetabular component as shown in Figure 10.1. The femoral component can be classified as cemented or uncemented depending on the mode of fixation. Cemented stems are fixed using a bone cement layer between the femur and the implant, while cementless stems are placed directly in contact with the bone. In this case, the biological fixation is promoted by the stem coating such as hydroxyapatite or in general a porous surface.

In the first 10 years after surgery, the THA success is greater than 80%; nevertheless, it is possible to indentify some causes of failure. Aseptic loosening is the main cause for failure, being responsible for approximately 60% of revisions [1, 4]. This failure can be due to mechanical problems such as excessive bone loss [5], inefficient initial stability [6], or cement mantle fatigue [7]. Thigh pain is another important mechanical cause for implant failure, with more than 6% of incidence [4]. On the other hand, deep infection is the major biological problem with more than 5% of prevalence [1, 4]. Actually, the majority of failures of the THA are related to mechanical factors.

Regarding the mode of fixation, cemented stems are globally more durable than cementless ones, with a 90% survival rate after 12 years. For uncemented implants this rate is less than 80% [1]. However, some cementless stems achieve more than 95% survival rate after 12 years. In fact, cementless stems with good initial stability and osseointegration exhibit a superior performance [8]. Cemented stems are preferred for elderly people, but, for patients less than 60 years old, the number of uncemented stems has increased, particularly, in the past few decades. In the early 1990s cementless stems represented only 20% of THA and in 2005 this value increased to approximately 45% [1].

As stated above, the long-term implant success is strongly related to mechanical factors. For cementless stems the initial mechanical conditions on bone–implant interface (primary stability) are very important for success of the prosthesis. In

Biomechanics of Hard Tissues: Modeling, Testing, and Materials.
Edited by Andreas Öchsner and Waqar Ahmed

ISBN: 978-3-527-32431-6

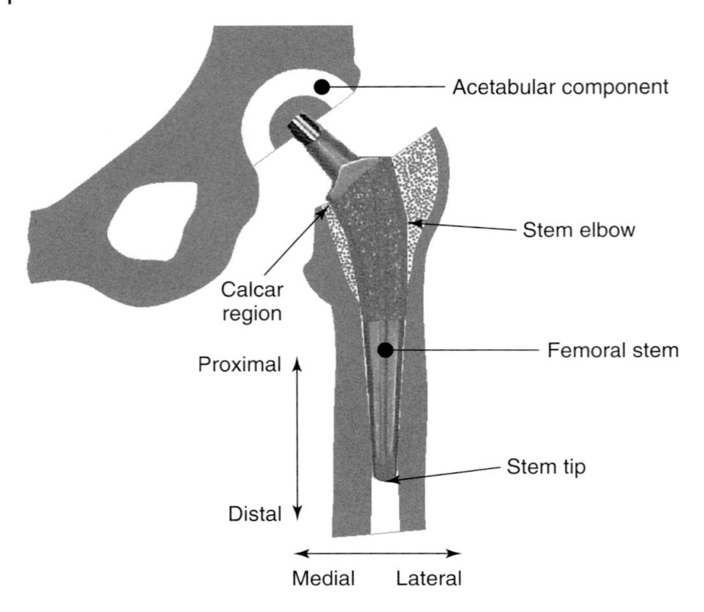

Figure 10.1 Schematic representation of a total hip replacement. (Adapted from Ruben *et al.* [2, 3].)

fact, "small" relative displacements between bone and stem and "small" contact stresses promote bone ingrowth into the stem porous coating, essential for biologic fixation [9]. Also, thigh pain is related to an inefficient initial stability and to excessive contact stresses [10]. In addition, stress shielding effect due to the presence of a metallic implant inside the femur leads to a proximal bone loss, which promotes implant loosening and reduces bone stock for a revision surgery.

Stem geometry plays an important role in the biomechanical behavior of the implant, since it determines the way the load is transferred to bone [11]. Although the stem design is subjected to several clinical requirements, there are stems with very different geometries in clinical use. From a biomechanical point of view, stem geometry. Actually, from a biomechanical point of view, stem geometry and porous coating length can be studied in order to improve initial stability and, thereby improving implant durability.

Computational mechanics tools are very attractive for analyzing and designing medical devices, with finite element method being the most commonly used. This chapter describes structural optimization methods used to succeed in bone implant. A multicriteria optimization procedure to determine the stem geometry that maximizes the initial stability and minimize the stress shielding effect is presented here. Although the model is developed for the femoral component of a hip prosthesis, similar models can be used to design other artificial joints.

A concurrent model for bone remodeling and osseointegration was also used in order to study the long-term effect of optimized stem shapes, and to confirm the relation between initial conditions and implant durability.

10.2 Optimization Methods for Implant Design

Rapid development of computational facilities and numerical tools to solve engineering problems resulted in the development of computational biomechanics. Nowadays, this field plays a key role in the modeling of living tissues, analysis and design of medical devices, and pre-clinical tests. The bone implant design has benefited from these tools, such as development of some shape optimization processes used to obtain geometries with better performance. Also, numerical bone remodeling and osseointegration models have been applied to study the stress shielding effects after a bone implant [12].

In the case of hip implant, some shape optimization models have been developed to study the relation between stem geometry and prosthesis performance for both cemented and uncemented stems.

10.2.1 Cemented Stems

Huiskes and Boeklagen [13] developed a model to optimize the shape of a cemented stem with the objective of minimizing the strain energy density in the cement mantle. This work was part of a research program with surgeons and bioengineers to obtain a new stem, which culminated with the scientific hip prostheses (Biomet Europe).

In the work of Yonn *et al.* [14] a two-dimensional model was used to obtain a stem shape that minimized the maximum stress in the cement mantle.

Katoozian and Davy presented two optimization models [15, 16]. Both of them considered a three-dimensional geometry of the bone with the implant. In the first work, the optimal shape is obtained in order to minimize the von Mises stress in the cement. In the second work, three single cost functions were considered. The first two are related to the stress field in the cement mantle, and the third is the cement strain energy density. In these studies, all optimized shapes have small distal sections and a stiffer proximal part.

Hedia and coworkers [17] minimized the fatigue in cement mantle with a constraint on the proximal bone stress.

Gross and Abel [5] optimized the thickness of a two-dimensional hollow stem. In order to do that, three single functions were considered: minimization of the cement mantle stress, maximization of the proximal bone stress with a constraint on cement stress, and maximization of the proximal bone stress without any constraint.

Tanino *et al.* [18] minimized the maximum principal stress on the distal half of the cement mantle. In this case, a three-dimensional model was considered with 15 design variables and some geometric constraints to obtain clinically admissible shapes.

These shape optimization models are important in developing a better understanding of the influence of stem shape on stress shielding effect and also on cement mantle stresses and fatigue. However, in all these models the interface was defined as fully bonded.

10.2.2 Uncemented Stems

In the studies mentioned above [15, 16], Katoozian and Davy also performed shape optimization for uncemented stems. In this case, they minimized the stress in the bone adjacent to the stem and the bone strain energy density. The finite element model considered that the bone and the stem are fully bonded. The optimized shapes presented a thin distal section and a large proximal part. These results are similar to the ones obtained for cemented stems.

In 2001, Chang *et al.* [19] developed a three-dimensional model with frictional contact at the bone–stem interface in order to minimize the difference in strain energy density of the intact femur and implanted bone. The relative tangential displacement between bone and implant was limited (to 50 μm) to avoid large displacements. A reduced midstem implant design is optimized, where the two design variables define the size of the middle part of the stem.

Kowalczyk [20] presented an optimization process to minimize the stress on bone–stem interface. In this three-dimensional model, the stem has a proximal collar and the interface is assumed to be fully bonded in coated regions. For the uncoated surface the interface condition is frictionless contact. The design variables define the coated region and stem axial length.

Fernandes *et al.* [21] presented a two-dimensional optimization procedure to obtain the shape of a hip stem to minimize the relative displacement and normal contact stress on stem–bone interface.

Ruben *et al.* [2] considered a multicriteria cost function in order to maximize initial stability. The 17 design variables are geometric parameters to define three-dimensional stem shape, and some constraints were considered to obtain clinically admissible implants. The multicriteria function permits the simultaneous minimization of relative tangential displacement and normal contact stress on stem–bone interface. Frictional contact was considered in coated regions and frictionless contact in uncoated ones. With this optimization process a set of non-dominated points were computed, and in all cases the initial stability is better than the initial shape defined based on a commercial prosthesis.

Besides these studies on shape optimization, other works on material optimization applied to implant design are available. Usually, such models assume as design variable the distribution of elastic modulus on the stem. An example of this approach is the model presented by Kuiper and Huiskes [22], where the cost function is the difference between the shear stress at stem–bone interface and a reference value. Hedia *et al.* [23] used a two-dimensional model to minimize the maximum shear stress value at the interface. In both cases, the optimized implants are stiffer in the proximal part and the modulus of elasticity decreases up to the distal part. Also, Katoozian *et al.* in 2001 presented a material optimization model for hip stem using fiber reinforced polymeric composites [24].

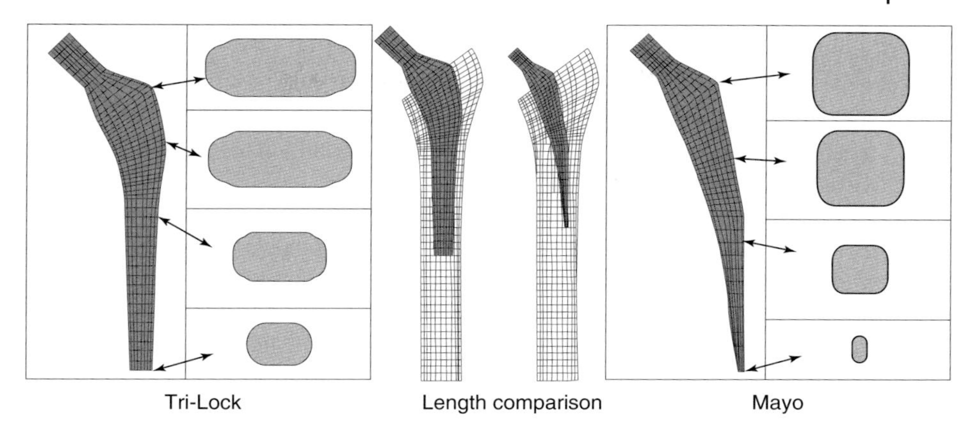

Figure 10.2 Finite element model of Tri-Lock and Mayo prostheses. (Adapted from Ruben *et al.* [3].)

In addition to the optimization models described above, some comparative analysis has also been reported in recent years. This type of work is useful in analyzing some particular shape characteristics. For instance, Mandell *et al.* [10] compared the influence of the collar size on the stability and on the stress shielding. In Sakai *et al.* [25] a new type of fixation was compared with the traditional ones. Goetzen *et al.* [26] presented computational results for a new distal design. The stem has a biodegradable pin to increase the distal diameter and thus to improve distal fixation. When the pin is completely biodegraded a gap is formed between femur and the distal part of the stem. The load-shift design tries to improve initial stability in proximal part and consequently, the osseointegration. Ruben *et al.* [3] compared the Tri-Lock stem from DePuy with the minimally invasive stem Mayo from Zimmer (see Figure 10.2).

This brief bibliographic review illustrates the relevance of the use of structural optimization methods to obtain new implant designs with better performance. In fact, these methodologies used in high-tech industries such as automotive and aerospace can also be a contribution for Biomedical Engineering, leading to innovative products made of new materials and using new manufacturing techniques in order to satisfy the tight design requirements. Moreover, shape optimization can be very useful to design new custom-made hip implants. In fact, each patient has his own characteristics and, in some cases, customized prosthesis is a better choice to improve the quality of life.

10.3 Design Requirements for a Cementless Hip Stem

As mentioned above, aseptic loosening is the main cause of THA failure, and, for cementless stems, it is essentially due to an inefficient primary stability and excessive bone loss. Thus, an uncemented hip prosthesis with a good initial stability and a moderate bone loss (reduced stress shielding effect) has a bigger probability to perform efficiently.

10.3.1 Implant Stability

The fixation of a cementless hip stem is biological. After surgery, the stem is press-fitted to the bone and fixation is achieved through an appropriate geometry and the friction interaction. Good primary stability is achieved when the bone starts to attach to the coating surface (osseointegration), establishing the necessary biological fixation. To have a good initial stability, "small" relative tangential displacements and "small" contact stresses at stem–bone interface are necessary [9]. On the other hand, thigh pain is related to excessive interface displacements and contact stresses [10]. In summary, initial stability is an essential requirement for the success of THA and can be quantified by the relative displacement and interface stress.

With respect to displacements, "large" relative tangential displacements at stem–bone interface lead to the formation of a soft fibrous tissue with reduced fixation capacity. Indeed, high values of displacements can lead to absence of bone ingrowth into porous surface. However, the threshold value of the displacement to obtain bone ingrowth is not precisely known. For instance, Rancourt *et al.* [27] suggested 28 μm as the limit value for relative tangential displacement. For Viceconti *et al.* [28], the limit is between 30 and 150 μm, and displacements between 150 and 220 μm lead to the formation of a fibrous tissue layer. Finally, to achieve a very fast osseointegration the displacement value should be less than 30 μm [28].

Contact stresses can also avoid bone ingrowth. In fact, for human femur, cortical and trabecular bone has an ultimate compressive strength of 170 MPa [29] and 7.89 MPa [30], respectively. Since proximal stem is in contact with trabecular bone (see Figure 10.1), osseointegration in this region is very sensitive to contact stress. On the contrary, the distal part of implant is in direct contact with cortical bone or is surrounded by marrow. This way, interface stresses can be greater, but not excessive to avoid thigh pain and stress shielding effect.

Besides the contact stress and relative displacement, the size of the interface gap is also an important parameter to obtain bone ingrowth. However, this parameter is not considered as a requirement in the design model described in this chapter since its relevance is relative. In fact, even with a gap of 150 μm between the bone and stem, osseointegration can occur [31].

10.3.2 Stress Shielding Effect

Another important requirement for a bone implant is the minimization of stress shielding effect.

Actually, bone is a living tissue in continuous adaptation, and its morphology depends on applied loads. This relation between loads and bone remodeling was first observed by Julius Wolff [32] at the end of the nineteenth century. From his observations he stated the law of bone remodeling (Wolffs law), which assumes that bone adapts to mechanical loading and that this adaptation follows mathematical rules. Indeed, it is possible to say that bone structure is regulated by cells that react to mechanical stimulus.

The stress shielding effect is a consequence of the mechanism of load transfer from prosthesis to femur [5]. Before THA, load transfer from pelvic bone to femur is carried out directly by the natural joint. After surgery, loads are transferred by interaction forces from the implant to femur. However, the implant is stiffer than bone and part of the total load is supported by the stem and the stress on bone tissue is globally reduced. Consequently, the femur loses mass and it becomes less dense. This stress shielding effect is greater for stiffer prostheses because fewer loads are transferred to bone tissue [5, 33].

After THA, the load is transferred from the stem to bone, from the interior to the outside part of the femur; thus there is formation of new bone next to the implant [34, 35]. This fact is more evident near the stem tip because of maximum distal stresses [33]. On the proximal region and away from the stem, the stress shielding effect is stronger and can lead to excessive bone loss.

10.4 Multicriteria Formulation for Hip Stem Design

The design of an implant has to take into account the requirements for a good primary stability (low stress and low relative displacement at the interface) and reduced stress shielding. In the case of the hip stem, these requirements lead to different geometries when they are considered alone. Thus, one needs to consider all the requirements simultaneously.

To address this problem, a multicriteria optimization process is presented in this chapter to obtain the three-dimensional femoral stem shape with better primary stability and less stress shielding. The multicriteria objective function combines three single cost functions. The first two are related to the primary stability, that is, they are a measure of the relative tangential displacement between the bone and implant and the normal contact stress. The third one is related to the stress shielding.

The optimization problem can be stated in a general way, as,

$$\begin{aligned}
&\min_{\mathrm{d}} f(\mathbf{d}) \\
&\text{such that} \\
&(d_i)_{\min} \leq d_i \leq (d_i)_{\max} \quad i = 1, 2, \ldots, 14 \\
&h_j(\mathbf{d}) \leq 0 \quad j = 1, 2, \ldots, 10
\end{aligned} \tag{10.1}$$

where $\mathbf{d}$ represents the geometric design variables with $(d_i)_{\min}$ and $(d_i)_{\max}$ being the lower and upper bound of these variables, and h_j is a set of geometric constraints necessary to obtain clinically admissible stem shapes.

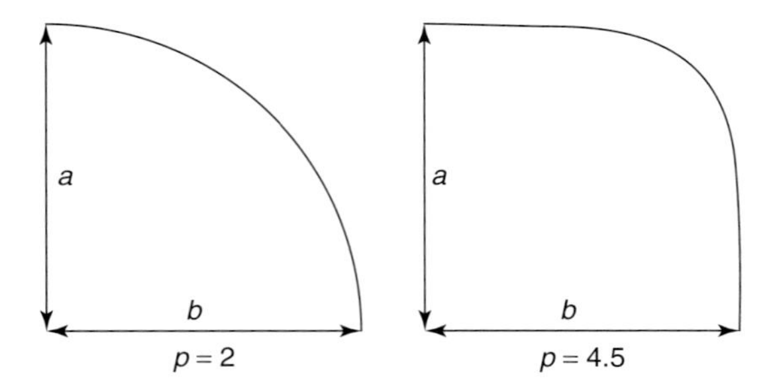

Figure 10.3 Influence of parameter p in Eq. (10.2), representation of first quadrant with $a = b$. (Adapted from Ruben *et al.* [2, 3].)

10.4.1
Design Variables and Geometry

To define the design variables, let us consider the stem transversal sections defined by,

$$\left(\frac{x}{a}\right)^p + \left(\frac{y}{b}\right)^p = 1 \tag{10.2}$$

in a local coordinates system xy. The parameters a, b, and p characterize the section geometry as shown in Figure 10.3. If $a = b$, then the section is circular for $p = 2$ and becomes a square when p increases. For $a \neq b$, circular and square sections become elliptical and rectangular, respectively. The 14 design variables (**d**) defining the four key sections are presented in Figure 10.4. These design variables are the parameters a, b, and p that are defined in Eq. (10.2), and an appropriate interpolation of the key sections is used to obtain the stem geometry. The initial geometry, observed in Figure 10.4, is based on the commercial Tri-Lock prosthesis from DePuy, a Müller-type stem with a high survival rate.

All 14 design variables have lower and upper bounds to maintain the prosthesis inside the bone. Additionally, 10 linear constraints are considered to achieve clinically admissible stem shapes:

$$\begin{aligned}
&h_1 = d_1 - d_4 \leq 0;\; h_2 = d_2 - d_5 \leq 0;\; h_3 = d_4 - d_7 + c_3 \leq 0\\
&h_4 = d_4 - d_8 \leq 0;\; h_5 = d_5 - d_9 \leq 0;\; h_6 = d_7 - d_{11} + c_6 \leq 0\\
&h_7 = d_{12} - d_8 \leq 0;\; h_8 = d_9 - d_{13} \leq 0\\
&h_9 = d_{11} - d_7 + c_9 \leq 0;\; h_{10} = d_7 - \frac{d_{11} + d_4}{2} \leq 0
\end{aligned} \tag{10.3}$$

Constraints h_3, h_6, and h_9 are necessary to always obtain b-splines with negative slope, as illustrated in Figure 10.5. The values for c_3, c_6, and c_9 depends on geometry and in this case are $c_3 = c_6 = 3.5$, and $c_9 = -9$. Finally, constraint h_{10} assures a convex b-spline, as shown also in Figure 10.5.

Without these constraints, the optimization process can lead to clinically unfeasible stem shapes that, at least, imply special insertion techniques or make the removal process difficult [13].

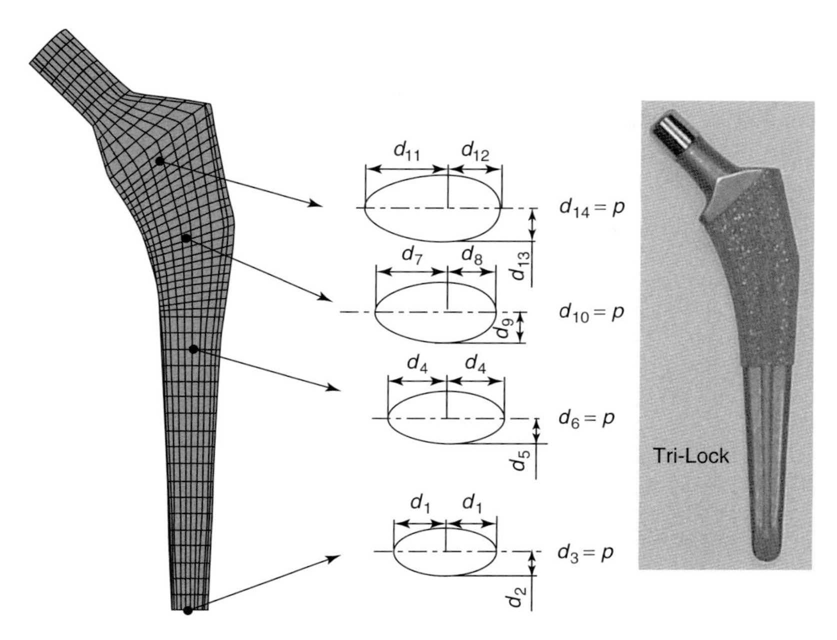

Figure 10.4 Definition of the initial geometry and design variables based on the Tri-Lock stem represented on the right image (section dimensions are not scaled).

10.4.2
Objective Function for Interface Displacement

One of the requirements for the hip stem is to minimize the interface relative displacements. With this objective, the single cost function f_d is considered:

$$f_\mathrm{d} = \sum_{P=1}^{NC} \left(\alpha_P \frac{C}{\Gamma_\mathrm{c}} \int_{\Gamma_\mathrm{c}} \left| \left(\mathbf{u}_t^\mathrm{rel} \right)_P \right|^2 \mathrm{d}\Gamma \right) \tag{10.4}$$

This function is a measure of the tangential displacement on the contact boundary, where NC is the number of applied loads, α_P are the load weight factors, with $\sum_{P=1}^{NC} \alpha_P = 1$, and $(\mathbf{u}_t^\mathrm{rel})_P$ is the relative tangential displacement on bone–stem interface Γ_c for load case P. The constant C is assumed equal to 10^5 to avoid numerical instabilities due to values of f_d close to zero.

10.4.3
Objective Function for Interface Stress

The control of interface contact stress is also crucial for prosthesis performance. It is necessary to avoid stress peaks and also to maintain a low stress level on all interfaces. To achieve this, a measure for the contact stress, given by f_t, is

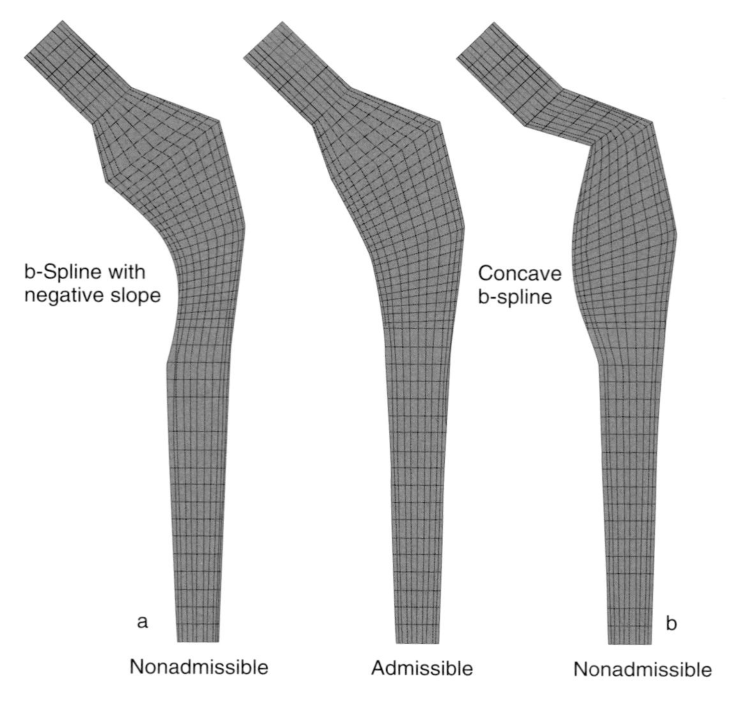

Figure 10.5 Two examples of nonadmissible stem shapes.

considered as the objective function,

$$f_t = \sum_{P=1}^{NC} \left(\alpha_P \frac{1}{\Gamma_c} \int_{\Gamma_c} |(\tau_n)_P|^m \, d\Gamma \right) \tag{10.5}$$

where $(\tau_n)_P$ is the normal contact stress on stem–bone interface Γ_c for load case P. Here, α_P is again the load weight factors, and m is constant taken equal to 2. It should be noted that the value of m can control the contribution of peak stress values for the function. When m increases the relevance of the peak values also increases.

For both functions (Eqs. 10.4 and 10.5), the state variables, tangential displacement and contact stress, are the solution of the equilibrium problem with contact conditions.

10.4.4 Objective Function for Bone Remodeling

After surgery bone starts a new remodeling process in order to adapt to the new load conditions. In fact, a smaller stress shielding effect promotes a smaller proximal bone loss. A femoral stem that minimizes the bone loss is the one that maintains the mechanical conditions of the intact bone. To achieve this, a common criterion for stem design is to minimize the difference between strain energy density before and after the implant (see for instance Chang *et al.* [19]). However, in this chapter another criterion is used, with the objective to minimize the bone loss in proximal femur. Since after surgery, the stiff stem supports most of the load, the strain energy in bone decreases too much. Thus, the objective is to increase the strain energy density in proximal femur. To reach that goal the remodeling function f_r was defined by,

$$f_r = \frac{D}{\sum_{P=1}^{NC} \alpha_P \left(\sum_{j=1}^{N_{bp}} U_j \right)_P} \tag{10.6}$$

where N_{bp} is the number of elements in proximal femur, that is, where the trabecular and cortical bone exists, U_j represents the strain energy density, and P is the load case. The constant D is taken equal to 10^3.

10.4.5 Multicriteria Objective Function

The three single cost functions are important to understand the influence of the stem shape on the remodeling and interface displacement and stress is in an individual manner. However, a multicriteria objective function is necessary to obtain simultaneously the implant geometries with less stress shielding and improved stability, that is, with better performances. To do that, a multicriteria objective function combining the three single cost functions was considered,

$$f_{mc} = \beta_d \frac{f_d - f_d^0}{f_d^i - f_d^0} + \beta_t \frac{f_t - f_t^0}{f_t^i - f_t^0} + \beta_r \frac{f_r - f_r^0}{f_r^i - f_r^0} \tag{10.7}$$

where f_d^0, f_t^0, and f_r^0 and f_d^i, f_t^i, and f_r^i are the minimums and the initial values of f_d, f_t, and f_r, respectively; and β_d, β_t, and β_r are the weighting coefficients with $\beta_d + \beta_t + \beta_r = 1$.

The multicriteria cost function is based on a weighting objective method as described in Osyczka [36], and Marler and Arora [37] reported that this approach to normalize objective functions is the most robust.

10.5 Computational Model

The problem described above is solved numerically using a suitable discretization by finite elements and appropriate optimization methods. Next, the optimization algorithm is described in detail and the finite element mesh is presented.

10.5.1
Optimization Algorithm

The shape optimization process is solved with a hybrid method that combines the method of moving asymptotes (MMA) [38] with a gradient projection (GP) method [2]. MMA is based on a convex approach for the cost function and this monotonous approach gives MMA a fast convergence property. However, MMA is not globally convergent and oscillations can appear near the optimal solution [39]. In fact, this behavior of the MMA is also present in the shape optimization process presented in this work. Thus, a hybrid method that combines the MMA with GP method) is used (see for instance [40]). Firstly, MMA approaches the solution to a point near the optimum, and then the GP starts to avoid numerical oscillations. In Figure 10.6, it is possible to observe MMA fast convergence and strong oscillation, and also the soft, but slower, GP global convergence. This hybrid method has proved to be a good strategy to solve this particular shape optimization problem.

Computationally, the optimization problem is solved following the flowchart presented in Figure 10.7. First, for the initial stem geometry, the objective function (f_d, f_t, f_r, or f_{mc}) is computed using the values of interface displacement, contact stress, and strain energy density, which are the solution of the equilibrium problem with contact conditions solved with the finite element program ABAQUS [41]. Next, the sensitivity derivatives are obtained using forward finite differences with the step size $\delta = 10^{-5}$ [2]. When the optimization method starts, the initial iterations

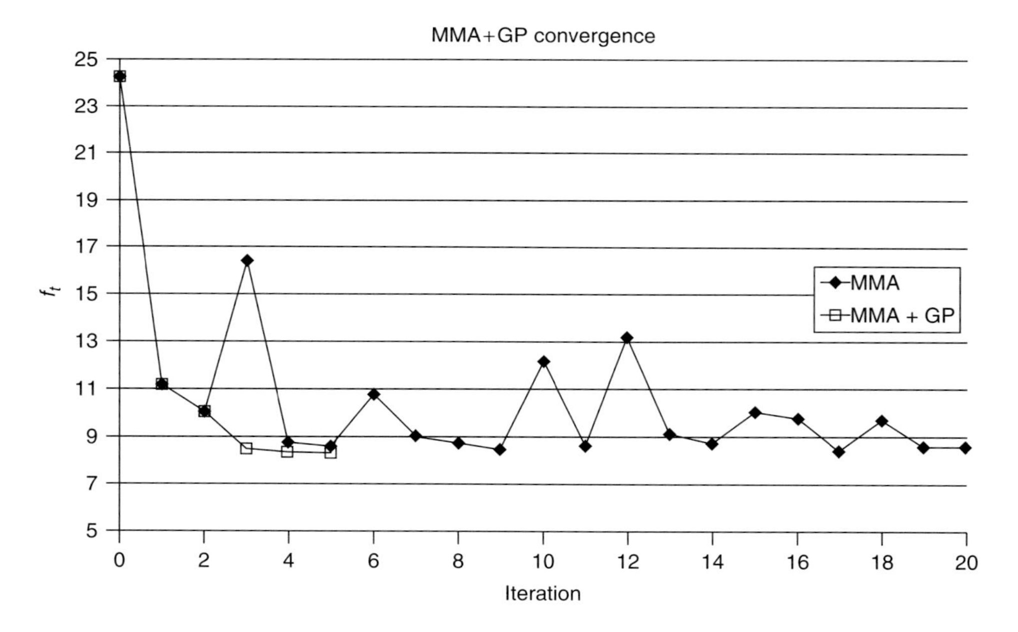

Figure 10.6 Convergence for MMA and MMA + GP hybrid method. Totally coated stem.

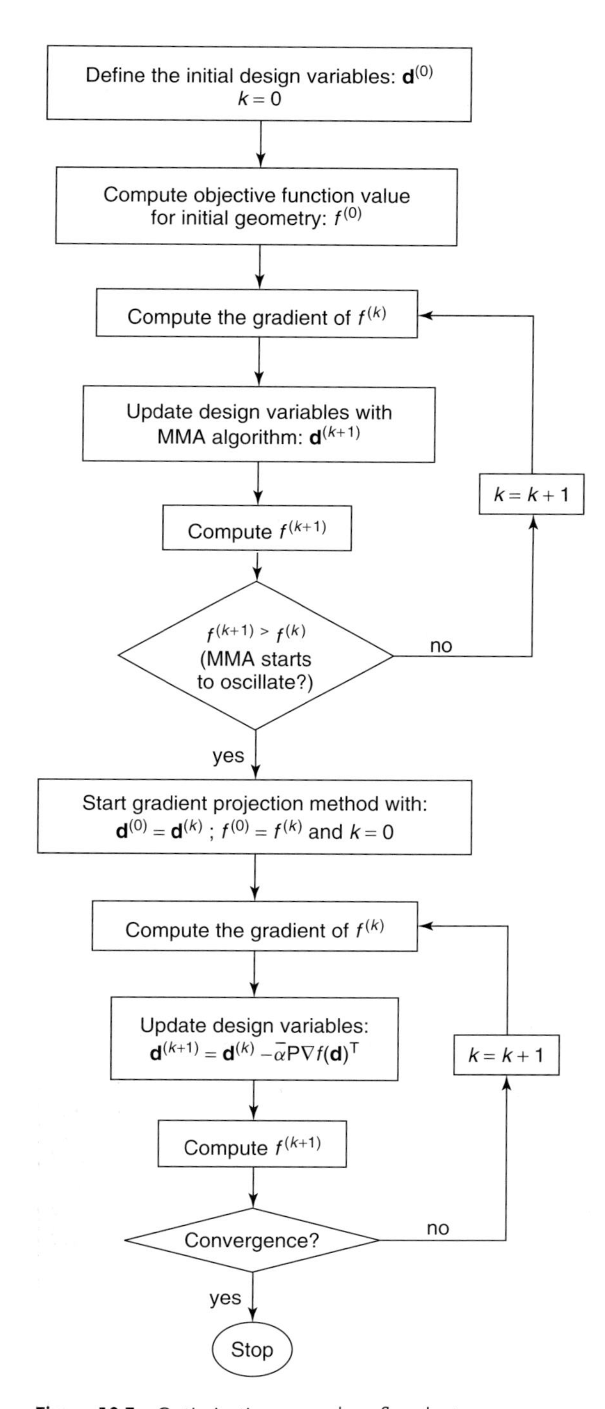

Figure 10.7 Optimization procedure flowchart.

are always produced by MMA. After the first oscillation the method changes to GP until convergence is reached.

10.5.2
Finite Element Model

The contact formulation used allows the study of several porous coating lengths. In the present work two coating lengths were considered, a totally coated stem and a half coated one (similar to the Tri-Lock prosthesis shown in Figure 10.4). A friction coefficient of $\vartheta = 0.6$ is assumed for coated surfaces while uncoated surfaces are modeled as frictionless contact.

To simulate several daily activities a multiple load case with three loads were considered [42]. With the multiple load case the three loads are applied sequentially in order to reproduce successive activities. In Figure 10.8 it is possible to see load directions and intensities.

To have a suitable finite element mesh of the stem–bone assemblage, it is necessary to have accurate geometries of femur and femoral stem. The femur mesh was built using a geometry based on the "standardized femur" [43]. Stem shape changes in all optimization iterations. In order to assure similar stem meshes in all iterations a meshing algorithm was developed. Therefore, the finite element mesh discretization has a total of 7176 eight-node brick elements, 5376 for the femur and 1800 elements for the femoral stem.

The bone is a nonhomogeneous structure where one can distinguish between trabecular and cortical bone. Additionally, long bones have a cavity in the bone shaft filled with marrow. Therefore, the femur model considers marrow, trabecular, and

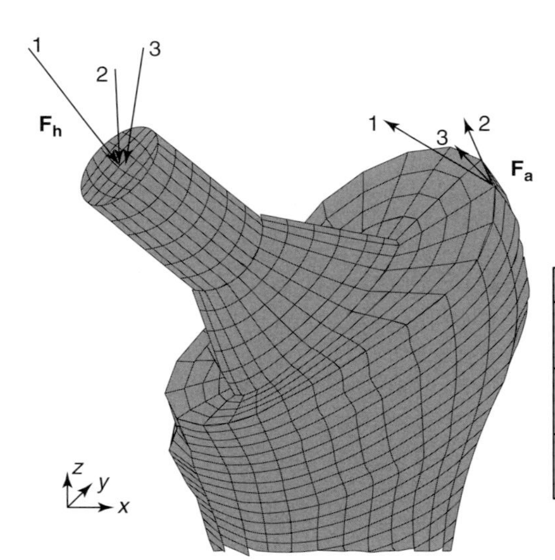

Load case		F_x (N)	F_y (N)	F_z (N)
1	F_h	224	972	−2246
	F_a	−768	−726	1210
2	F_h	−136	630	−1692
	F_a	−166	−382	957
3	F_h	−457	796	−1707
	F_a	−383	−669	547

Figure 10.8 Load cases.

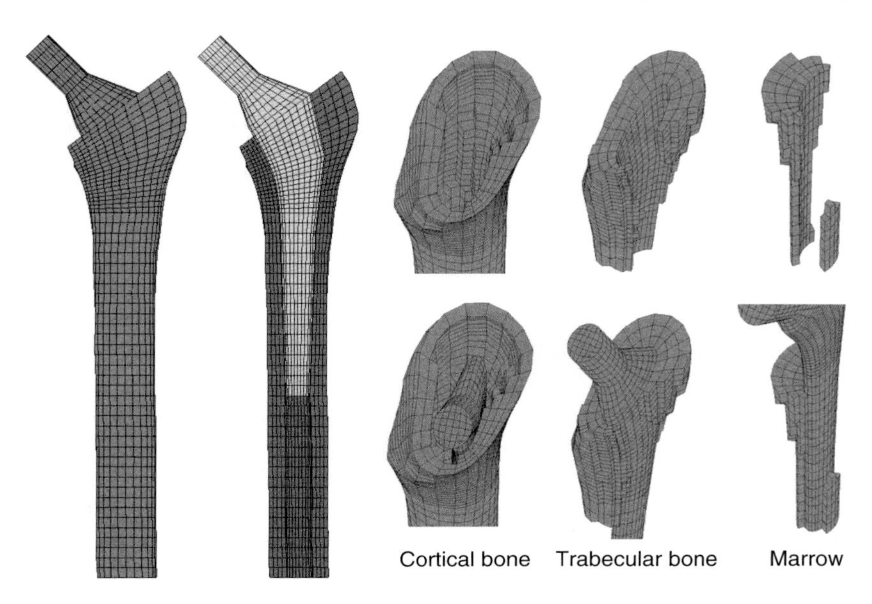

Figure 10.9 Finite element mesh for initial stem shape with the representation of different bone tissues.

Table 10.1 Material properties.

	E (GPa)	ν
Cortical bone	17	0.3
Trabecular bone	1	0.3
Bone marrow	10^{-7}	0.3
Femoral stem	115	0.3

cortical bone regions, as shown in Figure 10.9. Stems are considered to be made of titanium. All material properties are presented in Table 10.1.

10.6
Optimal Geometries Analysis

Four objective functions were considered in the shape optimization of uncemented hip femoral stems. Three of them are single cost functions, and the fourth one is a multicriteria combining the previous ones. In this section, the optimized stem shapes obtained for the single cost functions, followed by the multicriteria optimization results, is presented. For all analysis equal load weight factors were considered, that is, $\alpha_1 = \alpha_2 = \alpha_3 = 1/3$.

10.6.1
Optimal Geometry for Tangential Interfacial Displacement – f_d

Numerical results for the minimization of relative tangential displacement objective function are presented in Table 10.2. For the initial shape, half coated stems have greater function values than totally coated ones. For optimized shapes, totally coated implants also present lower f_d values. In Table 10.2 it is also possible to verify a reduction of approximately 50% in objective function values.

In Figure 10.10, it is possible to compare the optimized shapes with the initial solution. Section 2 (see section numbers in Figure 10.10) is rectangular (parameter *p*) and is larger (parameter *a*) than the initial one, improving rotational stability. It is also possible to observe a small section 1, which is big enough for a slight fixation on cortical bone. These two sections, 1 and 2, define a distal wedge design, important to improve axial stability. Section 3 is also larger to increase cortical fixation at elbow region (see Figure 10.1). Finally, section 4 is thinner (parameter *b*) and almost rectangular to improve rotational stability.

Although the characteristics mentioned above are similar for half and totally coated stems, there are some details that depend on coating length. For instance, for half coated stems section 2 and 3 are thin (parameter *b*) and the distal tip is larger (parameter *a*) than for the totally coated stems, to compensate the lack of friction in the distal part.

In Figure 10.11 it can be observed that the area with tangential displacements below 25 μm increases from the initial to the optimized shapes. Furthermore, for optimized geometries, displacements greater than 50 μm are concentrated in small

Table 10.2 Objective function values. Minimization of displacement function f_d.

	f_d – initial	f_d – final
Totally coated	40.18	21.58
Half coated	52.41	26.98

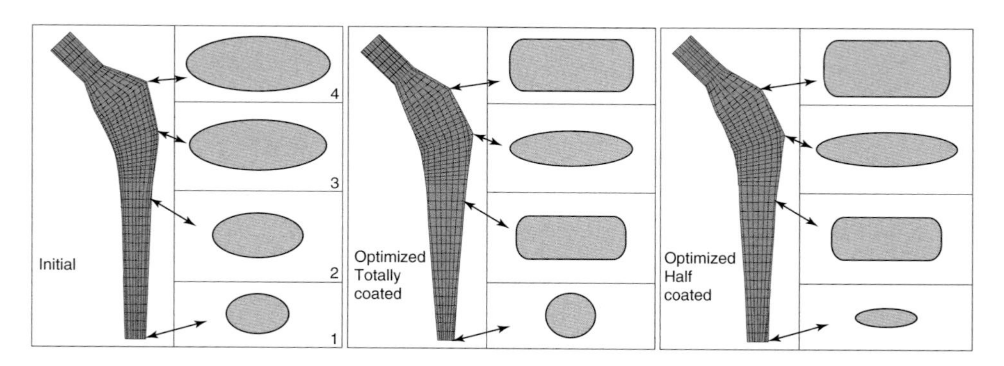

Figure 10.10 Initial and optimized stems. Minimization of displacement function f_d.

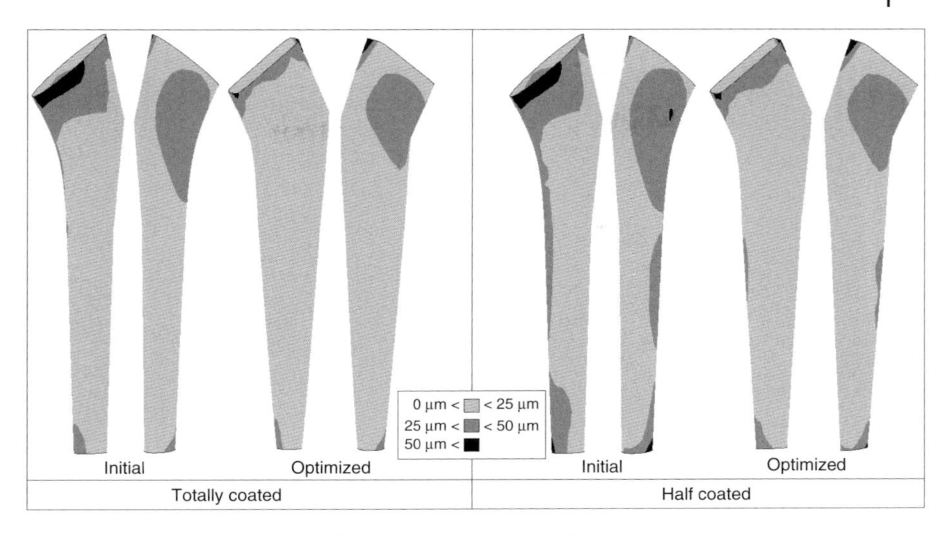

Figure 10.11 Relative tangential displacement for the initial and the optimized shape obtained for the minimization of displacement function f_d (load case 1).

areas. Assuming a bone ingrowth threshold of 25 μm, which is a conservative value, large portions of porous coated regions will be a candidate for efficient osseointegration. In Figure 10.11 the displacements are shown only for the most severe load case. For the other two load cases the tangential displacement at stem–bone interface is lower.

10.6.2 Optimal Geometry for Normal Contact Stress – f_t

Table 10.3 presents the numerical results for the minimization of normal contact stress cost function. In this case, it is also possible to verify that objective function values for totally coated stems are lower than those obtained for partially coated implants.

In Figure 10.12 it is possible to compare the initial shape with the optimized ones. Usually, the maximum normal contact stress is observed at the stem tip. To minimize the stress objective function it is necessary to decrease this peak stress. Consequently, a small and circular stem tip is the best solution (see Figure 10.12),

Table 10.3 Objective function values. Minimization of stress function f_t.

	f_t – initial	f_t – final
Totally coated	24.26	8.31
Half coated	39.03	9.25

Table 10.4 Maximum normal contact stress (MPa). Minimization of stress function f_t.

	Initial shape	Optimized
Totally coated	88.08	40.95
Half coated	135.1	40.36

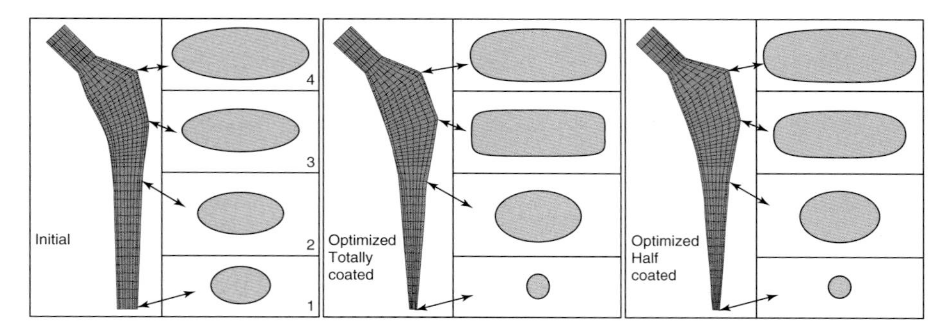

Figure 10.12 Initial and optimized stems. Minimization of stress function f_t.

because it stays surrounded by marrow and never directly touches the cortical bone. In Table 10.4 the maximum stress values for initial and optimized shape are presented. For initial shape the half coated stem has a greater peak stress than the totally coated prosthesis. However, optimized shapes have almost the same maximum contact stress, because the lack of friction is not relevant for small and circular stem tips. In Figure 10.12 it is also possible to verify that sections 2 for optimal shapes are almost circular in order to increase cortical bone contact at middle stem. Finally, sections 3 and 4 are thicker (parameter b) than the initial solution to increase the contact surface and thus to minimize stress at the medial region, particularly in calcar region (see Figure 10.1).

10.6.3
Optimal Geometry for Remodeling – f_r

In Table 10.5 numerical results for remodeling objective function are summarized. In this case, optimized shapes have the same function value for totally and half

Table 10.5 Objective function values. Minimization of remodeling function f_r.

	f_r – initial	f_r – final
Totally coated	50.34	17.93
Half coated	49.61	17.93

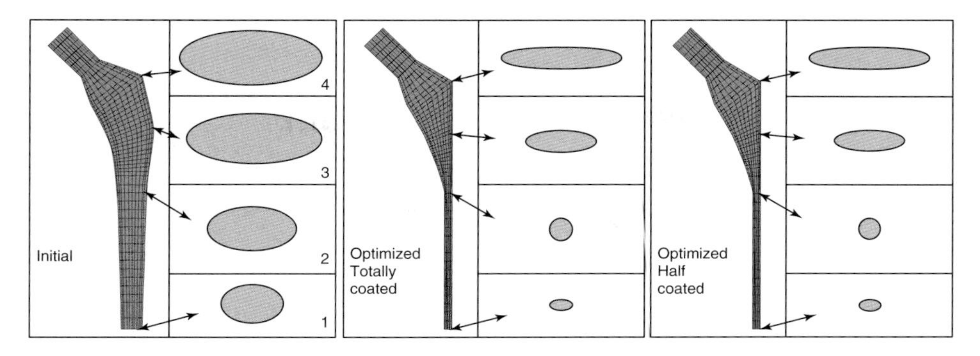

Figure 10.13 Initial and optimized stems. Minimization of remodeling function f_r.

coated prostheses. In Figure 10.13 it is possible to observe that optimized shapes are similar for both coating lengths.

The optimized shapes tend to be minimal invasive hip prostheses, as shown in Figure 10.13. In fact, minimal invasive prostheses are less stiff and reduce the stress shielding effect. In these results, the distal half stem is very small in order to increase proximal load transfer. In the proximal part, sections are thin (parameter b) and there is also no elbow to obtain a more favorable load transfer. Comparative contact stresses presented in Figure 10.14 show an increase in stress on proximal region. However, the stress in distal tip is almost zero for minimal invasive optimal shapes. In summary, with the remodeling cost function stem geometries with more proximal load transfer were obtained in order to reduce the stress shielding effect.

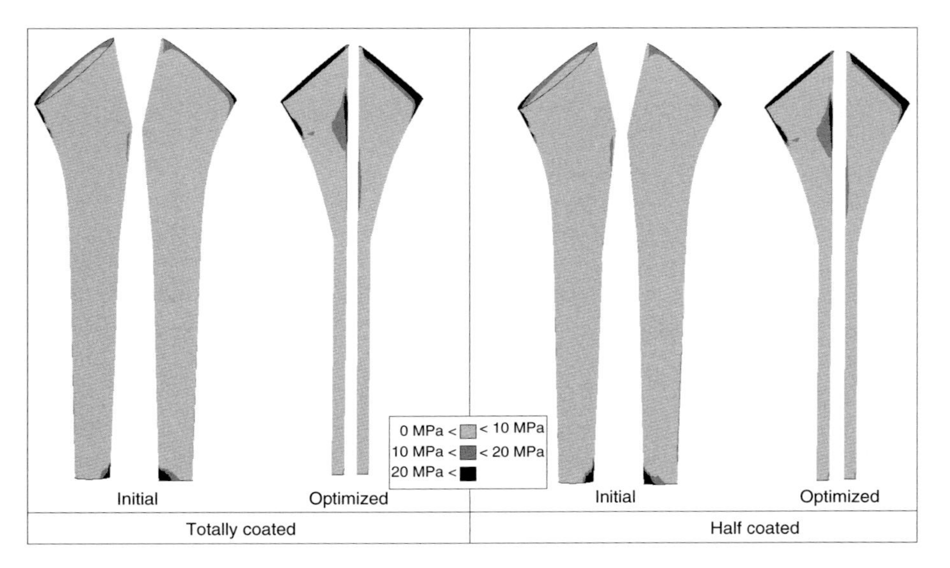

Figure 10.14 Normal contact stress in stem–bone interface for the initial and the optimized shape obtained for the minimization of remodeling function f_r (load case 1).

10.6.4
Multicriteria Optimal Geometries – f_{mc}

The geometric characteristics obtained for the single objective functions are, in many cases, opposite. However, to improve uncemented hip prostheses it is necessary to reduce, at the same time, relative tangential displacement, normal contact stress, and stress shielding effect. Therefore, a multicriteria optimization procedure allows us to simultaneously minimize the three single objectives.

With a multicriteria optimization algorithm it is possible to obtain a set of nondominated points [44]. In this case four nondominated points were computed for these weighting coefficients of Eq. (10.7): $(\beta_d, \beta_t, \beta_r) = (1/3,1/3,1/3)$, $(0.6,0.2,0.2)$, $(0.2,0.6,0.2)$, $(0.2,0.2,0.6)$. In Table 10.6, the function values for these nondominated points are presented and they are compared with the results obtained with single cost functions, that is, $(\beta_d, \beta_t, \beta_r) = (1,0,0)$, $(0,1,0)$, $(0,0,1)$. It is possible to conclude that totally coated stems have lower f_d and f_t function values, but for remodeling function f_r, half coated implants have the lower value. Almost all nondominated points have better values for the three objective functions compared to the initial shape, which is not obtained with the single cost functions.

Figures 10.15 and 10.16 show the nondominated shapes for totally and half coated stems, respectively. When all objectives have the same weight, that is, $\beta_d = \beta_t = \beta_r = 1/3$, a small and circular stem tip is observed in order to reduce maximum contact stress and also to improve proximal load transfer. On the other hand, section 2 is larger (parameter a) to increase rotational stability. Section 3 is also larger to improve elbow cortical fixation. Section 4 is thicker to increase contact surface and almost rectangular to give more rotational stability to the implant.

When the weight for the displacement objective is bigger ($\beta_d = 0.6$), more geometric characteristics to minimize interfacial displacement are observed. For instance, section 1 is large enough, having a slight fixation on cortical bone, section 2 is large and rectangular, and section 4 is also rectangular. However, section 4 is thicker to reduce contact stress in the calcar region.

Table 10.6 Objective function values. Minimization of multicriteria function f_{mc}.

$(\beta_d, \beta_t, \beta_r)$	Totally coated			Half coated		
	f_d	f_t	f_r	f_d	f_t	f_r
(1,0,0)	21.58	17.84	67.97	26.98	43.95	56.48
(0,1,0)	61.30	8.31	48.13	100.6	9.25	46.23
(0,0,1)	37.69	186.04	17.93	40.21	185.49	17.93
(1/3,1/3,1/3)	27.41	12.64	48.08	32.98	16.23	40.79
(0.6,0.2,0.2)	22.40	15.85	58.54	29.51	20.16	47.94
(0.2,0.6,0.2)	29.37	10.69	48.03	35.70	13.00	42.13
(0.2,0.2,0.6)	32.85	16.21	38.62	38.97	17.12	35.48
Initial shape	40.18	24.26	50.34	52.41	39.03	49.61

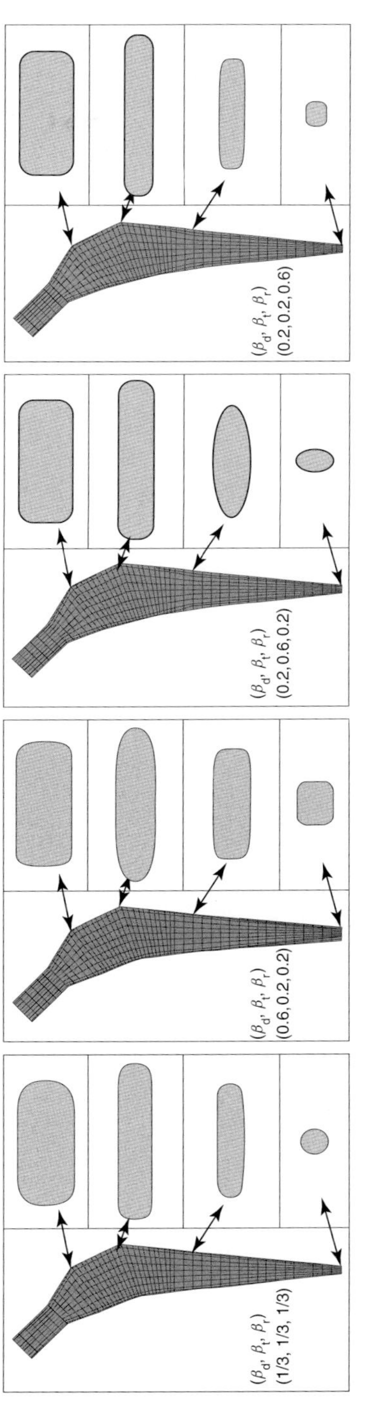

Figure 10.15 Nondominated shapes for totally coated stems. Minimization of multicriteria function f_{mc}.

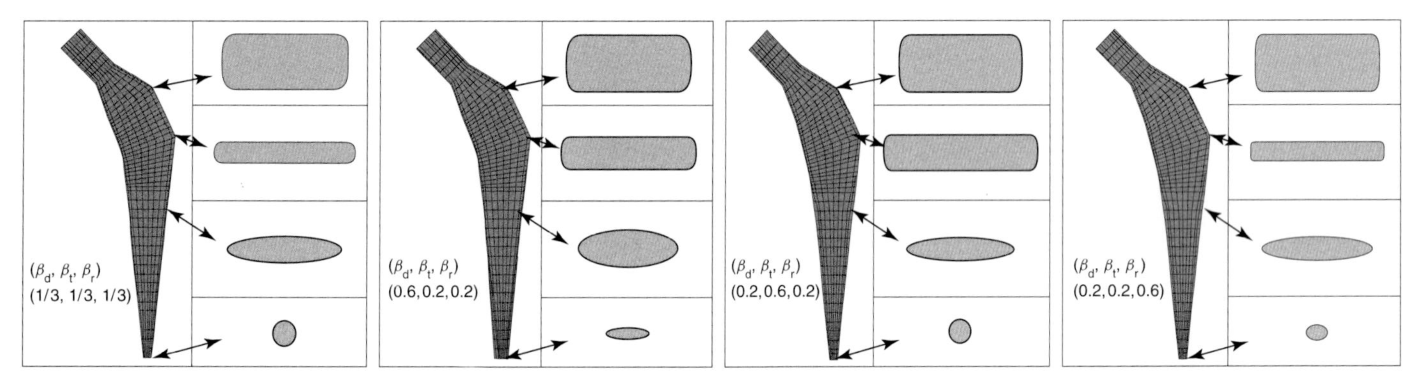

Figure 10.16 Nondominated shapes for half coated stems. Minimization of multicriteria function f_{mc}.

When the weight for the contact stress objective is bigger ($\beta_t = 0.6$), the stem tip is small to avoid direct contact with cortical bone, and section 4 is thick and almost rectangular, also to reduce stress objective function. But, in Figures 10.15 and 10.16 it is also possible to see that section 3 is thinner and larger to reduce both the displacement and the remodeling cost functions.

Finally, non-dominated shapes, when the weight for the remodeling objective is bigger ($\beta_r = 0.6$), have a very small stem tip, and sections 2 and 3 are very thin in order to improve proximal load transfer. However, section 4 is thicker to reduce contact stress and almost rectangular to improve rotational stability.

Finally, in the same Figures 10.15 and 10.16 a wedge design is observed in all nondominated points. In fact, the wedge design is important to improve axial stability and a small stem tip is also essential to reduce maximum contact stress and to increase proximal load transfer.

10.7 Long-Term Performance of Optimized Implants

With the multicriteria shape optimization process nondominated points with better primary stability and less stress shielding after surgery were obtained. However, the prediction of the long-term performance of optimized hip stems is necessary to confirm the relation between initial conditions and implant success. Therefore, an integrated model for bone remodeling and osseointegration was used in order to study the long-term effect of optimized stem shapes.

The remodeling model presented by Fernandes *et al.* was used [12, 45]. In this model bone tissue is considered a porous material with a periodic microstructure that is obtained by the repetition of cubic cells with prismatic holes with dimensions a_1, a_2, and a_3, as shown in Figure 10.17. The orthotropic elastic properties are obtained by the homogenization method [46]. Relative density at each point of femur is solution to an optimization problem, and depends on local porosity,

$$\mu = 1 - a_1 a_2 a_3 \tag{10.8}$$

with the extreme values $a_i = 0$ and $a_i = 1$ corresponding to cortical bone and marrow, respectively. Intermediate values for relative density correspond to trabecular bone. Assuming that bone adapts according to applied loads in order to maximize its stiffness, the law of bone remodeling is derived by considering an optimization process where the holes dimensions (**a**) at each bone finite element are the design variables, and the objective is the minimization of a linear combination of the compliance (inverse of the stiffness) and the metabolic cost to maintain bone mass. Considering a multiple load formulation the optimization function is,

$$f_{\text{remo}} = \sum_{P=1}^{NC} \alpha_P \left(\int_{\Gamma_f} (f_i)_P (u_i)_P \, d\Gamma \right) + \kappa \int_{\Omega_b} (\mu(\mathbf{a}))^2 d\Omega \tag{10.9}$$

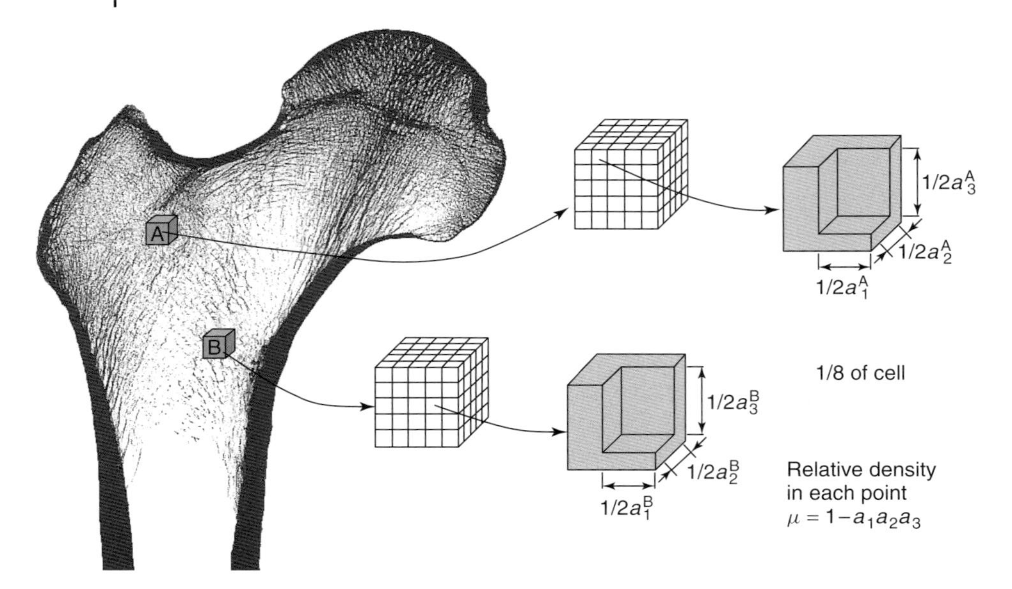

Figure 10.17 Bone material model. (Adapted from Fernandes *et al.* [12].)

In Eq. (10.9), $(f_i)_P$ is the applied load for the load case P, $(u_i)_P$ is the displacement field for load case P, κ is the metabolic cost to maintain bone mass, Γ_f is the surface where loads are applied, and Ω_b is the total bone volume.

This remodeling model is combined with an ingrowth algorithm to predict the ability of bone to attach to the stem surface. Actually, if a good primary stability is achieved bone starts to grow into the porous coating. The bone ingrowth depends on relative tangential displacement, contact stress, and the gap between bone and stem. Moreover, after bone ingrowth starts the connection between bone and implant can be disrupted if the mechanical conditions become adverse.

To take into account the mechanical factors mentioned above, an algorithm based on previous ingrowth models [12, 31, 47, 48] is used. In this algorithm there are five levels for the interface stiffness, from 0 to 30 N mm^{-1}, as shown in Table 10.7. In each interface point bone ingrowth occurs if the following five conditions are satisfied for every load cases:

Table 10.7 Stiffness and shear stress threshold in bone–stem interface.

Osseointegration level	Stiffness K (N mm^{-1})	Shear stress threshold (MPa)
1	0	4.5
2	7.5	18
3	15	33
4	22.5	35
5	30	35

- relative tangential displacement should be less than 25 μm;
- gap between bone and stem must be lower than 150 μm [31];
- contact tension stress limit is 0.8 MPa [31];
- contact compression stress limit is 7.89 MPa [30];
- shear stress should be inferior to the threshold values in Table 10.7, depending on stiffness level [49].

Immediately after surgery the interface stiffness is zero, that is, coated interface has contact with friction and uncoated surface has no friction. After each step (remodeling model iteration) interface conditions, contact stress and relative displacement, are computed at all interface nodes. If a certain point satisfies all osseointegration conditions then interface stiffness increases by one level until level 5 is reached. If at least one condition is not satisfied then, in that point, interface stiffness goes back to level one (zero stiffness).

The remodeling model combined with the ingrowth algorithm is used to predict the long-term bone adaptation and osseointegration of the optimized stems. Results for this analysis are summarized in Tables 10.8 and 10.9.

In Table 10.8 the percentages of bone ingrowth are presented, comparing the percentage of surface that verifies the conditions for bone ingrowth after the THA and the long-term prediction of ingrowth. Note that 100% corresponds to bone ingrowth on whole coated surface. It is possible to verify that all nondominated points have more osseointegration just after the surgery than the initial shape because of a more efficient primary stability. The differences between "after surgery" and "long-term" illustrates the influence of bone remodeling on the interface conditions and how these two processes are coupled. Thus, although all nondominated points have more osseointegration just after surgery than initial shape, in the long-term analysis it changes. In fact, the bone remodeling could increase contact stresses to values above the limit. On the other hand, bigger amounts of osseointegration promote a better load transfer and reduce the stress shielding.

Table 10.8 Osseointegration percentage after surgery and at long-term.

$(\beta_d, \beta_t, \beta_r)$	Totally coated		Half coated	
	After surgery (%)	Long-term (%)	After surgery (%)	Long-term (%)
(1,0,0)	73.6	85.8	62.1	78.4
(0,1,0)	57.2	75.4	19.6	55.1
(0,0,1)	54.3	44.0	33.7	66.7
(1/3,1/3,1/3)	71.3	85.5	47.8	58.9
(0.6,0.2,0.2)	73.3	85.8	57.6	62.9
(0.2,0.6,0.2)	70.5	81.3	50.4	60.3
(0.2,0.2,0.6)	69.0	79.3	42.9	57.4
Initial shape	61.6	81.0	39.5	70.8

Table 10.9 Bone mass variation.

$(\beta_d, \beta_t, \beta_r)$	Totally coated			Half coated		
	Zone P (%)	Zone PC (%)	Zone E (%)	Zone P (%)	Zone PC (%)	Zone E (%)
(1,0,0)	−23.32	−29.57	−56.48	−18.48	−26.25	−54.61
(0,1,0)	−21.44	−33.98	−61.92	−18.67	−32.36	−62.31
(0,0,1)	−5.55	−21.20	−55.94	+10.43	+2.76	−50.16
(1/3,1/3,1/3)	−14.72	−24.72	−54.62	−12.15	−23.49	−53.31
(0.6,0.2,0.2)	−20.03	−27.57	−55.16	−16.74	−27.01	−55.12
(0.2,0.6,0.2)	−18.46	−26.90	−56.51	−14.04	−26.38	−54.96
(0.2,0.2,0.6)	−9.78	−19.06	−51.77	−9.97	−21.53	−52.69
Initial shape	−14.24	−24.87	−58.20	−14.01	−24.75	−58.09

Table 10.9 presents the bone loss variation in proximal and exterior femoral regions defined in Figure 10.18. The percentage variation is referred to as the total amount of bone immediately after the surgery and the minus sign means loss of mass. In this case, half coated stems have less bone resorption than totally coated ones. In fact, the lack of porous coating in the distal part implies a proximal fixation, which is useful to minimize the bone resorption. Nevertheless, for all nondominated points and both coating lengths, a smaller amount of bone resorption is achieved in the exterior region (zone E) compared to the initial shape. In zones P and PC, nondominated solutions with high weight for the remodeling function (β_r) also presents less bone resorption than initial geometry. In Figure 10.19 some examples of long-term bone density distribution are presented,

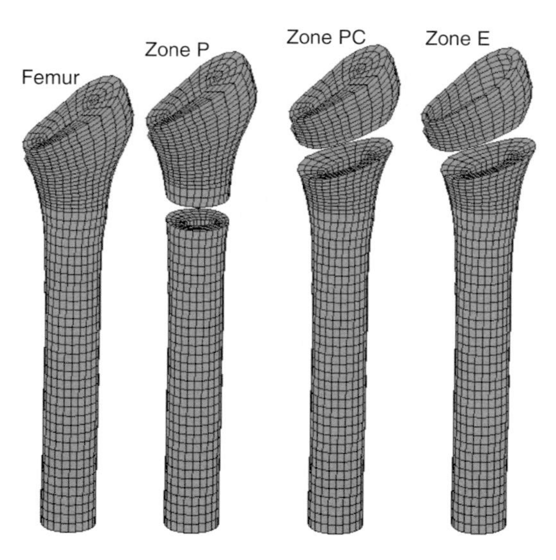

Figure 10.18 Definition of the zones to compare bone remodeling results.

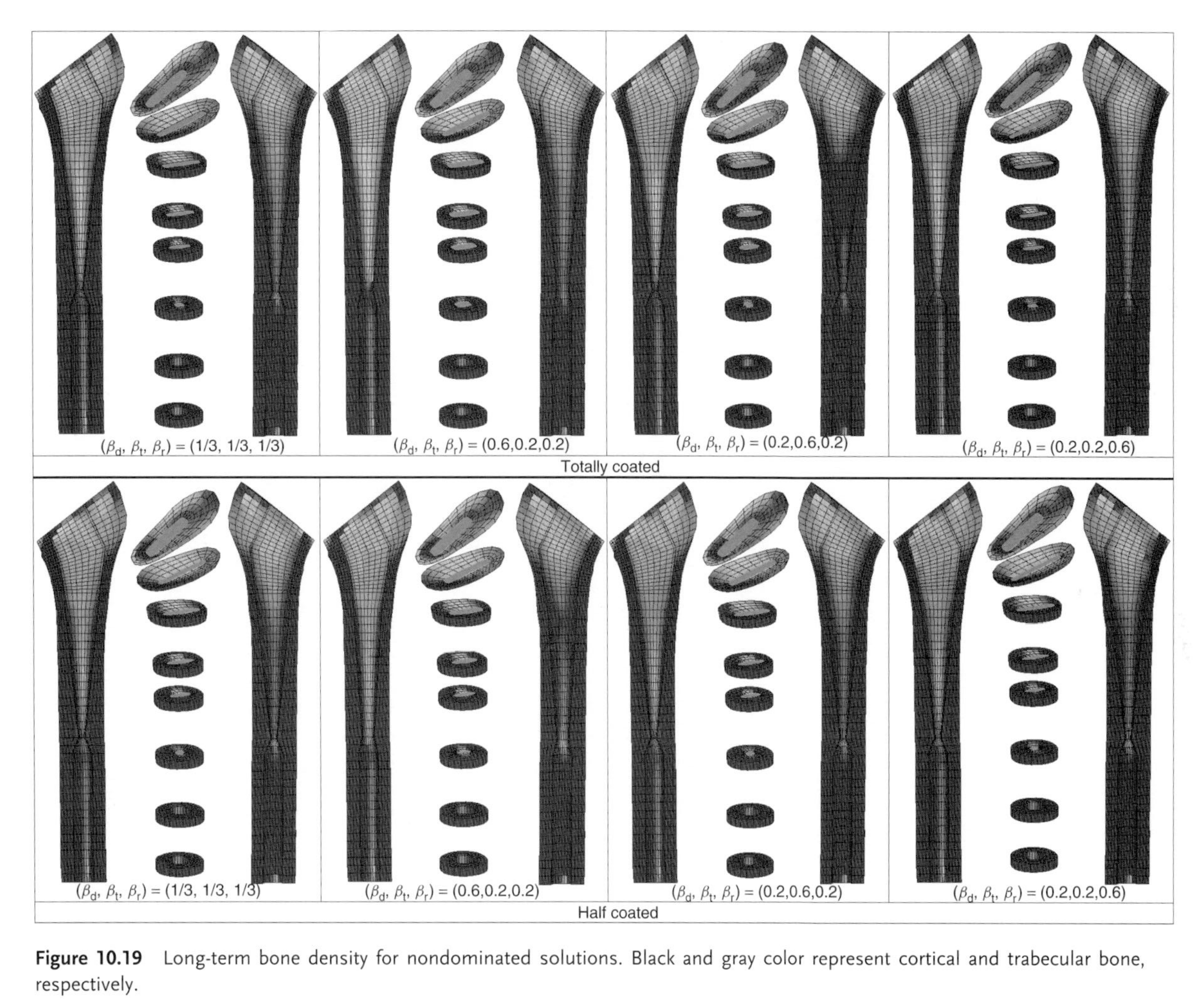

Figure 10.19 Long-term bone density for nondominated solutions. Black and gray color represent cortical and trabecular bone, respectively.

and it is possible to verify less bone resorption for half coated stems and also the hypertrophy near stem tip due to distal peak on contact stress.

This concurrent model for remodeling and ingrowth confirmed the good performance of optimized shapes. Some nondominated points have more bone ingrowth and less bone resorption than initial shape. Additionally, all nondominated solutions promote less bone loss in the proximal outside region of femur (zone E) compared to the initial shape. Notice, that bone resorption in this particular region is one of the major reasons for THA failure [33].

10.8 Concluding Remarks

A new multicriteria shape optimization procedure has been developed in order to obtain cementless hip prostheses with improved performance. The multicriteria objective function combines three major causes of implant failure, the relative tangential displacements, normal contact stress, and stress shielding effect.

A concurrent model for bone remodeling and osseointegration was also used to study the long-term performance of initial shape and optimized geometries. The remodeling process used was developed by Fernandes *et al.* [45] and assumes that bone adapts to the mechanical loads in order to maximize the stiffness and minimize the metabolic cost to maintain bone tissue. An algorithm for bone ingrowth is proposed, taking into account the five interfacial conditions that must be satisfied to obtain osseointegration into the porous coating.

Stem shapes obtained with the three single cost functions are contradictory. Nevertheless, several geometric characteristics obtained are in accordance with clinical observations. The minimization of displacement function f_d leads to stems with wedge design to improve axial stability, and some sections are almost rectangular and the elbow is more accentuated to improve rotational stability [50, 51]. With the minimization of contact stress function f_t stem tip is small and circular to avoid direct contact with the cortical bone [52] and proximal sections are thicker to reduce stress at calcar region. The minimization of remodeling objective function f_r leads to minimally invasive stems without elbow [33]. The multicriteria optimization is highly efficient in dealing with these differences, since nondominated solutions obtained have a combination of geometric characteristics depending on weight coefficients: β_d, β_t, and β_r. In fact, from a computational point of view, nondominate solutions have a better performance than the initial prosthesis, that is, the three single functions are simultaneously reduced.

Results obtained with the model for remodeling and bone ingrowth illustrates the dependence of ingrowth on bone remodeling and vice versa. This model confirms the importance of stem stability in long-term success of uncemented hip implants. The long-term prediction obtained with this model is also in agreement with clinical observations. For instance, Niinimäki *et al.* [33] and Sluimer *et al.* [34] also verified that totally coated stems are more stable but partially coated stems lead to less bone resorption. In addition, minimally invasive stems have less stability but have less stress shielding effect compared to implants with stiffer distal parts [33, 34]. Finally, the distal hypertrophy is slighter for minimally invasive stems, since stem tip stress is smaller.

It should be noticed that in this study all optimization processes considered a standard femur. However, the optimization model can be applied to a patient-specific analysis where different clinical situations can be considered. For instance, bones with osteoporosis can originate a more difficult initial stability and a fast adverse remodeling process.

Notwithstanding some efforts to refine the computational models, this work shows that shape optimization is a powerful tool for implant design. In fact, the model presented in this chapter leads to useful conclusions on the relationship between shape, porous coating, stem stability, and stress shielding. This information is important for new prosthesis design and for surgeons who have to decide from among numerous commercial stem shapes. Finally, it should be noted that although the model was applied to uncemented hip prostheses, it can be easily extended to cemented ones and also to other bone joints and bone implants.

References

1. Kärrholm, J., Garellick, G., and Herberts, P. (2006) The Swedish Hip Arthroplasty Register – Annual Report 2005, Joint replacement unit, Sahlgren Hospital.
2. Ruben, R.B., Folgado, J., and Fernandes, P.R. (2007) *Struct. Multi. Optim.*, **34**, 261–275.
3. Ruben, R.B., Folgado, J., and Fernandes, P.R. (2006) *Virt. Phys. Prototyp.*, **1**, 147–158.
4. Furnes, O., Havelin, L.I., Espehaug, B., Steindal, K., and Sørås, T.E. (2006) Report 2006, Centre of Excellence of Joint Replacements, Haukeland University Hospital.
5. Gross, S. and Abel, E.W. (2001) *J. Biomech.*, **34**, 995–1003.
6. Viceconti, M., Muccini, R., Bernakiewicz, M., Baleani, M., and Cristofolini, L. (2000) *J. Biomech.*, **33**, 1611–1618.
7. Lennon, A.B., McCormack, B.A.O., and Prendergast, P.J. (2003) *Med. Eng. Phys.*, **25**, 8333–8841.
8. Emerson, R.H., Head, W.C., Emerson, C.B., Rosenfeldt, W., and Higgins, L.L. (2002) *J. Arthroplasty*, **17**, 584–591.
9. Herzwurm, P.J., Simpson, S., Duffin, S., Oswald, S.G., and Ebert, F.R. (1997) *Clin. Orthop. Relat. Res.*, **336**, 156–161.
10. Mandell, J.A., Carter, D.R., Goodman, S.B., Schurman, D.J., and Breaupré, G.S. (2004) *Clin. Biomech.*, **19**, 695–703.
11. Bernakiewicz, M., Viceconti, M., and Toni, A. (1999) Investigation of the influence of periprosthetic fibrous tissue on the primary stability on uncemented hip prostheses, in *Computer Methods in Biomechanics and Biomedical Engineering – 3* (eds J. Middleton, M.L. Jones, N.G. Shrive, and G.N. Pande), Gordon and Breach, New York, p. 21–26.
12. Fernandes, P.R., Folgado, J., Jacobs, C., and Pellegrini, V. (2002) *J. Biomech.*, **35**, 167–176.
13. Huiskes, R. and Boeklagen, R. (1989) *J. Biomech.*, **22**, 793–804.
14. Yonn, Y.S., Jang, G.H., and Kim, Y.Y. (1989) *J. Biomech.*, **22**, 1279–1284.
15. Katoozian, H. and Davy, D.T. (1993) *Bioeng. Conf. ASME*, **24**, 552–555.
16. Katoozian, H. and Davy, D.T. (2000) *Med. Eng. Phys.*, **22**, 243–251.
17. Hedia, H.S., Barton, D.C., Fisher, J., and Elmidany, T.T. (1996) *Med. Eng. Phys.*, **18**, 299–654.

18. Tanino, H., Ito, H., Higa, M., Omizu, N., Nishimura, I., Matsuda, K., Mitamura, Y., and Matsuno, T. (2006) *J. Biomech.*, **39**, 1948–1953.
19. Chang, P.B., Williams, B.J., Bhalla, K.S.B., Belknap, T.W., Santner, T.J., Notz, W.I., and Bartel, D.L. (2001) *J. Biomech. Eng.*, **123**, 239–246.
20. Kowalczyk, P. (2001) *J. Biomech. Eng.*, **123**, 396–402.
21. Fernandes, P.R., Folgado, J., and Ruben, R.B. (2004) *Comput. Methods Biomech. Biomed. Eng.*, **7**, 51–61.
22. Kuiper, J.H. and Huiskes, R. (1997) *J. Biomech. Eng.*, **119**, 166–173.
23. Hedia, H.S., Shabara, M.A.N., El-Midany, T.T., and Fouda, N. (2004) *Int.J. Mech. Mat. Des.*, **1**, 329–346.
24. Katoozian, H., Davy, D.T., Arshi, A., and Saadati, U. (2001) *Med. Eng. Phys.*, **23**, 503–509.
25. Sakai, R., Itoman, M., and Mabuchi, K. (2006) *Clin. Biomech.*, **21**, 826–833.
26. Goetzen, N., Lampe, F., Nassut, R., and Morlock, M.M. (2005) *J. Biomech.*, **38**, 595–604.
27. Rancourt, D., Shirazi-Adl, A., Drouin, G., and Paiment, G. (1990) *J. Biomed. Mater. Res.*, **24**, 1503–1519.
28. Viceconti, M., Monti, L., Muccini, R., Bernakiewicz, M., and Toni, A. (2001) *Clin. Biomech.*, **16**, 765–775.
29. Fung, Y.C. (1993) *Biomechanics – Mechanical Properties of Living Tissues*, 2nd edn, Springer, Berlin.
30. Kaneko, T.S., Bell, J.S., Pejcic, M.R., Tehranzade, J., and Keyak, J.H. (2004) *J. Biomech.*, **37**, 523–530.
31. Viceconti, M., Pancati, A., Dotti, M., Traina, F., and Cristofolini, L. (2004) *Med. Biol. Eng. Comput.*, **42**, 222–229.
32. Wolff, J. (1986) *The Law of Bone Remodelling* (Das Gesetz der Transformation der Knochen, Verlag von August Hirschwald, 1892), (Translated by P. Maquet and R. Furlong), Springer, Berlin.
33. Niinimäki, T., Junila, J., and Jalovaara, O. (2001) *Int. Orthop. (SICOT)*, **25**, 85–88.
34. Sluimer, J.D., Hoefnagels, N.H.M., Emans, P.J., Kuijer, R., and Geesink, R.G.T. (2006) *J. Arthroplasty*, **21**, 344–352.
35. Gabbar, O.A., Rajan, R.A., Londhe, S., and Hyde, I.D. (2008) *J. Arthroplasty*, **23**, 413–417.
36. Osyczka, A. (1992) *Computer Aided Multicriterion Optimization System (CAMOS) Software Package in Fortran*, International Software Publishers, Cracow.
37. Marler, R.T. and Arora, J.S. (2004) *Struct. Multi. Optim.*, **26**, 369–395.
38. Svanberg, K. (1987) *Int.J. Numer. Meth. Eng.*, **24**, 359–373.
39. Bruyneel, M., Duysinx, P., and Fleury, C. (2002) *Struct. Multi. Optim.*, **24**, 263–276.
40. Luenberger, D.G. (1989) *Linear and Nonlinear Programming*, 2nd edn, Addison-Wesley Publishing, Reading, MA.
41. ABAQUS (2005) *ABAQUS, Version 6.5*, ABAQUS, RI.
42. Kuiper, J.H. (1993) Numerical optimization of artificial hip joint designs. PhD thesis. University of Nijmegen.
43. Viceconti, M., Casali, M., Massari, B., Cristofolini, L., Bassani, S., and Toni, A. (1996) *J. Biomech.*, **29**, 1241.
44. Cohon, J.L. (1978) *Multiobjective Programming and Planning*, Academic Press, New York.
45. Fernandes, P.R., Rodrigues, H., and Jacobs, C. (1999) *Comput. Method Biomech. Biomed. Eng.*, **2**, 125–138.
46. Guedes, J.M. and Kikuchi, N. (1990) *Comput. Methods Appl. Mech. Eng.*, **83**, 143–198.
47. Moreo, P., Pérez, M.A., García-Aznar, J.M., and Doblaré, M. (2007) *Comput. Methods Appl. Mech. Eng.*, **196**, 3300–3314.
48. Folgado, J., Fernandes, P.R., and Rodrigues, H.C. (2008) *Int.J. Comput. Vision Biomech.*, **1**, 97–106.
49. Svehla, M., Morberg, P., Zicat, B., Bruce, W., Sonnabend, D., and Walsh, W.R. (2000) *J. Biomed. Mater. Res.*, **51**, 15–22.
50. Swanson, T.V. (2005) *J. Arthroplasty*, **20** (Suppl. 2), 63–67.
51. Min, B.W., Song, K.S., Bae, K.C., Cho, C.H., Kang, C.H., and Kim, S.Y. (2008) *J. Arthroplasty*, **23**, 418–423.
52. Romagnoli, S. (2002) *J. Arthroplasty*, **1**, 108–112.

Index

Biomechanics of Hard Tissues: Modeling, Testing, and Materials.
Edited by Andreas Öchsner and Waqar Ahmed

ISBN: 978-3-527-32431-6